The Automobile and Urban Transit

The Formation of Public Policy in Chicago,
1900–1930

Technology and Urban Growth

A Series Edited by

Blaine A. Brownell,
Mark S. Foster,
Zane L. Miller,
Mark Rose,
and
Howard J. Sumka

The Automobile and Urban Transit

The Formation of
Public Policy
in Chicago,
1900–1930

Paul Barrett

TEMPLE UNIVERSITY PRESS

PHILADELPHIA

Library of Congress Cataloging in Publication Data

Barrett, Paul.
    The Automobile and Urban Transit.

    (Technology and urban growth)
    Bibliography: p.
    Includes index.
    1. Urban transportation policy—Illinois—Chicago—
History—20th century.   2. Automobiles—Illinois—
Chicago—History—20th century.   3. Traffic engineering—
Illinois—Chicago—History—20th century.   4. Chica-
go (Ill.)—Transit systems—History—20th century.
I. Title.   II. Series.
HE310.C45B37   1983   388.4′068   82-19371
ISBN 0-87722-294-0

Temple University Press, Philadelphia 19122
© 1983 by Temple University. All rights reserved
Published 1983
Printed in the United States of America

Contents

Series Preface

The design, construction, and consequences of trolley and highway networks have engaged the attention of a vast number of historical scholars. The earliest studies of urban transportation tended to portray the central actors in local dramas regarding transport innovation in Manichean terms, as heroes or as villains. Trolley operators, for example, were often presented as heartless monopolists squeezing the last pennies out of widows and orphans who had little choice but to ride their rickety, dangerous cars. Or else they came across as enlightened public servants, struggling valiantly to preserve indispensable mass transit systems against concerted attacks by predatory automobile and highway interests and suburban developers bent upon selfish ends. Although emotion-charged works continue to be published, a new generation of studies has provided more balanced perspectives concerning the complex development of urban transportation. Following the lead of Sam Bass Warner, Jr., whose study of late nineteenth-century trolley suburbs in Boston provided a model of careful, innovative scholarship, several historians have examined the effect of decision-making and changing transportation technology upon American cities. Their studies have assessed the relationship between transportation innovations and such factors as suburban growth, demographic trends, local economic development, politics and topography, and the role of federal funding.

Paul Barrett's analysis of trolley and highway planning in Chicago between 1900 and 1930 is presented against the dramatic backdrop of local politics; yet by avoiding facile labels and refusing to editorialize, his study fits squarely into the emerging school of sophisticated, transportation history. Barrett highlights the role of decision-making upon local transportation, carefully unraveling the perspectives and ideologies of conflicting interest groups active in Chicago's tumultuous politics. Several themes are prominent in Barrett's book. The leaders of Chicago's competing neighborhood, ethnic, business, and political factions never achieved consensus regarding such basic transit issues as whether mass transit or the auto should serve certain areas and where major routes should be located. Political leaders could agree that public transit must be convenient, cheap, and reliable, but virtually all of the decisions affecting transportation in Chicago were based ultimately on factors that had little to do with improving mobility for the masses. Barrett argues that local transportation experts prepared some excellent plans; unfortunately, none enjoyed the clout necessary to put even the most modest into effect. Nobody was in charge. According to Barrett, even expediency fails to explain transportation decisions in Chicago; as often as not, final outcomes were inadvertent.

However chaotic the politics of transportation in Chicago was, its long-term effect was clear-cut, in that decision-making was subverted mass transit. Barrett suggests that this outcome was not simply due to the mean-spiritedness of public officials. From 1880 to about 1906, mass transit operators brought on many of their own problems. After 1900, though, urban politicians imposed expensive conditions on operators, such as snow removal, street repair, and requiring service to low-density and thus unprofitable areas near the city's outskirts. In effect, trolley operators had to subsidize the construction and maintenance of clear roads for motorists. Under these conditions, even the most conscientious, public-spirited mass transit officials would have experienced serious difficulty. As a regulated public utility, mass transit endured close scrutiny or neglect, depending upon its public image at the moment. Ultimately, a vacillating public policy severely hampered long-range planning and the ultimate viability of mass transit.

Barrett argues, however, that negative responses by politicians to mass transit were less important in shaping the city than the undeniable fact of having to accommodate the automobile. The private car, in vivid contrast to public transit, was depoliticized. Because virtually every politician and most of his constituents believed that motorists had an inherent right to use the streets without interference, the only question was how best to keep the automobile moving. Everyone benefited from improved streets, or so ran the reasoning. At the center of the push for easy circulation of automobiles was the downtown merchant, who demanded wide avenues to the central city and curbside parking for his customers. Little did he imagine in the 1920s that in a few more years improved boulevards and highways would carry many potential patrons to streamlined shopping centers located along Chicago's periphery.

Barrett's examination of public transportation policy in Chicago has broad, national implications. Although several cities experimented with municipal ownership and still others provided huge subsidies for mass transit, the results in these American cities generally resembled those in cities that provided little or no public assistance. Throughout the United States public and private transportation encouraged the rapid outmigration of households, retailers, and even factories.

Barrett carefully avoids claiming too much influence for transportation policy. Decisions concerning but a single area of technological innovations could not effect rapid spatial growth by themselves. In fact, a wide range of urban decision-makers—including land developers, highway planners, high-rise office promoters, sports complex and convention center visionaries—all harnessed emerging technological advances and helped shape American cities.

The editors of this series are interested in exploring the role of technological advances in the urbanization process. Such changes are usually physically obvious, and their impact can be measured with varying degrees of precision. The processes by which city-dwellers apply particular technologies in individual urban settings remain both controversial and elusive. Yet too many social critics ritualistically parrot clichés about these processes. We believe Barrett's work guards carefully against oversimplification, while elaborating changes in mass transit technology and illuminating the politics of traffic and city planning during the early twentieth century. This is the third volume of the series.

The Editors, Technology and Urban Growth Series

Preface

Local transportation is a perennial urban crisis. The composition of the traffic jam has changed from generation to generation; the scene has shifted from trolley tracks to expressway, but the problem has remained unsolved. Popular explanations for the difficulties of moving about the city have altered. The auto industry, city bureaucrats, and city planners have replaced horse-drawn wagons, utility owners, and absence of planning as the favorite villains. Villains, however, there have always been, and the traffic jam remains.

This study began as a search for villains. As a child I saw a villain in the municipal agency which operated the transit lines, forever raising the fare charged for rides on slow, growling, creaking cars. The auto too was a handy villain: even a child could see how a single misplaced motorist might block a line of clanging trolleys. As I grew to adulthood, I watched men clearing the way for a superhighway as they cut a wide swath through my neighborhood, to the benefit of I knew not who. The road they built was immediately jammed with cars. Planners were at once added to the list of villains.

At the University of Illinois, Chicago, I produced a doctoral disertation on the history of Chicago's urban transportation problem. In it, public policy emerged as the villain: favoring private transportation and destroying public transit through misdirected and politically motivated regulation. Further thought, research, and experience have modified that judgment in many ways, and a complex of causes has replaced the list of villains.

The urban transportation problem now appears to me as the product of an evolution, characterized chiefly by the absence of a coherent transportation policy. Public and private transportation were long treated separately, and each system was shaped by its users as well as by its own technology and by the public policy within which it operated. In a city in which both planning and utility regulation were for the most part the results of compromises among organized interests, straphangers could influence the treatment of the transportation problem only by the way in which they used—and then abandoned—public transit. Ultimately private choices—how and whether to use mass transit or the automobile—did much to shape the outcomes of public policy and planning. The choices made by the poor had far less impact than those made by men and women who could afford automobiles, but every user helped to shape the transit system.

The crucial fact, however, was that local government possessed the power and the mandate to accommodate the new automotive technology. Automotive technology was flexible, but so was local government when dealing with the private car. Mass transit proved inflexible not merely because streetcars ran on rails but because its only conceivable model was, and continues to be, the profit-seeking corporation which moved people in masses.

The present study has a long history, in the course of which many debts have been incurred. Space permits the acknowledgment of only a few of these. Among the scholars who have given important advice and encouragement since the inception of this work, none has been more important than my graduate school mentor, Peter d'A. Jones, of the University of Illinois, Chicago. Along with Melvin Holli, Perry Duis, Glen Holt, and many others, he helped to give breadth and perspective to my work. From Mel Holli I learned much about the meaning of urban reform, from Perry Duis I gained a realization of the importance of the early twentieth-century debate over the distinction between public and private space, and from Glen Holt I began to learn the importance of the social implications of mass transit.

Several scholars have given invaluable advice and encouragement during the course of my work on this book. Foremost among these are Mark S. Foster and Mark Rose, who have given invaluable editorial advice. A number of other scholars have provided important advice and encouragement. Among these are Joel Tarr, Blaine Brownell, Michael Ebner, Harold Platt, Howard Rosen, John Root, Steven Meyer, Wilbur Applebaum, and many others. Length limitations, the pressures of time, and the limitations of the focus of the book did not always permit me to follow the suggestions I received from these men. Thus while each has contributed much to this work, its shortcomings are my responsibility alone.

Many persons outside the academic world have helped to make this book possible. Harold Hirsch, General Manager of the Chicago Transit Authority, and George Krambles, his immediate predecessor, have been extremely generous with their time. Leroy Dutton, Louis Traiser, and Evan McIlraith, all traffic and scheduling experts with the Chicago Surface Lines during the 1920–1947 period, were also most helpful and forthcoming. Whatever understanding of the operations of a transit system may be reflected in this book is due largely to the efforts of these men. My errors are, of course, my own responsibility.

Many librarians have also given invaluable aid. Especially knowledgeable and helpful were Archie Motley of the Chicago Historical Society and Carolyn Moore and her staff at the Chicago Municipal Reference Library. The staffs of the Northwestern University Transportation Library, the Cudahey Memorial Library of Loyola University, Chicago, and the John Crerar Library, along with that of the University of Chicago's Regenstein Library, have also been most helpful.

Cartographer Toby Roberts has succeeded in cramming a great deal of information into the maps in this volume. His fast and skillful work has done much to make this book more readily understandable. George Krambles has kindly provided the cover illustration, and Kevin Harrington has assisted in its preparation. Several heroic typists, most notably Kate Carroll and Barbara Lezinski, have survived encounters with this manuscript in its various manifestations. Greg Baiocchi and Dr. Deborah Holdstein provided invaluable assistance with the production of the final version.

Finally, personal friends have helped to make this book possible. Bill Abler, inventor, sculptor, linguist, and author, gave helpful advice on the writing of the introduction. Bill Lauria, Frank Gutierrez, and other former fellow workers in the taxi business have given constant encouragement and the not entirely irrelevant assurance that I do have an alternative career. From first to last, Mary Carroll, a life-long friend, has contributed ideas, encouragement, editorial skill, and unfailing faith. Most important of all, my wife, Annette, contributed not only her advice and typing skill but a sane perspective on an insanely competitive world.

To these and many others—including Temple's skillful production editor Candy Hawley, and editors-in-chief Ken Arnold and Mike Ames—my heartfelt thanks are due. Once again, the shortcomings of this work are the responsibility of the author. For its merits, those named above deserve much credit.

Chicago, Illinois
November 1, 1982

The Automobile
and Urban Transit

The Formation of Public Policy in Chicago,
1900–1930

Introduction

The policy definitions of mass transit and the automobile which developed in the American city between 1900 and 1930 did much to determine the roles which these two modes of transportation would play in urban life. In essence, mass transportation was defined as a regulated private business, while the accommodation of the automobile became an undisputed public responsibility. This anomaly—public responsibility for personal transportation and mixed, public and private control of mass transit—prevailed in Chicago and other American cities during the years in which the automobile achieved its position of dominance in urban transportation.

Public policy did not cause the triumph of the automobile in Chicago or anywhere else. Yet, by placing the auto and mass transit in artificially separated categories, public policy *did* cripple mass transit and confuse public thinking on local transportation. The transit systems created by regulated private enterprise became the models for all subsequent solutions to the "transportation problem." What may be called "accommodative planning" became the standard approach to dilemmas created by the growing numbers of private cars. These facts in turn continue to prevent planners and policy-makers from dealing with urban transportation as the single, integrated phenomenon it is.[1]

In the period from 1900 to 1930 these policy definitions largely determined the focus of city planning in Chicago. They conditioned the pace and direction of innovation in transit system design and highway planning and helped determine the role of engineers in public life. In addition, public policy influenced the development of urban bureaucracies that regulated transit, traffic, and highway planning. It also influenced the internal organization of transit corporations. It created a largely destructive form of regulation and helped to alter, even as it reflected within itself, political relationships among interests in the city.

While it remains a fundamentally local event, conditioned by local circumstances, this Chicago story is an essential part of the national saga of the rise of the automobile. This seeming contradiction deserves a moment's consideration before we begin. Until the coming of massive federal intervention in urban affairs, the public policies of cities had essentially local roots; yet,

3

since so many of the circumstances cities had to deal with were similar across the nation, the story of one city's policy decisions and their consequences tells us much about what was going on in other cities as well.

On the surface, the mass transit policies of most American cities "look alike." Regulated capitalism was the basic approach to mass transit almost everywhere, despite the existence of strong movements for "municipal ownership." Just as important, cities almost never became involved in the financing of transit improvements until all other alternatives had been exhausted. The crowded streets of New York, Boston, and Philadelphia made expensive off-street rapid transit a necessity in those cities. In San Francisco the desire to pierce Twin Peaks helped bring the city into the transit business. In the eastern cities at least, this need for capital intensive improvements helped assure that transit could not be self-supporting. Still, as Charles Cheape has shown, differing local traditions and circumstances produced different outcomes in each city.[2]

In Chicago, despite the existence of a ninety-seven-mile elevated system, the great majority of commuters traveled like those of most other cities: they rode the surface streetcars. Through the 1920s, the reasonably good surface system *was* self-supporting. Thus, in Chicago, the idea that mass transit should "pay for itself" became firmly established and assumed a defining role with respect to local transit policy. In this respect, Chicago was more typical of other American cities than were the eastern "big three."

The central role of surface mass transit also meant that the city's street and traffic policies were of paramount importance for the future of public and private transportation alike. Because most commuters traveled on the street, the contest between public and private transportation came to turn as much on the regulation of street traffic as on transit policy or the inherent attractions of the private car. The battle for the streets took place in public view, with neither subways nor superhighways offering escape from the confrontation. Traffic segregation was the major panacea advanced by Chicago's planners, whether they focused on mass transit or on street traffic as such. But regulated capitalism could not provide subways for mass transit, and separate, high-speed streets for the automobile evolved slowly. As a result, the emerging definition of the motorist's rights and responsibilities became the most important element in determining how the streets would be used—whether to the advantage of motorist, straphanger, or both.

Most American cities faced similar situations. Yet important differences emerge when one looks closely at the way each city responded to these common problems. Chicago's transit system became, to an unusual degree, an adjunct of neighborhood business and community life. By 1915, local businessmen looked to the streetcar to serve their interests, and patrons expected it to fill a niche similar to that later carved out for the automobile. Public policy, combined with the city's geography and demography, created a surface system—the nation's largest—which went as far toward filling these expectations as regulated, privately owned mass transit could.

In city planning, too, Chicago differed in some respects from the national pattern. As Mark Foster has shown, there was no single planning response to the automobile, nor did planners possess the power to implement their disparate visions.[3] In Chicago, the implementation of Danial Burnham's monumental 1909 *Plan of Chicago* quickly became the special province of a business-dominated, "nonpolitical" body, the Chicago Plan Commission. Mass transit planning was divorced from city planning at this point, to a degree and with a decisiveness not common elsewhere. The thinking of planning experts and traffic engineers outside Chicago continued to have some impact on the city's efforts to deal with the automobile: innovations like the limited-access interchange, for example, were clearly borrowed. But local needs, as understood by the business-political policymaking community, determined how and whether such innovations were adopted in Chicago.

Even with all the parallels and lines of influence between Chicago and other cities, it is clear that each metropolis developed its transportation network and policy in a local context and according to a rationale which arose out of its own needs and history.

Local transportation evolved and had its impact almost imperceptibly and over a long period of time. Mass transit and automobile policy alike were made primarily by men whose chief concerns lay elsewhere—in political influence or survival, in channeling the city's economic growth, or, on occasion, in one or another program of reform. When mass transit or, more rarely, the automobile, *did* become the focus of public discussion, that discussion employed language and definitions derived from the city's long and poorly understood history of transportation policymaking. While policy ideas from "outside" were utilized in Chicago, they were acceptable only in so far as they fit the local traditions and definitions of transportation problems.

In short, the development and impact of early twentieth-century transportation policy must be studied in detail and in its local context in order to be understood. Despite the common problems faced by American cities during this period, each city made its own transportation policy. The constraints of state law, the national love affair with the automobile, the national commitment to private ownership of public utilities, all helped shape transportation policy in every city, but policy itself remained embedded in local history. The words used to describe aspects of the transportation problem might be the same from city to city, but they derived their meanings from previous local history. At the same time, transportation policy decisions were, for those participating in them, a part of local, short-term struggles—struggles against an immediate traffic or transportation crisis or, more often, battles over political power or influence.

To motorists, straphangers, and policymakers alike, mass transit and the automobile were essential facts of daily life, not self-contained issues or historical phenomena. Policy was formed in the context of many concerns which may seem to us irrelevant. Its impact was often obscured by politics, rhetoric, and the constantly changing cityscape. Often public and policy-maker alike were unable to see the effects of past policy. Most important, both auto and mass transit policy were based on assumptions so fundamental that they were not even addressed as such. It is these assumptions—the policy definitions of public and private transportation—which are the focus of this book.

Throughout our period, politicians, reformers, courts and the public agreed that good transit service could and should be profitable. Most thought it should be privately owned, and those who favored public ownership con-curred that transit should be run as a business. The relief of traffic congestion, on the other hand, came to be understood on grounds of safety, commercial advantage, and political and legal necessity as a matter of public responsibil-ity. These differing policy definitions—mass transit a regulated, privately owned utility and streets and traffic control fundamental charges of urban government—had far-reaching consequences for the treatment of these two aspects of the city's single local transportation problem.

The fundamental difference was, of course, that facilities for the auto-mobile were publically subsidized while mass transit was regulated and taxed. This difference in treatment is of great importance, but it is only one of several. The corporate ownership of mass transit made it the subject of persistent political controversy. In the course of this controversy, issues of social policy, environmental protection, and the distribution of political and economic power became attached to transit policy formation. Individual auto ownership helped to insulate street-building and traffic policy from such questions during our period.[4] Transit regulation, the product of many political and economic interests, was often destructive, despite its positive intent.[5] Traffic regulation, initially negative, soon became positive and accommoda-tive, because it dealt directly with a large and growing group of citizens, and because the mutual adjustment of the city and the automobile presented the city with an urgent problem of pressing concern to important interest groups.

The battle between centralizing, CBD-oriented business interests and decentralizing neighborhood business organizations affected the two modes differently. Both the technology of mass transit and its business definition caused transit policy to become a battleground between decentralizing neigh-borhood interests, whose concerns were geographically defined, and central-izing citywide business organizations, dominated by the CBD. Engineers favored centralization and professionalization of policymaking functions in their own hands, but their plans could not in themselves resolve the differ-ences between CBD-citywide and neighborhood-oriented business and real estate interests. Transit engineers helped shape and legitimize transportation policy, but CBD business and some politicians took the lead in centralizing transit policymaking and removing it from the arena of public debate.

The auto, on the other hand, escaped this controversy. The private car's flexibility and depoliticized status, together with the emerging planning tradition (which developed together with the auto's urban role) made street improvements and accommodative traffic regulation appear to be in the interests of all sections. The auto and the streets planned for it implied freedom of movement. So long as power divisions in the city were conceived of as geographically based, the auto seemed neutral. While real power centralized, the auto offered to redistribute urban functions in ways which could not at once be forseen. Its rise seemed potentially compatible with the interests of all.[6]

Street planning and traffic control were for the most part in the hands of the CBD business community from early in our period. There was no need here for the role of the engineer or planner as centralizer. Transit planners in particular contributed heavily to the body of skills and knowledge upon which city planners built, and the Chicago Plan Commission did much to enlarge the city's power to execute plans.[7] But city planning in this period was in reality a function of the CBD business community—to so great a degree that there is little possibility of distinguishing the real impact of planners and engineers as such, or of discerning the motives behind their plans or their professionalization.

Their policy definitions caused mass transit and the automobile to be treated differently in many ways, yet they were alike in surprising ways as well. Most notable, perhaps, is the fact that mass transit, like the private car, was used in ways which facilitated the separation of the city's different social classes and ethnic groups. Routing and fare structures sometimes accommodated this reflection of social distinctions in the journey to work, but rider choices and the spatial segregation of the city itself were largely responsible. The spatial ordering of the city also assured, as none should be surprised to learn, that inner-city straphangers experienced more crowding and effectively received less in return for their fare than did other classes of riders. Regulatory policies ostensibly designed to alleviate this condition sometimes exacerbated it instead.

In most respects, however, mass transit and the automobile were treated differently because they were differently defined. It is well to stress from the outset that this difference in treatment was not the result of a conspiracy: the private car was not foisted upon Chicagoans by automakers, planners, or anyone else. Nor was the simple fact of regulation to blame for the decline of public transportation: in some East Coast cities, where regulation was accompanied by public responsibility for mass transit, the nation's best and most comprehensive transportation systems resulted.[8] Rather the assumptions that mass transit must be treated fundamentally as a business and planning for the auto as a duty of local government helped to promote and to shape the urban transition from public to private transportation.

Operating below the surface of discussion, these assumptions helped to subvert the planning process, retard mass transit innovation, and create an anomalous role in policymaking for the engineer. Our theme, then, is the impact of policy definitions on planning, innovation, and the role of the engineer. Our story is that of one city's attempts to adjust to technological and demographic changes while using a conceptual framework which became increasingly inapplicable.

Chapter 1:

The Decision for Regulation

Section I:
The Context of 1907

> In the last analysis it is a question of public endurance. . . .
> Whether for the sake of peace they will . . . transfer the fight to
> another generation.
>
> *Clarence Darrow, 1905*[1]

Between 1905 and 1907, Chicago made the decision to define mass transportation as a regulated private enterprise. During the following quarter-century, in consequence of this decision, mass transit planning became divorced from the rest of the city planning movement, innovation in mass transit was held back, and a curious, ultimately untenable public role was created for the transportation engineer.

Around the nation, other cities were reaching the same decision. Many in the transit industry were committed against regulation. Many cities were bound by political tradition, sometimes even by their charters, to city-owned mass transit. In some states, companies were already seeking to escape city politics through the creation of state regulatory commissions. Everywhere, however, the trend was toward the definition of mass transit as a mixed, public and private responsibility.[2]

In Chicago, this decision was embodied in a franchise agreement with the city's street railway companies—an agreement popularly known as the 1907 ordinances. This settlement formed the legal and (for lack of a less pretentious term) the ideological basis of the city's mass transit policy for the coming two generations. It enshrined late nineteenth-century beliefs about the profitability of mass transit and the necessity for separating "traction" (the current term for all mass transit) from "politics."

Unfortunate though its consequences were, the decision was based in political and legal necessities of the time. The principles which long governed Chicago's local transportation policy evolved from and were expected to promote the resolution of an immediate political and transportation dilemma. Further, the policymakers of that era—like those of later times—responded not only to problems of engineering, politics, and finance but to contemporary popular and expert *perceptions* of these problems as well. In short, the crucial 1907 settlement arose out of turn-of-the-century understandings of the history as well as the realities of urban transportation.

Two sets of aspirations stand out among the issues and perceptions attached to mass transit at the beginning of the century. One may be called the Planner's Dream, though it predated the city planning movement as such in Chicago. This aspiration, shared by reformers and business people with different specific aims, saw in mass transit a means of transforming the urban environment; it was to thin out the slums and promote one or another tendency in the city's spatial arrangements. The other aspiration represents the Straphanger's Dream—the simple wish for swift, safe, convenient, and economical transportation. Between 1900 and 1907, both dreams were attached alternately to programs for public or private ownership of mass transit. Ultimately, after 1907, they were transferred to the automobile and the form of city planning which developed around it.

Yet these aspirations do not appear in clear focus in the history of transit policy formation between 1900 and 1907. One sees instead a confusing picture of political calculation and compromise, rhetoric about recent transit history, reform and other interest group pressures, legal decision, and apparently genuine popular outrage. All this was dramatized almost daily by the city's press. For most of the city's 670,000 daily commuters, whether or not they read the English-language press, this storm of controversy held at least some relevance. By all reliable accounts, the daily experience of Chicago's mass transit riders was inconvenient, uncomfortable, and not infrequently, outright dangerous.[3]

To gain any real sense of the complex of issues and interests involved in the creation of the pivotal 1907 settlement, we must retrieve something of the atmosphere from which that settlement emerged. We can begin in the autumn of 1906 with an incident which, in other times, might have been on only passing interest to the average Chicagoan.

      •   •   •

Frank Murphey saw what happened that late November morning on Chicago's near west side. Of all the accounts surviving, the young machinist's seems least calculated to prove a point.

Murphey was late for work that Monday morning. It was already 7 A.M.

when he ran up the last few steps to the elevated platform. A train stood there in the cold, late autumn mist, "packed to the guards" with people. Murphey ran up to it and, as any turn-of-the-century Chicagoan might have done, kicked at the nearest of its metal gates. The gate, already broken, swung ajar, and Murphey jumped on, followed by a girl.

Turning, he saw an inch or two between the edge of the girl's brown-checked skirt and the end of the car's open platform. "Plenty of room to stand," he thought, and grabbed hold of the metal latticework. But the platform guard was scowling; the gate would not re-close. "You'll have to get off," he shouted to them both. Already, before either of them had arrived, he had rung the rope-operated bell which was the signal for the motorman, who was several cars away, to proceed. Now the guards in the cars ahead passed the signal on; the train lurched and started to roll.

Murphey just clung tightly to his handhold on the car, but the girl, frightened and confused, looked first at the guard, then at Murphey. Suddenly she stepped back onto the station platform, reeling from the car's momentum. Two men standing at the corner of the platform reached for her. One seemed to get a hold, but the slick material of her dark red jacket slipped through his hands. The train ground to a stop in time for Murphey and perhaps a hundred others to see the girl fall screaming to the muddy alley thirty feet below. Pedestrians rushed in from the corner of Grand and Paulina to find that the city's transit system had claimed another life—the 131st of 1906.

For twelve hours no one knew who had died in this latest tragedy. As the news spread, some parents journeyed to the morgue, but most had to stay on their jobs and could only wait anxiously for the end of the day. Then, at home, they would known whether their daughters had returned from work.[4]

But if no one knew who had died, almost everyone thought he knew why the accident had happened. A spokesman for the Metropolitan Elevated Company, from whose platform the girl had fallen, explained that the victim was a hooligan trying to hitch a ride on the outside of a moving car. Two eyewitnesses reported that the elevated guard had shoved the girl to her death by slamming the gate in her face.[5] In the evening, the *Chicago Record-Herald* went to Clarence Knight, attorney for another elevated company, to see if the story could be milked any further. Knight, who had long been a spokesman for the now-departed traction magnate Charles Tyson Yerkes, stated the company's position. He drew attention to the stubborn complexity of any issue related to mass transportation.

"How could the tragedy have been prevented?" the *Record-Herald*'s reporter wanted to know. "It could not have been prevented," Knight replied. Crowding caused the girl to die, and crowding—as well as slow, undependable service on all the city's surface and rapid transit lines—was the result of two things. One was, as Knight pointed out, "the fact that three quarters of the people of the city had to come downtown between 7 and 9

A.M.," and the other was the fact that, for reasons which were more political and economic than physical, the number of vehicles terminating downtown far exceed the space available to turn those vehicles around.[6]

Knight moved on quickly from his first point; there was favorable publicity to be had from the second. But he had touched on one of the elements that made mass transit in Chicago what it was. By 1906, mass transit officials were eager to cooperate in engineering plans for spreading out the city's central business district and, before long, a few would be asking whether it was really necessary for all businessess to open and close within two hours of one another.[7] But the city's transit lines, from the first horsecar of 1859, had been built to serve and feed upon the central district and to foster the way of life which centered on the clock.

This was so because of a discovery made by all American cities in the decade before the Civil War. While people had to walk to work, the desirable districts of a commercial-industrial city could not extend more than two to three miles from its center. Once it became apparent, however, that large numbers of people could be moved across a city by horsecar at relatively little expense—but at great profit—a remarkable vista opened. Every city of any size was besieged by capitalists asking permission to start horsecar lines which would more than double the city's potential size and population, raise the value of real estate, and expand the market for the city's commerce. In addition, transit entrepreneurs offered to take the entire risk of this venture upon themselves. It was an offer no city could reasonably refuse—even had it not been so often accompanied by a timely bribe.[8]

In return for their beneficence, the promoters usually asked only their choice of routes and a guarantee—usually in the form of a franchise—that they could continue to operate long enough to pay off their investment and accumulate a fitting profit.

Mass transit, then, began as an adjunct of commerce and a profit-making business in itself; it had to behave accordingly. By the end of the Civil War, Chicago had had three major horsecar systems, but each sought quite naturally to serve the most profitable market: traffic to and from the Central Business District.[9]

For the most part, industrial laborers in the nineteenth century appear to have walked to work. Even workers who could afford to spend a dime each day on carfare commonly settled within walking distance of their factories or shops. Ethnic solidarity and poor transit service had as much to do with this as frugality. Communities like Chicago's Packingtown, Pilsen, and the near northwest-side Polonia were almost self-sufficient and, in any case, their residents could not conveniently travel crosstown on the carlines. So the horsecar systems—and the cable and electric lines which replaced them in the 1880s and 1890s—were laid out and developed as middle-class utilities, serving the more affluent segments of the population which worked and shopped downtown.[10]

Since there was thought to be little profit in crosstown service, each company (there were seventeen in Chicago at the height of their proliferation in 1896) had stayed within one of the three divisions of the city demarcated by the Chicago River. Many of these lines were strictly local affairs, but those which could sought some form of access to the richest territory marked out first by the Loop described by the cablecar lines and later by the elevated structure which went into full operation in 1898. Each feasted on the daily rush hours, sending as many cars as possible to squeeze into the two-track elevated Loop or the twelve-streetcar thoroughfares of the central district. Each ran as few cars as it dared at other hours of the day, and with few exceptions, none at all at night.[11]

There were advantages to this compact use of space and time. Chicago's Loop district provided a variety of commercial and (at its western and southern edges) light industrial services rarely found in so small an area. In an age when personal contact was still highly valued by executives, the Loop put many of the city's business leaders within walking distance of one another.

But the district was as rigid as it was compact. Hemmed in on the south and west by railyards, with but four bridges for access on the north, the Loop was a victim of its own advantages. By the time the first good traffic count was taken in 1907, almost 57,000 vehicles a day were streaming in and out of the Loop from the west and north districts alone.[12] Inside the iron fence of the elevated, these vehicles often became hopelessly entangled. Heavy four-horse wagons transferring cargo shipments from one rail terminal to another crept down the trolley tracks in front of the clanging cars. Construction materials were piled halfway out into the streets, and delivery carts parked so close to the tracks that they routinely grooved the sides of passing streetcars.

Four years before the tragic accident that kept Clarence Knight in his office declaiming on congestion to the man from the *Record-Herald*, a nationally known transit engineer had begun his report to Chicago's city council on street railway service with a description of the crowds of people who filled the streets at the edges of the center of town; these people were walking to work as their grandparents had half a century before. The downtown streets had become so crowded that, as pedestrians, they could easily outpace the trolleys.[13]

•   •   •

It was here that the second reason for poor transit service—the second to which Clarence Knight would admit—came into play. There was far too little space in the elevated Loop to turn the number of cars required by the rush hour throngs. Every elevated line in the city funneled into the Loop, but most of the Loop's platforms could accommodate only one train at a time. Why couldn't the platforms be extended? Why did the city refuse its permission and back the small group of landowners who resisted this public benefit because it might lower property values? "We have ordered 40 new cars,"

Knight told the reporter, "but it will do little good. Already our trains are backed up every morning from Lake Street to Chicago Avenue [nearly a mile north of the Loop]." What was true of the elevated on its private right-of-way applied doubly to the surface cars which were slowed and delayed by competing traffic. The business of taking as many people as possible to the same place at the same time was foundering amid the results of its success.

Here, then, were two major reasons for congestion and poor transit service. First, the quest for maximum economic advantage on the part of downtown business and the transportation companies had led to the creation of a city so geographically centralized and a workday which concentrated traffic in so few hours that good transportation had become physically impossible. In theory this problem could have been dealt with in a variety of ways. Business hours could be staggered; transit companies could through-route their vehicles or turn some of them back at the edge of the CBD; more facilities—subways or elevated lines—could be built. But in 1906, each of these potential solutions faced obstacles which were, for the moment, insurmountable. Through-routing and staggered hours called for cooperation among businesses not accustomed to such teamwork and seemed to threaten profits. Engineering solutions, such as subways, required cooperation not only among transit companies, but between the companies and the city as well.[14]

Thus the second major reason for the miseries of Chicago's commuters was the adversary relationship between the city and the transportation companies. The point which Clarence Knight had wanted to make to reporters was that the city itself, in its efforts to force the companies to accept regulation, was actually making service worse.

In this particular case, the city council hoped to trade permission for the extension of downtown elevated station platforms for an agreement by the three rapid transit companies to through-route their cars instead of turning each one back onto the line from which it came. But if the cars were through-routed, each company feared it would lose the fares passengers traveling through the Loop had to pay when they changed from the cars of one company to another. Each side, city and companies, stood its ground, and as a result, service was worse than it would have been had either side given way.[15]

This standoff was not a unique occurrence. It was typical of the counter-productive battles which took place throughout the 1900–1930 period, whenever the city administration and the mass transit companies were at odds. Regulation, the obvious intent of which should have been the improvement of service, often made service worse. The whole complex history and mythology of city–company relations stood behind this anomaly of destructive regulation.

• • •

Many of the same events that brought about congestion helped to bring the transit companies into conflict (often only apparent, but sometimes very real) with members of the city government and some parts of the business community. Among these factors were the growth and concentration of the city's population, the rapid expansion of its boundaries during the late 1880s, a series of technological innovations which, together with other factors, led to a sharp rise in mass transit ridership, and a drive within the transit industry toward concentration of ownership and maximization of profits.

This nineteenth-century background has been covered and debated by historians and journalists for three quarters of a century.[16] Here it can be touched upon only briefly; but the brevity of this treatment should not disguise the period's importance. The actual events of the late 1880s and the 1890s created the twentieth-century mass transit system and the dilemmas which accompanied it. Popular, reform, and political perceptions of what happened during this time shaped the policymaking of the 1905–1907 period, which in turn was a controlling force in the development of twentieth-century transportation policy.

The reality of the late nineteenth century was a rapidly expanding city with the nation's most dramatic population growth and a corrupt and volatile political system. Between 1887 and 1893 the City of Chicago tripled its geographical size with the annexation of well over 120 square miles. The 1890 population was nearly 1,100,000, almost double the 1880 figure. Annexation accounted for much of this increase, but the old, preannexation city alone added 289,000 during that decade. The whole city added another 700,000 by 1900 and had reached a population of 2,185,283 by 1910, about 1,000,000 of whom lived within four miles of the center of the CBD. The total increase for the 1880–1910 period amounted to some 350 percent.[17]

The city's mass transit system, like others around the country, went through an equally rapid transformation. Powered exclusively by the horse in 1880, the system had become predominantly cable-powered by 1890 and, in November 1906, had just completed the final conversion to complete electric power. This rapid technological change had both helped to cause and been made possible by a tremendous increase in the rate of transit use.

In 1890—the first year for which reliable data are available—mass transit use stood at an average of 164 paid rides per year for every man, woman, and child in the city. By 1900 the system averaged 215 rides per capita, and in 1905 the figure stood at 253. With 320 paid rides per capita in 1910, the city's transit system had nearly doubled its rate of use. Indeed, transit use had risen almost twice as fast as population. If the free rides given policemen, firemen, company employees, and small children could be counted, the rate of increase might be even greater.[18]

Chicago's transit system grew and improved with this increased ridership. Car speeds increased as electric trolleys replaced horse and cable cars. Over eighty miles of elevated railway were built between 1892 and 1906, and the street railways doubled their mileage.[19] Nevertheless, while improved mass transit facilitated the city's growth, it did not keep pace with public expectations.

Much of the problem could be traced to the city's growth and centralization, and the resulting congestion. Longer rides meant more discomfort, and
congestion meant slow, unreliable service and crowded cars. Complaints
about these conditions were constant throughout the 1890s.[20]

The adversary relationship that eventually developed between city and
companies was not, however, a result of the difficulties caused by the city's
rapid growth. It came about because of the way those difficulties were
interpreted. The public saw, not the growing pains of a burgeoning city, but
rather the actions of villainous "traction barons" and "boodling" politicans.

This view, though it may have missed the main point (that the city had
been growing and changing at a rate to which mass transit could not adjust)
was not without its merits. Politicans and traction owners contributed significantly to their own poor reputations. This was so in part because late
nineteenth-century laws almost invited politicians and transit businessmen to
interfere in one another's affairs.

For practical purpose, before 1907, the city had the power to grant
street railways permission to operate but could not effectively regulate them.
It could insert what amounted to regulatory clauses into such "franchise"
ordinances and did successfully require streetcar companies to sweep the
streets between their tracks, remove snow in the winter, and limit their fares
to 5¢ per ride. But this sort of arrangement, called a "contract ordinance,"
could not provide for day-to-day regulation or the setting of service standards.
With the elevated companies (legally regarded as railroads and therefore
subject only to the state government) the city was in an even weaker position.
Only when an "el" company wished to infringe technically on city property, as
by extending its platforms, could the council attempt to bargain with it.[21]

Politicians, then, could not do much to improve the quality of local mass
transit. However they could use the occasion of granting street railway
franchises as an opportunity for extracting bribes from traction promoters. As
late as 1897, Chicago had fifteen ostensibly separate transit companies, and
each of the surface companies had twenty-year franchises. Further, each
request from an existing company for permission to extend its lines provided
yet another occasion for graft. Not surprisingly, then, transit companies were
perceived by reformers as a major source of funds for city council corruptionists.[22]

Reformers, as will be seen, made much of this relationship. But they
also made much of the doings of the traction companies themselves. Complaints of crowding and poor service were brought to a head most dramatically
by the efforts of one man, Charles Tyson Yerkes, to consolidate and rationalize
the city's transit system.

•   •   •

Yerkes, who arrived in Chicago in 1887, had by 1897 gained control of
two thirds of the city's surface cars and about half of its elevated trackage. This

villains, rather than a single villain. Henry Demarest Lloyd, a publicist who led the nationwide movement against trusts, declared that "monopoly grows too rich, too concentrated, too keen, too unscrupulous to be regulated."[28] Others, somewhat later, would question whether, with even the most public-spirited of managements, mass transportation could yield both high quality service capable of keeping pace with the city's dispersion and a reasonable profit for investors. Most subscribers to this school of thought believed Chicago's els and streetcars must be owned and operated by the city. Only then, they thought, could the transit system give the people its full and undivided efforts.

There was much to be said for the first point of view—that bad men made bad service—and Yerkes himself had inadvertently provided the best arguments. For one thing, he overcapitalized his lines. By 1906, when the average American street railway carried about $115,000 of securities per operated mile of track, the largest of the companies Yerkes had built was capitalized at over $250,000 per mile.[29] It was on this gigantic watered investment that the straphangers of the north and west sides paid interest and dividends.

Yerkes' questionable financial practices—of which his less-notorious contemporaries were not entirely innocent—led to exposés which helped to fix in the public consciousness the image of utility-as-thief.[30] His personal character, as portrayed by the press, gave an aura of villainy to practices which might otherwise have been attributed to simple greed. Yerke's apparent contempt for reformers and for the self-righteous press, his seeming disregard for the comfort and safety of his passengers, his reportedly cavalier treatment of business and community leaders—all these combined with the traction baron's drive to control every situation of which he was a part to create a symbol that was, perhaps, too easy to hate.[31] Yerkes unknowingly came to personify the "traction problem" and, in so doing, made it all the harder to resolve.

The image of the traction company as villain was, by 1906, all-pervasive. Any politician with a healthy instinct for survival made a point of subscribing to it publicly and with fanfare. On the evening of November 26, 1906—before an inquest had been held into the morning's accident—even before the victim had been identified—Chicago's city council met to pass a strict and certainly unenforceable anticrowding ordinance and to condemn the transit companies. Big, affable "Bathhouse" John Coughlin, who coaxed the bums and tavern hangers-on to vote in the interest of the traction companies whenever necessary and profitable, voted for the ordinance. So did his first ward comrade, Michael "Hinky Dink" Kenna. John Powers, the perennial Irish boss of the polyglot nineteenth ward, cast his vote against the companies, too. William Dever, the progressive reform alderman from the near northwest side, and John Derpa, who liked to call himself a "labor man"—both staunch advocates of municipal ownership—backed the new measure to regulate the companies. Milton Foreman, who had spent the last six years fighting for a "fair and reasonable settlement" with the private companies, voted for it, too. Even Alderman Adolph Larson, who had told the *Chicago Tribune* that "I get tired whenever I hear about traction," roused himself for a yes vote. To a man, the council members gladly voted another condemnation of the companies and deplored another needless death because, by 1906, no politician dared do otherwise. A Chicago alderman—to be even plausible as an honest man— could do or say nothing which might be construed as favorable to the "traction interests." This was the legacy of Charles Tyson Yerkes.[32]

•   •   •

The public's negative attitude toward aldermen and traction companies alike was of major significance. As will be seen, planners and opinion makers of the 1906 era did subscribe to the nationally pervasive belief in the power of good transportation to improve urban life. But by 1906 neither transit companies nor politicians had any major reservoir of public trust upon which to draw.

The difficult situation was made even more unwiedy by the fact that, in 1906, the city council lacked any clear, positive focus. It was a curious creature: many of its aldermen were adherents to factions of the internally fragmented Democratic or Republican machines, but most had been elected with the endoresement and aid of reformers. This strange situation was in large part the result of the efforts of a group of civic-minded professionals known as the Municipal Voters' League.

The League had come into its glory in the battle against Yerkes' 1897 effort to secure a fifty-year franchise, first from the state and then from the city council. In an effort to gain the election of aldermen who would fight the "franchise steal," the MVL's first secretary, George Cole, made up a pledge for aspiring candidates to sign. The pledge condemned the fifty-year franchise and called for the state to grant the city both the power of regulation and the right to buy the traction conpanies outright. A staff of young attorneys pored over the records of aldermen and candidates, and Cole, having considered all this information, produced a list of endorsements. The press cooperated enthusiastically by informing the English-reading public as to which aldermen were the most notorious grafters and which candidates were the most likely prospects for a "clean" city council.

The MVL was, for Chicago, a moderate organization, but it subscribed to the creed held by the most thoroughgoing reformers in other cities—that political corruption was primarily the fruit of corporate greed. Thus, the League's early platforms, its neighborhood mass meetings, and its torchlight parades combined to infuse in the alert Chicago voter the idea that a dishonest alderman and an alderman who voted for legislation favored by street railway companies were likely to be one and the same man. Applying research techniques learned in Chicago aldermanic contests, the MVL jumped into the campaign against Yerkes' legislative package, the "Humphrey Bills."[33]

The MVL achieved its greatest single victory in the aldermanic election of 1898: as a result of that election, forty-two of the council's sixty-eight aldermen had subscribed to the League's anti-Yerkes platform. The reformers went into the final battle in the council chambers with all the city's newspapers (save the one owned by Yerkes) on their side. They had two comprehensive reports on the companies' financial practices to impress any doubters. They had the mayor, too—Carter H. Harrison II—happily tagging along and proclaiming himself at the head of the movement. The morality play ended in the council chambers on the night of December 19, 1898, with Yerkes in defeat. Even the archvillain himself did not seem surprised.[34]

The crusade of 1898 and the years of wrangling which followed left their impression on a wide range of citizens. Muckraking journalist Ida Tarbell declared in 1906 that "even humble shopgirls" knew something of the rights and wrongs of traction.[35] The difficulty was that organization and propaganda of the sort which had defeated Yerkes could not solve the city's transportation problems. The MVL could fight crooks, but it could not deal with confusion, or with the mistrust of both government and corporations which it had helped to create.

        •   •   •

By 1906, the council which met and, in the words of the mayor's transit expert, "trembled to act" in response to the crowding accident at Grand and Paulina was probably as honest a legislative body as ever sat in Chicago's city hall. Between 1901 and 1911, 85 percent of the candidates elected to city office bore the Municipal Votes League's endorsement. Of course, not all these men had halos under their derbies: the MVL would often endorse a lesser crook in the hope of unseating a great one. But each MVL alderman had signed a pledge that called for strict dealing with the traction companies in the public interest, and most of them were relatively clean. "The people of Chicago . . . have beaten boodle [graft]," muckraker Lincoln Steffens exulted in 1902. "That is about all they have done so far, but it is about all they have tried deliberately and systematically to do."[36]

Steffens summed up the matter rather well. It was corruption from any source—but particularly that which originated in public utility corporations—which the League was created to fight. Consequently, the MVL was supremely indifferent to political partes, since corruptionists were to be found in all of them. The parties were of roughly equal strength, so there was no single great party "machine" on which the MVL could focus its attack. Instead each part contained a number of small machines, each based in patronage and ethnic tradition as well as party loyalty. The MVL's tactic was to bargain with these machines and attempt to bring about the nomination of more honest candidates. It would then back these candidates with all the conventional nineteenth-century campaign devices.[37]

The result was a city council which lacked cohesion on controversial issues. Though the League's president told the *Tribune* that "there must be a machine and a boss," the MVL's "machine" functioned effectively only at election time. Having built its reputation on nonpartisanship, it dared not endorse specific proposals (such as a referendum on public ownership or a franchise to a private company) which might be tied to the interests of one party or another. Nor, for this reason, would the MVL endorse a candidate for mayor.[38]

The man who was mayor between 1897 and 1905, Carter Henry Harrison II, could not fill the leadership vacuum. Dapper, smooth talking, and personally honest, Harrison nonetheless earned the mistrust of reformers because he tolerated a measure of graft and vice. More important, Harrison had no clear position on the transportation issue, aside from a steadfast opposition to Yerkes. By his own admission, the mayor knew nothing of the "traction question" when he took office in 1897. Over the years, he stressed public ownership or regulation as the political need arose, and he seems not to have kept up with the details of policymaking even in his later terms. In November 1902 the council received an engineering report it had commissioned, which rapidly became the Bible of the city's transit planning. Three months later, the mayor of Chicago still had not read the 236-page document.[39]

The city council of 1906, then, was in a poor position to decide local transportation policy for the coming century. The council had achieved a great deal between 1897 and 1906. In defeating Yerkes' bid for a fifty-year franchise, it had taken an important first step toward gaining the power to regulate mass transit. It had thereafter made significant progress (of which more will be said later) toward creating a transportation policy. But the council had never managed to form a lasting consensus for either regulated private ownership or municipalization.

The absence of effective leadership was not the only reason for this impasse. Aldermen had to seek not only a policy which would give the city planning power over transit in the long run, but one which would produce immediate service improvements. They were subject to public pressure which manifested itself in aldermanic elections as well as in referenda. This pressure for a quick solution to the "traction problem," when combined with the business community's objections in principle to public ownership and the transit companies' resistance to regulation, pushed the city back and forth between the public and private alternatives.

Between 1900 and 1905, moderates favoring private ownership had dominated policymaking. They had commissioned two important studies, one on the legal and policy aspects of local transit, the other on the problem's engineering dimensions. Both studies helped structure policy and thinking in the years thereafter. During this time the council also created a permanent Committe on Local Transportation, which soon became the closest thing to a transportation planning and policymaking body Chicago would develop during our period. Finally, council moderates oversaw the drafting, promotion, and passage of a state law which set the terms for any future transit "settlement," public or private. Yet by November 1906, this approach had produced no final solution. In consequence, a new mayor, Edward F. Dunne, sat in city hall.

Dunne and his council allies were firmly committed to immediate municipal ownership of mass transit. They used many arguments, but for Dunne and his chief spokesman, Alderman William Dever, public ownership clearly represented more than just the quickest solution to the transit problem. Dever, whose whole political career was bound up with municipalization, made this expecially clear.

MVL or no MVL, an alderman stayed in office in 1906 by showing his constituents that they needed him. William Dever was in the council chambers on the night of November 26 to vote for the new crowding ordinance, but the next afternoon he would be with the parents of the accident victim—asking questions of Frank Murphey and the Metropolitan Elevated representative and going through the motions of another useless inquest. Dever represented the seventeenth ward, where the accident had occurred. It was the nature of his ward which, combined with Dever's unusual background, helped make him city council's most articulate advocate of municipal ownership.[40]

In the rhetoric of Dever and others like him, urban transportation was more than a mere utility, a facilitator of commerce and developer of real estate; it was a remarkable tool for social change as well. To reach its full potential, however, mass transit would have to be managed by people responsible, not to stockholders, but to the voter. That was why Dever would tell a friendly audience that "there is something more involved here than the right of a citizen to get on a streetcar and ride downtown for 5¢ . . ." or that "the real traction issue is the more equal distribution of the great wealth of this country; . . . the great movement for a higher standard of living . . . for higher morals. . . ."[41]

The idea that mass transit could be used to change the quality of people's lives was not new with municipal ownership advocates nor was it unique to them. Around the nation reformers had log argued that better transit meant better homes for workers. Yerkes himself had been fond of boasting that his car lines improved the living standards of the working class.[42]

A group of Chicago businessmen reporting on "tenement house" conditions in 1884 analyzed the links connecting congestion, high rents, poor living conditions. Their report explained "the extravagantly high proportion of rent to their wages paid by workmen" as well as the reason for "the almost entire impossibility of their being able to relieve their condition by their own efforts,

. . . why they show little if any desire for social improvement, . . . why 4,000 saloons flourish [which is that] their patrons have no better place for meeting and recreation."[43]

High rents were a result of congestion. So was the lack of recreation space. Chicago's densest tenement districts were more congested than London's most notorious slums. In the area around Division and Noble, just north of Dever's ward along the Chicago River, over six hundred persons per acre crowded into three- and four-story tenements and flat buildings.[44]

To the minds of many among the late nineteenth-century middle class, these immigrants and native-born Americans were prisoners, not voluntary citizens of ethnic communities. Adna Weber, the first great student of the American city, placed responsibility for freeing them squarely upon street railways. So did Delos F. Wilcox, a nationally recognized expert on the politics of urban transportation. New York and Boston staked public capital for the building of subways when private capital shied away. There was more than one reason for this commitment, but the chairman of New York's Public Service Commision stressed his belief that without good mass transit "the moral and physical well being of a large proportion of the population is menanced."[45]

The theory behind this widespread belief was simple enough. Good mass transit would develop outlying areas and thus allow the more affluent citizens to move out. Their move would cut congestion—and rents—in the inner city. Alternatively, good crosstown service would enable workers to take the best jobs available and to live in the best neighborhoods they could afford without undue time lost in the journey to work. And a single fare throughout the city would make all districts equally accessible to anyone with a nickel. "The zone fare builds tenements," a Scottish transit expert told the Western Society of Engineers. But a single low fare and good service all over the city might even out population, open new jobs to the poorest workers, and break down ethnic barriers.[46]

Perhaps these hopes now seem naive and exaggerated. By 1911, Delos Wilcox had come to recognize that what he called "slum life" had attractions of its own. It would take very good transportation indeed, he felt, to break down the "tenement habit." The Tenement House Commision of 1884 knew that the central (and ultimately unsolved) urban problem was finding a way for private enterprise to build livable housing for the very poor.[47] Yet there was much truth to the argument Dever and his settlement house allies asserted with such a passion, and the argument was believed, to some degree, by everyone concerned with mass transit.

"It takes me 20¢ and a long time to get to work," one west sider told the *Chicago Daily News* in 1893. A decade later, the trip from the "slums" around Polk and Halsted Streets to the lakefront—a two-mile journey—required two fares, three car changes, and, often enough, the better part of an hour.[48] At the turn of the century, 80 percent of the city's streetcar passengers traveled to or from the Central Business District. There was no way to reach the north or west division of Chicago from the south side's Packingtown or Back-of-the-Yards neighborhoods without either going downtown and paying another fare to go west or north, or riding the Halsted car (one mile west of the Loop) to the Chicago River, walking across a bridge, and handing another nickel to a new conductor. A west side night worker would have to wait eighty minutes for a car on major radial streets like Milwaukee or Blue Island Avenue.[49] And on the west side grew the city's most outrageous slums, with their crowded, windowless rooms and subsidewalk privies.

Special handicaps like these for people whose living and working patterns did not match those of daytime CBD commuters particularly concerned reformers. Jane Addams, who realized that more things than transportation tied immigrants to their homes, favored municipal ownership and unified operation of the streetcare companies "so that workers can change their jobs without moving." Municipal ownership was widely thought to mean lower fares as well; the Chicago Federation of Labor often supported public ownership on this ground.[50]

Five cents in 1906 was the price of a loaf of bread. In that year, a seamstress would work almost an hour to earn her daily carfare. Many packinghouse workers labored over half an hour for the same amount in 1903.[51] Lower fares were probably not as essential as reformers in Detroit and Cleveland believed them to be, but for the very poor, the young, the aged, and the unemployed, they could mean the difference between riding and walking.

•　•　•

Transit shaped and reflected people's lives in other ways as well. By the end of the nineteenth century, Chicago's transit companies had developed an unplanned system of class segregation. In 1886 a south side woman who signed herself "Frozen Feet" complained in the *Chicago Tribune* that her numb toes were a side effect of social class, since "the Rich have their [Illinois Central] steam cars to ride on." Commuter railroads like the Illinois Central served the upper middle class of every section. By 1906, electric transit expert Bion J. Arnold—faced with the choice for his own journey downtown of an in-city commuter train or an elevated line, part of whose modern equipment he had invented—selected the steam railroad.[52]

The elevated lines themselves had their own class connotations. They had been built specifically to tap developing sections that were often really suburbs within the city limits, and elevated lines with the trackage to do so concentrated their efforts on express trains between major outlying centers like Uptown or Englewood and the Loop. Without such express service, the "nice class of people" (as a company spokesman described them) would have to stand and then might complain to the press and the city council.[53] Of course more express trains meant fewer locals. Laborers from factories in the industrial belt around the city's heart could not squeeze aboard the trains which did stop near where they worked. By 1906 it appears that factory workers did ride mass transit when they could, but the el was difficult—even dangerous—to use in the inner city factory districts.[54]

The routes of the streetcar lines helped define the world of the working class: they limited its possibilities and reflected its ethnic conflicts. When the State Street cable car made its last run on the night of July 21, 1906, a *Chicago Tribune* reporter rode along. As the car moved southward toward 12th Street, he observed how "negroes, always more negroes" got on. The reporter apparently didn't thinking of speaking to the passengers, but the white conductor, who had worked the line since 1882, told him how the "Black Belt" had moved southward since that time—from 25th to 63rd, always following the cable line.[55]

Neighborhoods of all sorts communicated chiefly with other areas directly accessible by streetcar. "We don't want to be carried down Wentworth," Dever told the city engineer during one discussion of through-routing to the south side for the northwest side Milwaukee Avenue car. "The character of this neighborhood and the Northwest side is such . . . that the people living here will very seldom want to go up there and vice versa. Put it on State Street [three blocks east of Wentworth], and then we can use it."[56]

To municipal ownership advocates, the power of mass transit was limited chiefly by one thing—the profit motive. Elevated lines were indisputably the best openers of new territory. On the north side of Chicago they developed great new districts of middle-class housing.[57] But the very spread of population made elevated lines less profitable. Yerkes had known this. By 1904, the South Side Elevated was struggling to pay a 4 percent return on its relatively lean capital account. The problem was that the more the elevated lines succeeded in dispersing population, the less well they paid. The els carried most of their passengers more than four miles for a nickel.[58] Zone fares would have helped the companies, but they would have negated the social benefits of transit according to contemporary thinking. And there would have been resistance: in 1911 workers bound for Maywood, a western suburb, simply folded their arms and refused to pay or get off when conductors on the Lake Street Elevated tried to collect an extra charge on trains that crossed the city border.[59]

After the inner city, it was the suburbs which were most severely affected by the need of the transit companies to make a profit. Thin population made the trips between towns expensive for companies and rides alike. While Chicagoans could travel from the Loop to 63rd Street (eight miles) for 5¢ residents of suburban Cicero paid 8¢ to reach neighboring Oak Park.[60]

But the transportation companies resisted through-routing and universally valid transfers even within the central city. "It is against every business principle that a man should throw open his business to be operated by another company," E. R. Bliss told the city council's Committee on Local Transportation in explaining why the Chicago City Railway opposed through-routes. Transfers in the downtown district, he argued, were simply out of the question. "It is an attempt to run this company's business by those who have no pecuniary interest in it," Bliss declared. "These people [who] have got their money here want to make some money, and if they can't . . . they want to get out of the business and you take it and you run it."[61]

Dever himself could not have stated the argument for public ownership more clearly. Transit planning and the needs of profit-making corporations seemed incompatible. A wide variety of people backed public ownership, for varied reasons. Some notorious grafters supported it, perhaps with a view to a wider field of patronage. But ex-Governor John Peter Altgeld, Clarence Darrow, and H. D. Lloyd also sought municipalization. Labor oganizers, settlement workers, and socialists found their way into the movement, each with their own set of broader goals.[62] But public ownership drew general support on two main grounds: it seemed to offer both immediate service improvement and an opportunity for the city to plan transit development, and so, its own growth.

Yet municipalizationists had no monopoly on either of these promises. Those who favored regulation simply sought to gain indirect city control of mass transit and thereby to facilitate both service improvement and planning. Hindsight reveals that the policy of regulation could achieve neither. American urban history provides no clear examples of what municipal ownership could or could not have done. For the moment, we can only seek to understand why Chicago chose to deal with mass transportation indirectly, as a regulated, privately owned utility.

• • •

Three factors combined to divorce the idea of public ownership from the dreams of planners and straphangers alike, and to assure that the city would choose regulated private ownership as the basis for its mass transit policy. First, the pragmatic approach followed by the city council between 1900 and 1907 created legal, institutional, and what may be called conceptual circumstances which favored private ownership. Second, irrelevant issues—political personalities and the fear of "municipal socialism"—became attached to the concept of public ownership and helped to discredit it. Third, and most important, voters consistently sought the realization of the Straphanger's Dream—speedy, safe, convenient service. Neither public nor private ownership attracted a consistent following, since voters regularly endorsed whichever solution seemed most likely to bring immediate service improvements.

The first factor, the development of a legal, institutional, and conceptual tradition favoring private ownership, is extremely complex. Only the highlights can be touched upon here. They consist of the policy and engineering studies, state legislation, and bureaucratic developments already alluded to.

The first landmark in council policymaking was the report of the Street Railway Commission, created by Mayor Harrison at council request in 1900. Headed by moderate Republican alderman Milton Foremen, the Commission studied transit policy developments around the United States, took testimony from advocates of public and private ownership, and produced a report which drew national attention and set forth an important preliminary definition of the city's policy goals.[63]

The Commission began its report with the significant complaint that Chicago's street railways had for the most part "been permitted to develop . . . in a haphazard manner, without regard to any well defined and comprehensive plan or policy." The Commission traced most of the city's transportation ills to this defect.[64] Accordingly, it called for coordination of streetcar, elevated, and steam railroad service; unified management of all surface lines; investigation of the possibility of unified service throughout the city at a single fare; and, most significantly, public ownership of street railway tracks "and of whatever may form a part of the public street" at the earliest possible time.[65]

Thus the Commission took an advanced position on the city's proper role in local transportation—one which, had it been carried to fruition, would have given Chicago a degree of planning power for public transit not developed anywhere else. Despite this clear linkage of transit planning and public ownership, however, the Commission's general approach was pragmatic.

The Commission recommended that the city proceed at once to seek from the state the power to own and operate street railways. It made clear, however, that this power was to be sought ". . . because the having of such power in reserve, to be used in case of need, would put the city in a much more advantageous position in its dealings with private corporations." Indeed the Commission expressed opposition to the idea of municipal *operation* of street railways and declared that the whole question was ". . . one, not of principle, but of policy or expediency."[66]

The Commission of 1900, then, called for public ownership of streetcar tracks, but prepared the city to fight for the power of regulation. It suggested the creation of a permanent city council committee with expert staff to see to the enforcement of regulations, and it discussed at length what form of franchise should be granted private companies.[67]

This seeming confusion reflected the fluid state of transit policy. The Commission borrowed ideas from eastern cities; for example, it favored indefinite-term franchises (a policy which would not find public favor in Chicago for over a quarter-century). Yet the Commission tried to adapt "outside" ideas to Chicago prejudices, especially to the tradition of mistrust of both city and companies. It suggested only minor steps toward the new type of franchise and insisted that any city securities issued for the purchase of transit facilities be separated from the rest of the city's debt.[68]

The one clear position which emerges from the report is the determination to unify mass transit and to gain regulatory control over its development and operation. In the years after 1900 the behavior of the council, and particularly of Alderman Milton Foreman (chairman of the 1900 Commission), made it evident that these were the only goals on which a meaningful consensus had been reached.

The city council's most important achievement between 1900 and 1905 was the commissioning of an engineering study. The report of 1900 had called for "an appropriation . . . for investigation and preparation of plans" for Chicago's first subway.[69] What the city council commissioned at the beginning of 1902 was a report, not only on subways, but on the "engineering and operating features of the Chicago transportation problem." For the then substantial sum of $10,000, the city retained Bion Joseph Arnold, an electrical engineer, inventor, transportation expert, and businessman, to look at Chicago's local transit problem with the limitations of politics and economics set aside.[70]

Arnold's work made several important contributions to the city's transportation policy. It brought the engineer into the process of policy formation and, for a time, placed him in the camp of the reformers. Just as important, Arnold stressed the power of transit to change the city for the better, and purported to show that this could be done under private ownership and at a reasonable profit.

Arnold began his report in his letter of transmittal by placing himself outside the arena of political controversy. His plans, he declared, ". . . would be the same whether the system be owned and operated by a private corporation or . . . by the city."[71] After this disclaimer, however, he proceeded to tell the city that it could expect more of its private transportation companies than they had hitherto been willing to give.

Chicago, Arnold declared in his report, was really five cities. There were the three great divisions marked out by the Chicago River, and two outlying districts—the northwest and far south sides. The latter two sections were suburbs by any but a political definition. By 1902, the former Yerkes companies had consolidated into two groups, with one—Chicago Union Traction—serving the inner north and west divisions, while another served outlying areas. On the south side, the Chicago City Railway dominated, while three small companies existed in the territory south of 63rd Street. Routes, fares, and transfer rules reflected this division.

Like the 1900 Street Railway Commission, Arnold found this unnatural division to be at the heart of the city's transportation problems. "Chicago," wrote the engineer, "is one city. . . . The citizens have the right to expect and demand that they be transported in . . . the whole district in one general direction for one fare. . . ."[72] He showed in detail how this could be done—by means of through-routing and free transfers outside the downtown area—

without unduly cutting the companies' profits. The 1900 Commission asked whether a "one city–one fare" policy was possible; Arnold showed that it could be implemented at a profit.[73]

Still, Arnold expected that most riders would want to go to the central district. In the CBD, he saw no way of making good mass transit compatible with private vehicular traffic. His ultimate solution was a streetcar subway combining the relative flexibility of the trolley with the exclusive right-of-way of rapid transit. For suburbanites, Arnold proposed a subway that would link the railroad stations so that the commuter's fast journey to the Loop would not end in a traffic jam. And for all riders, there was Arnold's insistence that transit vehicles have the right-of-way on the streets.[74]

The Arnold report had several unintended results. The document at once made a national reputation for itself; locally, it became a sort of Bible for everyone interested in the transit problem. H. D. Lloyd made it the basis of his "Peoples' plan," Alderman Milton Foreman recurred to it, and the street-car companies interpreted its recommendations as statements of what the city would expect of them.[75]

This was one of the drawbacks of the Arnold report. For the next ten years, the 1902 recommendations seemed to set the limits of what the city could bargain for in its negotiations with the companies. Alderman Foreman pressed for transfers within the Loop, but Arnold had opposed them, so the companies successfully resisted Foreman's pressure.[76] Still it was in the raising hopes—not in setting limits—that Arnold was most effective.

Arnold, then, contributed several elements to Chicago's transit policy debate. He purported to show that the best of mass transit could be had under regulated private ownership. Good service was, indeed, compatible with the reasonable profits. And he affirmed that the problems of congestion could be dealt with through engineering and a limited degree of traffic regulation, as opposed to planned dispersion (Arnold maintained throughout his career that cities were naturally and inevitably centralized on a single nucleus).[77]

Finally, the report itself and the national attention it attracted placed the engineer in the spotlight. Almost magically, it must have seemed, Arnold had shown that the conflict between the city and the companies was more apparent than real. The neutral, rational expert had shown the city how to overcome the problems of chaotic traffic, greedy traction barons, and feuding politicians.

This impression lasted long after the report itself was published. On the day of the accident with which this chapter opened, November 26, 1906, the *Chicago Daily News* carried an editorial declaring that the "impartiality of engineers" provided the sole key to the solution of the traction dilemma. Engineers, not politicians, said the editorialist, should be given the power to plan and regulate transportation service. Good engineers, he went on, were worth "whatever they cost." This view was echoed by a variety of sources during the early twentieth century.[78]

Engineers—specifically Arnold—helped move the city toward the 1907 compromise by developing a specific physical plan, by demonstrating the ostensible compatibility of a rational, comprehensive transit system with private ownership, and by serving as neutral sources of information about what could or could not be done. They made regulation more plausible because they themselves might become the neutral, expert regulators.

The 1900 Street Railway Commission set a pragmatic tone for city transportation policy. Arnold's report seemed to show that regulation could work. Two more council actions tilted the scales in favor of regulation: the promotion and passage of a state law governing local transit policy and the creation of a standing council Committee on Local Transportation.

The state local transit policy act—named the Meuller Law after its legislative sponsor—gravely circumscribed the city's options. It limited franchise grants to private companies to twenty years and set debilitating conditions upon municipal ownership. The law came into existence in a way which showed the constraints on the city's own policymaking process.

Briefly, despite the 1900 Commission's call for a law allowing the city to buy street railways, little effort was made toward gaining such power until 1902. At the aldermanic elections in spring 1902, however, voters over-whelmingly endorsed a nonbinding advisory proposition for public own-ership, even though the referendum had the backing of no major political faction. Only after this did an active movement emerge among powerful interests within the city for a municipal ownership law.[79]

MVL secretary Walter Fisher, who helped the council draft the law, explained why. The Mueller Law, he told the Creditmen's Association, was not the product of "theorists or long haired reformers." It was, instead, a tool to be used in bargaining with the companies and an attempt to control public ownership sentiment, which Fisher rightly believed to have arisen from frustration with the city's inability to get better service from the companies. "If you delay [in using the Mueller Law to get better terms from the com-panies]," he told them, "this sentiment, which is now led by the conservative and practical men in this city, will break out under the leadership of demagogues."[80]

The Mueller Law passed the state legislature with the support of moder-ate reformers and some business elements only after a long battle which culminated when pro-Mueller legislators stormed the speaker's platform in the Assembly, armed with planks, cuspidors, and inkstands. Despite the drama, however, the new law was an essentially conservative piece of legislation.[81]

The law permitted the city to buy street railways if a majority of voters approved at a special referendum. To operate a streetcar line, however, the city needed a two-thirds majority on a separate proposition. City-owned street railways were to be purchased with special certificates, not part of the city's debt, which would be a lien only against the street railways themselves.[82]

In some ways the Mueller program was similar to the plans under which

East Coast cities contributed to the financing of subways. It might give the city some planning power, but (since the certificate holders would almost certainly be the holders of the old traction company securities) it would at best make the city, in effect, the managing company for the old private investors or their successors. They would be guaranteed a good return on the certificates and, if they did not get it, could recapture the property. All this resulted from the law's provision separating debt for mass transit improvements from the rest of the city's debt.[83]

The Mueller Law made public ownership a difficult proposition; the creationof a council Committee on Local Transportation laid the groundwork for regulation. Between 1903 and 1905 this committee engaged in a persistent effort to get the transit companies to accept some form of regulation. The companies variously cut service, went into friendly receiverships, and made half-hearted agreements. The policy did not work. The Mueller Law did not intimidate the companies into accepting regulation, because no plausible threat of municipalization existed.[84] But the failure of the CLTs quest for regulatory power was, in the long run, secondary.

Throughout these three years of fruitless negotiations, the Committee built up a tradition. It helped draft the Mueller Law and produced franchise ordinances, ultimately rejected, for the South Side company. It was to the CLT that Bion Arnold, correctly, addressed his report. The Committee on Local Transportation, in short, became the repository of aldermanic expertise on the legal and financial aspects of the traction question. Between 1903 and 1905, it made transit policy.[85] In the following quarter-century, every major transit plan which went before the voters had originated in the CLT. It remained the most important political body for transit policymaking and, despite a shifting membership, it almost always followed a moderate, pragmatic stance.

All these council actions "stacked the deck" in favor of regulation as the means by which Chicago would pursue the dreams of planners and straphangers. The second great factor working against public ownership was the nature of the municipal ownership movement in Chicago. Between 1902 and 1905, the results of referenda on public ownership seemed to show increasing popular support. But this support probably reflected more than anything else the failure of the city to gain the power of regulation. As political scientist Arthur Bentley long ago observed, middle-class voters—overwhelmingly straphangers in this period—voted for public ownership when regulation seemed impossible, and for regulation when municipalization appeared unachievable. Working-class voters appeared to follow the lead of their local machines.[86]

But if the public's interest in municipal ownership was pragmatic, the idea of municipalization itself aroused nationwide debate on what may be called ideological grounds. In Chicago and elsewhere, the idea was labeled municipal socialism, and it was rejected as such.[87] Just as important, the long drive against political corruption—corruption often associated with street railway franchising—gave rise to the argument that politicians were too corrupt and too incompetent to be trusted with the mine of jobs and plunder that a major street railway would provide. Franklin McVeagh, a prominent merchant who later pressed for greater city spending in connection with the *Plan of Chicago*, put it most succinctly: "The city government hasn't enough executive ability at present to run an inkstand. . . ."[88]

.   .   .

When the city council met on the evening of November 26, 1906, to vent its rage over the second transit death in as many days, the mayor presiding was not the evasive Carter Harrison II but a man committed firmly to municipal ownership: Edward F. Dunne. Dunne could—and did—declaim against crowding with much authority. He had sued the street railways for overcrowding, had met with the president of one of them, and had assigned police to prevent dangerous overloading of elevated cars. But such regulatory approaches were, to Dunne, mere temporary expedients.

A stern, uncompromising jurist, Dunne had campaigned for and won the mayoralty in 1905 on a platform promising "municipal ownership before the snow flies." He said, and apparently believed, that public ownership was both the answer to the Straphanger's Dream and a proper extension of the role of urban government. Ironically, Dunne was elected to preside over the process by which previous history and the public's demand for quick service improvements combined to make public ownership of transit, in the sense Dunne had imagined it, impossible.[89]

Dunne's background had something to do with this. Born in New England, he had been educated in Ireland—in the same class at Trinity College, Dublin, with Oscar Wilde. He was converted to municipal ownership, not by theories, but by seeing how well European street railways worked. Dunne also knew Tom Johnson, the burly Cleveland mayor who had once been a "traction baron" himself, but who now worked sincerely and effectively for municipal ownership and 3¢ fares.[90]

But Dunne also associated with people on the fringe of "respectable progressivism," people the *Tribune* called "the whole crew of female politicians and long-haired freaks." He listened to Margaret Haley, organizer of the Chicago Teachers Union, who seemed to believe quite literally that nothing good could come from a corporation, and to Louis Post, editor of *The Public* and constant enemy of corporate power.[91] From people like these, from his reading and experience, Dunne formed an idea of what municipal ownership could do.

As he had while still a judge, the Democratic candidate promised Chicagoans that they were in the vanguard of a world movement. Municipal ownership worked, he told them. Private ownership of utilities was "dying out" in Europe, and publicly owned water and gas plants in American cities operated more efficiently and charged lower rates than had their private predecessors.[92] Once the profit motive was removed, fares dropped, the product improved, and wages rose. Dunne was particularly fond of examples. Ignoring the differences between compact old European cities with zone fares and the sprawling young prairie metropolis which aimed at a single-fare ride from one end to another, Dunne gave the people statistics which were hard to resist. In Liverpool, he told them, fares had been cut in half yet receipts were up 86 percent (the lines, as it happened, were electrified at the same time they were municipalized).[93] In short, instead of dealing with public ownership as a necessary addition to the responsibilities of city government, he described it as something of a free ride.

Despite—or because of—his exaggerations, a solid majority of Chicago voters backed Dunne's position in 1905. At the same election, 152,135 Chicago voters cast referendum ballots against giving any franchise to any private company.[94] But votes alone could not translate Dunne's promises into fact.

To be successful, Dunne had to get a program through quickly. Republicans, Democrats loyal to Harrison and other old party leaders, the majority of the city's newspapers, and the companies themselves worked hard during Dunne's administration to convince Chicagoans that they could not wait any longer. But Dunne, who had told the voters that his solution was the quickest, found himself blocked at every turn by events which had nothing to do with mass transit or public ownership.

Dunne's ideas and personality made him highly vulnerable on issues which might have been mere annoyances to another mayor. His labor ties caused Dunne no end of trouble. At the beginning of his administration, the city's teamsters walked out in sympathy with striking garment workers. The conservative press blamed Dunne for the strike, while many unionists blamed him for its failure. Then the city was hit with a crime wave, real or imagined. The usual sorts of crime increased, and a new one emerged; "kodak fiends" began roaming the fashionable streets and "snapping" embarrassed young ladies. All this the press blamed on Dunne: how could personal privacy or property be safe in a city whose mayor attacked the sanctity of the utility companies' property rights?[95]

Saloons were also blamed for crime, and the crime wave brought a drive to cut their numbers by raising the license fee. Dever joined the movement; so did Foreman and other "honest alderman." But Dunne tried to remain aloof. To large numbers of working-class people, especially to some recent immigrants, the saloon was the family and neighborhood gathering place. The head of a party of immigrants could scarcely afford to attack it. Yet Dunne ultimately signed a "high-license" ordinance—too late to avoid being branded a friend of saloons by the people who thought that liquor made men bad, but soon enough to alienate the owners and patrons of bars. "He has closed the saloons and dance halls and everything," one angry worker cried out at a labor meeting. "He has sacrificed everything for municipal ownership."[96]

Despite the saloon tax, Dunne gained few friends on the "right." His appointments to the school board (Margaret Haley, Louis Post, and Rayomond Robbins, a founder of lodging houses for the poor and husband of labor organizer Margaret Dryer) frightened conservative parents. One active member of the conservative clean government movement told the City Club that Dunne was driving the opening wedge for socialism. "As the spirit of 1861 fought negro slavery," Eugene Prussing told the assembly, "the spirit of 1906 must fight industrial slavery by majority rule . . . which reduces all to mediocrity."[97]

Reformers and professional politicians alike found Dunne difficult. He worked easily with Hinky Dink Kenna, and went to the little grafter's ward meetings, as Harrison had never dared to do. But one of his department heads drove Dunne's more politically savvy allies to near distraction when he tried to implement merit appointment at the expense of the ward bosses' loyalty.[98]

Dunne's personal character proved an easy target for his enemies. Some charged that he was personally dishonest: at the beginning of his administration, he had asked the head of Glasgow's public tramways for an opinion on the Chicago situation and, when the Scot responded by advising against municipal ownership, Dunne suppressed the report.[99] Or perhaps he was daft. Dunne could look like a fanatic. The *Tribune*'s McCutcheon delighted in caricaturing the mayor's dome-shaped skull and oddly combed hair. Imaginative reporters enjoyed portraying municipal ownership advocates as amusing cranks. One, they told their readers, had invented an electric roller skate and proposed to end private ownership by making "Every Man His Own Trolley Car." Another fictitious Dunne follower discovered that bivalves could generate electricity and came out for "Immediate Municipal Oyster." Even the Mayor of the Midgets at White City Amusement Park got into the act by demanding graded fares: 2¢ for people under 3'2'' up to 25¢ for 6 footers.[100]

Dunne was not insane, corrupt, or even a socialist. He was, however, seriously overconfident of the public's commitment to public ownership and hopelessly intolerant of opposition to his most cherished goal. As a result he was unable to work effectively with the Committee on Local Transportation. The debate over transit policy became increasingly bitter, and more irrelevancies entered into the inherently difficult process of transit policy formation.

As soon as the crisis created by the teamsters' strike of 1905 had passed, Dunne sent two municipal ownership plans—one of them prepared by his traction counsel, Clarence Darrow—to the city council. There, the Committee on Local Transportation—which had been struggling for half a decade with the traction problem—treated his proposals with thinly disguised contempt. Dunne would not relent. He submitted the plans again and again and charged the city council with defying the people's will as repeatedly expressed in referenda. Alderman Dever pressed the issue in the council's Committee on Local Transportation in tense meetings which, on at least one occasion, degenerated into a session of angry name-calling. Dunne was adamant. "The traction companies," he told a Denver audience, "have succeeded in taking over 2/3 of the city council."[101]

Mayor Dunne left no room for honest disagreement, and Republicans and Harrison Democrats alike took umbrage at his slurs on their integrity. Staking all his hopes on another referendum vote, the mayor accepted the support of some of the city council's most notorious corruptionists in January 1906 in order to obtain authorization of a public vote on a plan to buy the street railways with Mueller certificates.[102] This only further alienated Walter Fisher of the MVL as well as others like him.

When the complex municipal ownership proposition went before the voters on April 4, 1906, the issues were more confused than ever. Harrison Democrats and some other Democrats who—unlike Harrison and Dunne— had ties to utility interests opposed the measure. The saloon issue and other irrelevancies entered the debate. The proposition for municipal ownership and operation failed to win the required two-thirds of the votes cast; the number voting for the plan barely exceeded the number opposing it.[103]

After April 1906, therefore, neither public nor private ownership could claim a mandate. City council actions since 1900, the irrelevant issues attached to public ownership, and the style of Mayor Dunne and his followers all "stacked the deck" in favor of private ownership. So did public impatience, now that immediate municipal ownership seemed out of the question. Negotiations with the companies began again, but in the autumn of 1906 no solution was in sight. Frustration with the impasse between city and companies was evident in the reaction of the press and the political community to events like the accident of November 26.

●　　●　　●

The victim's body lay unidentified in the morgue all day. Hundreds of parents worried, but Frank DeTomaso apparently did not. His first information on the accident came from the evening newspaper, which he read while hanging from a strap on the homebound trolley.

DeTomaso was on his way up to his flat before the city council downtown had begun its indignation meeting. A neighbor girl caught him in the hall. His daughter Maggie had not come to work. The father must have put the possibilities together in an instant. He did what half the Chicago fathers in his generation might have done: he headed straight for his church, St. Columbkille's. The priest did what any turn-of-the-century priest would have done: he got on the trolley with Frank DeTomaso and rode across to the morgue.

Down at city hall the politicians were getting under way, "stirred by the sacrifice of another life," as the *Record-Herald* reported. They asked the predictable questions. Why did straphangers have to pay the dividends—sometimes at the cost of their lives? Did the girl die for profits, or for politics? Why hadn't the council acted before? Why hadn't the mayor acted? Alderman Werno summed up the spirit of the evening. "Why don't *he* do something?" Werno asked, indicating someone else.

At the morgue, Frank DeTomaso didn't say anything; he just fainted across the body. The little group of reporters got what they needed from the priest. "Magdaline DeTomaso, garment factory operative, aged 18"—they wrote it all down. (The man from the *News*, a born Yankee for sure, wrote "Madeline.") Right or wrong, the city had a name. They could read it and forget it. [104]

Section II: The Ordinances of 1907

> Straps will continue to exist as long as we insist on the impossible.
> *Milton Forman, 1906* [105]

Between May and December 1906, Walter Fisher, Dunne's new "traction councel," managed a series of negotiations between the council Committee on Local Transportation and street railway officials which led to the settlement of 1907. Fisher, a man of considerable talents who went on to become U.S. secretary of the interior under President Taft, slowly brought about a compromise which the companies and the council moderates could accept. [106]

These were the remarkable ordinances of 1907—one for each of the major street railway companies. They were remarkable because they appeared to give all major parties to the long-standing "traction wars" a large portion of what each had wanted, and to do so without departing in any important way from the reform tradition which held that neither corporations nor politicians could be trusted. To achieve this unlikely feat, the ordinances simply put off the resolution of some important issues and "solved" others by attempting to create a balance of power in transit planning and operation which pitted companies, politicians, and engineers against one another. The settlement, which seemed to fulfill the dreams of planners and straphangers alike to the benefit of all, in reality assured that only the automobile could effectively be used for the realization of either dream.

The ordinances passed the city council after protracted debate on

February 5, 1907. But they passed only because supporters had reluctantly agreed to make their implementation subject to a referendum vote.[107] The result was an all-out campaign for their ratification on the part of the business community and some political forces. This campaign was as much an innovation as were the ordinances themselves, and we can best begin consideration of these complex documents by looking at the drive for their public ratification.

• • •

Some time in mid-March 1907, every Chicago voter received a slick cardboard flier from the Citizens' Non-Partisan Traction Settlement Association. Its message was simple. A referendum would be held at the upcoming mayoral election. If the voter marked his ballot Yes in that referendum, he would be guaranteed several things. First, he would have the best transit service in the world, provided by no less than 2,000 of the most modern streetcars anywhere. Second, he would be guaranteed a seat on any car he rode and would pay no more than 5¢, no matter where or how far he traveled. Third, there was to be municipal ownership any time the city wanted it. Finally, the streetcar companies would pay ample "compensation" to the city, thus (implicitly) keeping property taxes low. In case the voter lacked the time or ability to read, the flier carried a large color picture on its cover which made the message unmistakably clear: the picture showed a hand inserting a ballot marked Yes into a ballotbox and a huge, new, shiny, red streetcar leaping forth from the box in response.[108] The "traction question," the flier seemed to say, had finally been resolved. The fulfillment of the Straphanger's Dream was just a ballot away.

The organization which mailed this flier was itself something of an innovation. The CNPTSA was a creature of the Chicago Real Estate Board and banker D. R. Forgan of the Chicago Commercial Association—the same business group which had, a year earlier, commissioned the Chicago Plan. Formed in February 1907 after a flurry of statements from the Real Estate Board and the Commercial Association on the essentiality of the new streetcar ordinances of the city's commercial growth and financial prosperity, the CNPTSA constituted the first in what would be a series of attempts by downtown business interests to organize citywide support for specific plans dealing with mass transit and, later, the accommodation of the automobile.[109]

The CNPTSA was not alone in it support of this new solution to the dilemma which had haunted Chicago since the 1890s. Lined up behind the ordinances were the Republican Party and its mayoral candidate, the rotund ex-saloonkeeper Fred A. Busse; many Democratic followers of Carter Henry Harrison II; and all the city's newspapers save the two controlled by that lover of controversy, William Randolph Hearst.[110] Among those signing CNPTSA literature was a young reformer from the University of Chicago's Hyde Park neighborhood, Harold Ickes. Amos Alonzo Stagg signed the literature, too, as did dozens of Chicagoans whose names proved less nearly immortal.[111]

Yet there was dissent. The Chicago Federation of Labor made a sarcastic appropriation for "10,000 feet of rope" with which to deal with pro-ordinance aldermen, and signs reading "*Tribune* steals candy from babies" and "Fisher is Armour's goat" appeared at union rallies condemning the settlement and its backers. (The unions were not unanimous: the head of the carpenters' union lashed out at Margaret Haley and "feminine domination in the central labor body.") Mayor Dunne, running for reelection, opposed the 1907 ordinances, as did Dever and some other progressive aldermen. So did George Hooker and a few others from Hull House.[112] Their task, however, was difficult in the extreme.

The impossibility of campaigning against the ordinances must have come home to Dunne one cold March night in Dever's seventeenth ward. After trying for days in halls all across the city to explain that the ordinances did not really guarantee good service, that they prevented public ownership instead of encouraging it, the mayor's voice finally gave out. Earlier that day, a large assembly of west side businessmen had applauded him warmly when he spoke of his own campaign, but had fallen silent when he turned to the traction issue. When Dunne gave the platform up to his corporation counsel and the attorney began to speak, an Italian band carrying a Dever banner came tooting and stomping up to the rostrum and nearly drove him to despair. During the remainder of the campaign, Dunne rarely spoke of traction.[113] The truth was that, even for a voter with the time and inclination to study them, the 1907 ordinances were complex and ambiguous. After fourteen years of incessant agitation over "traction," it could hardly have been otherwise.

The principle of mistrust governed the ordinances. At the heart of the settlement was the decision to give the companies, not a franchise, but a "license to operate" for twenty years, theoretically terminable by the city at any time. This approach, originated by Charles Francis Adams' Massachusetts Street Railway Commission of 1897, appeared to combine safeguards for every interest. Bad service could lead, in theory, to termination at any time—through either municipal purchase or a city order to sell to a more cooperative company. Yet the streetcar companies were protected against displacement for frivolous or political causes by provisions guaranteeing them a 20 percent bonus for any forced sale to a private company, and by the referendum requirement of the Mueller Law. Thus, these ordinances, which supporters promoted as guaranteeing public ownership whenever the city wished it, were interpreted by company attorneys as being, at a minimum, twenty-year franchises.[114]

The principle of mistrust was further evidenced in the ordinances' provisions for service improvements and regulation. A thorough rehabilitation program for the first five years was laid out in detail. The companies

committed themselves to an expenditure of over $40 million for precisely
specified improvements in the first three years under the ordinances and to
the building of at least six miles of extensions per year thereafter. They further
pledged to contribute $5 million to the construction of any city-owned street-
car subway which might be built for their use.[115] To ensure fulfillment of these
commitments and to provide for continuing regulation, the negotiating parties hit upon a scheme which attempted to assuage the perpetual suspicion
among city, companies, and the public.

Under the settlement, the city gained the power of regulation—something the companies had not recognized before. But all regulations drafted by
the city council were to be subject to review by a Board of Supervising
Engineers, consisting of one representative each for the city and companies,
and headed by a neutral of unquestioned impartiality—Bion J. Arnold. This
board, to which Fisher and the companies had wished to give control of
extension requests and the actual making of regulations, came to be known as
the BOSE. It was to pass on the reasonableness of city council demands and
determine whether the companies were living up to the complex requirements of the ordinances' rehabilitation provisions.[116]

The creation of the BOSE, therefore, placed service regulation in limbo:
no single agency was fully responsible for it. The companies could appeal the
BOSE's decisions to the courts, and the city was free to ignore them. In effect,
the BOSE was to be a referee, though one without the power to make final
decisions. It was presented as a neutral force, but this neutrality hinged on the
impartiality of the chairman, Arnold, who was both a city consultant and a
traction businessman. Nevertheless, propaganda for the ordinances insisted
that the BOSE, because of its neutrality and expertise, could assure good
service.[117]

Some of the most involved and ultimately self-defeating provisions of
the ordinances arose from the beliefs that mass transit could and should make
substantial profits and that money spent on new equipment was the surest
guarantee of good service. Inevitably these provisions centered about the
obscure but essential issue of valuation.

The question of what the traction companies were actually worth was an
extremely vexatious one, and the 1907 settlement did not resolve it. The
traction wars of the 1890s left behind a large body of literature purporting to
show that the face value of the companies' outstanding securities was far
higher than the amount actually invested in the properties. The longest and
most frequently cited of these documents estimated that actual investment in
the three major companies in 1898 was but $39.5 million, while the combined
capital accounts of these companies showed an amount in excess of $63
million.[118]

But were traction companies worth even the amount actually invested
in them? Equipment decayed and was replaced, yet the companies did not
retire securities representing it. In 1902, Bion Arnold valued the plant and

equipment of the surface companies at $27 million. By 1906, with $108 million of securities outstanding, the companies were estimated by Arnold to have a "reproduction value" of $46.6 million. Ultimately the ordinances set the value of the major surface lines at $50 million. Arnold endorsed this figure, but it clearly represented a political compromise rather than a precise engineering judgment. [119]

All of this was important because the ordinances attempted to link the companies' profits to their valuations and enticed them into making service improvements by permitting them to pad their valuations each time they made an important improvement or extension.

Each year, each company under the ordinances was to set aside from its gross receipts an amount equivalent to 5 percent of its capital account. This amount—in addition to taxes, operating expenses, and salaries—had to be subtracted before any compensation was paid to the city. If the 5 percent could not be paid in a given year, it was deducted from the gross in subsequent years. Furthermore, no extensions could be required which might "unreasonably" reduce profits the companies would make over and above this 5 percent. [120]

This arrangement, together with the fact that the companies were to give the city 55 percent of their net profits, seemed to some a major step toward making transit a genuine public utility: both its profits and its risks were limited. Further, it seemed a necessary step if the companies were to raise capital for improvements; traction securities, especially those of companies with twenty-year franchises like those offered in Chicago, were already falling into some disfavor. [121]

On the other hand, this arrangement had serious drawbacks which were evident to some contemporaries. It gave the companies a positive incentive to keep their capital accounts high, since the higher the "book value" of the company, the greater would be the amount of the guaranteed 5 percent return. This was all the more important since the ordinances allowed all new construction during the first three years as well as a substantial portion of the cost of later renewals and extensions to be added to the companies' capital accounts with a bonus of 10 percent "construction profit" and 5 percent "brokerage" on the new money involved. Thus the companies had an incentive to buy new equipment, but they also had good reason to keep old equipment on the books and to refrain from amortizing any of their securities. If their valuations could be raised high enough, they could not only increase the amount of their 5 percent guaranteed profit, but they could "cover" worthless old securities, greatly reduce the amount of compensation paid to the city, and (because the valuation price was also their agreed purchase price) make municipal ownership even less likely. [122]

In return for these concessions, the city received a degree of planning power, as well as the assumption by the companies of a number of what would otherwise have been city responsibilities.

The city achieved a major transit planning goal when the companies agreed to a degree of through-routing of cars across the city and the issuance of transfers valid anywhere outside the five-square-mile CBD. These concessions were granted by the companies only on the basis of Bion Arnold's argument that the concessions would not lower profits or substantially affect riding patterns—an argument that reflected the contemporary belief that transit would continue to operate in a naturally centralized city.[123] Enticed by capitalization concessions, the companies also agreed to provide night service and allowed the city council authority to mandate six miles of extensions per year—a power which they had originally wanted reserved for the BOSE. The companies also agreed to contribute to the cost of downtown streetcar subways, a long-time city goal which reflected the belief both in centralized transit and the accommodative approach to street traffic problems.[124]

Other company concessions in the 1907 ordinances took their justification not from planning but from nineteenth-century ideas about the grievances against transit companies. The companies agreed to pave the center sections of the streets through which their tracks ran, to sweep and wash the streets in summer, and to plow the snow from them in winter. These were traditional obligations of traction companies and rested on the assumption that transit, as a business, should somehow return part of its profits to the city whose streets it used for gain. In fact, as will be seen, they merely added to the cost of service and encouraged private vehicles to drive on the streetcar tracks.[125]

A final important concession was the agreement to pay the city 55 percent of all profits above the 5 percent of each company's valuation deducted from the gross. Under the ordinances, this direct "compensation"—a long-time goal of reformers—could be used for any transportation purpose: municipal purchase, the building of subways, or, through rebates to the companies, fare cuts. Fifty million dollars eventually accumulated in this fund, but it was never enough to accomplish any of these ends.[126]

Chicago's 1907 ordinances contained many other complex provisions, but the essence of the compromise lay in the effort to harness private capital to public service through regulation while assuring that it would yield a competitive return to investors. With their franchises running out and capital hard to get, the companies took the best offer they could elicit. The city, lacking a political consensus, could hardly have held out for more. In essence, the ordinances were a compromise between the financial community, which would ultimately determine the viability of any settlement reached, and the council's business-minded reformers. Indeed, it was the CBD business and banking interests, together with the traction companies, real estate interests, and most moderate and conservative politicians, that worked to assure public acceptance of the 1907 compromise.[127]

As the referendum neared, large new cars appeared on the tracks of the south side's Chicago City Railway. The company's president, who had just purchased a large block of Chicago City stock, assured the public that more would follow if financing was not complicated by political considerations. The *Chicago Tribune* sponsored a poll in which thousands of readers expressed the opinion that the ordinances meant good service, while any alternative meant only delay. "It's almost impossible to get standing room," averred one Miss Yaseen in the poll. "It's all the same to me who gets the nickle." Council leaders voiced much the same opinion. Alderman Milton Foreman announced that "there is no legitimate opposition."[128]

Indeed the only argument against the ordinances, aside from their obscure financial provisions, was that service was not the essential issue at stake. George Hooker of Hull House wrote to the *Chicago Record Herald* urging that the real issue facing Chicago voters was not a seat on a trolley tomorrow but honest, active government for years to come. The editors printed Hooker's letter and then replied to it: "Mr. Hooker means well, but he writes flubdub."[129] Their response reflected a major theme of the campaign: immediate improvement of service was the real issue; all else was "flubdub." Cards were distributed showing Mayor Dunne smoking an outsized pipe and dreaming of better streetcars. Fisher pointed out that the only solid opposition to his compromise came from eminently impractical University of Chicago intellectuals, who lived where "municipal ownership is a little god."[130]

• • •

The results of the 1907 referendum support the position that service was the primary issue to those voters who responded to something other than the pressure of their precinct captains. The day of the crucial election, April 2, 1907, found 73,851 new voters on the rolls. Over 75 percent of the registered voters went to the polls, and nearly 90 percent of those voted in the referendum—by far the largest number yet to vote on any traction proposition. The morning of April 3, 1907, showed that 167,367 voters had endorsed the compromise, while 134,281 held out for public ownership or a better compromise. The ordinances won by a margin almost three times as large as that by which Mayor Dunne, who opposed them, was defeated.[131]

Interpretations followed. Steadfast proponents of municipal ownership cried corruption. Moderate progressives blamed Dunne, who was always erratic and now, some thought, quite thoroughly unbalanced. The *Tribune* simply sighed with relief—the nation's "most radical city" had finally come to its senses.[132]

None of these explanations accounts for the referendum's outcome. Company influence was certainly present: Chicago City Railway alone spent $350,000 on publicity for the ordinances—an amount which it later tried to charge off as an operating expense. Yet the ordinances did best in middle-class areas where simple corruption was least likely to settle the issue. Dunne was not daft; he contributed to the debacle of municipal ownership, but went on to become an effective and respected governor of Illinois. Chicago did not "come to its senses" in 1907; the city had not gone wild for public ownership in the first place. Middle-class voters favored private ownership in 1907, as they had favored municipalization in 1905, presumably because it seemed the fastest route to better service. Machine ward voters largely followed the lead of their local political leaders—for Dunne and public ownership in 1905, against both in 1907. Solidly Democratic wards which were neither "outlying" nor "slum," according to Arthur Bentley, tended to stay with the public ownership faith. Aside from this, the only clear pattern showed Dunne doing better than municipal ownership in almost every ward. But it should be remembered that the 1907 ordinances were promoted as being, among other things, the fastest route to municipal ownership.[133]

If the referendum did not produce a definitive statement of public opinion on the ownership of mass transit, it was nonetheless of far-reaching significance. The referendum itself witnessed the first full-scale effort on the part of the press and CBD business interests to "put across" a referendum proposition on a public policy issue. Techniques developed here would be used again and again, to good effect, on behalf of the Chicago Plan.

The ordinances themselves were genuine landmarks of local transportation policy. They froze into law the adversary relationship between the city and its transportation companies which had emerged from the traction wars of the 1890s. By effectively committing the city to regulated private ownership, they limited the role of transit in future city planning. By placing a premium on overcapitalization, they promoted innovation in the short run and discouraged it in the long run. By creating the BOSE they unwittingly involved transit engineers in political controversy and limited their role in the planning process. Finally—inevitably, given the city's policy history—the ordinances left the rapid transit lines out of the transit policy framework and attempted to deal with street congestion by putting streetcars in subways.

In 1907 the implications of the traction ordinances could not have been evident to the average voter or, indeed, to policymakers. What seemed apparent was the imminent realization of the Straphanger's Dream: fast, safe, convenient service with seats for all. Those who wished to see mass transit used in shaping the city's growth could look to the extension and regulatory provisions of the ordinances, to the promise of subways, and to the assurance that crosstown service and intercompany transfers would be soon be implemented.

Meanwhile the other half of the city's transportation policy was taking shape in the streets. Chicago tried to formulate a street and traffic policy centered around mass transit but, despite a good deal of legislating and a great deal of deliberating, it failed. The policy of regulated private ownership contributed to this failure. Thus street and traffic policy—essential elements of transportation policy—were for the most part shaped separately from mass transit policy.

In 1906, while the negotiations which led to the settlement ordinances were underway, some of the same CBD business elements which backed the traction settlement commissioned Danial Burnham to prepare his *Plan of Chicago*. With time this document, which focused on street and park improvements rather than public transportation as such, became the centerpiece of the Planners' Dream. Slowly, between 1907 and 1917, the street replaced mass transit as the tool-of-choice for those who wished to reshape the city. The automobile helped make this change of focus possible. Once planning had shifted to accommodate the automobile, the private car could begin to replace mass transit as the fulfillment of the Straphanger's Dream.

Chapter 2:

The
Battle for the Streets

First make the very best possible use of the surface of the street. When that proves insufficient, go under the street. When that proves insufficient, go above the street.

Bion J. Arnold, c. 1906[1]

. . . the greater the facilities for transportation, the greater the congestion becomes.

Cassier's Magazine, *1907*[2]

The ordinances of 1907 attempted to deal with the political, financial, and technological aspects of the urban transportation problem. Traffic was the other half of the city's transportation dilemma. Indeed, it could seriously hinder the city's efforts to deal with the mass transit problem as such. The 1907 ordinances led to the purchase of over 1,000 new electric streetcars, but street traffic, centralized routing, and synchronous hours of business determined what effect the new equipment would have. When 30 mph electric cars replaced 10 mph cable cars on one major route in 1906, the *Tribune* found that the new vehicles crept downtown at the same dismal rate as the old. The irrationality of traffic canceled out some of the effects of improved technology.[3]

In dealing with street traffic, the city had two basic, complementary options: it could regulate traffic and it could alter and improve the street. Each option could be used in a variety of ways. Regulation could be used to restrict specific uses of specific kinds of vehicles, to channel different sorts of traffic into different thoroughfares, or to give certain portions of the street over to specific functions—for example, by keeping private vehicles off the streetcar tracks. Street improvements likewise could channel traffic, accommodate its existing patterns, or create separate rights-of-way for different types of traffic.

Chicago used both options in all of the ways described, but the most important factor behind the city's street and traffic policies between 1900 and 1917 was the set of constraints and pressures within which the city operated. These constraints and pressures determined that the city's approach to street congestion would be fundamentally accommodative, and that its efforts would be centered on the CBD.

Chicago was in a far better position in dealing with the street and traffic aspects of the transportation problem than it was when grappling with the legal and financial "traction question." This was so for two reasons. First, street building and traffic control were recognized public responsibilities from the beginning of our period. Powerful interests accepted the fact of this city power and sought to increase it. In contrast to the determined opposition to public ownership of mass transit on the part of some members of the business and press communities, business sought to *add* to the city's responsibility for and power over the street. Street building and traffic control, though they offered plenty of opportunities for graft, were not "businesses'; by 1900 they were considered part of the necessary costs of running a city. Reformers, city planners, and "public spirited" business groups labored to spread the gospel of good streets and ordered traffic.

Second, traffic and street-building policy could be made *ad hoc*. Transit policy required a contract with private corporations. Every change in policy, every new plan, meant a new series of negotiations. Street and traffic policy-making were by no means easy. Established interests and patterns of thought as well as limited city powers constricted policy options. But street and traffic plans and policies could be changed. Compared with transit policy, street and traffic policy was extremely flexible.

There were, however, important limitations to what the city could do. Even though street modernization was politically easy when compared with the building of a subway, there were still regional jealousies to be dealt with. In consequence, improvements downtown made improvements in other parts of the city politically necessary. This meant that street improvements were subject to a political bargaining process, and thus could become extremely expensive.

Traffic regulation was an even more difficult matter. When the city sought to regulate heavy wagon traffic, it met with some success because it was dealing for the most part with "teaming" companies over which, as with the traction companies, it had a certain "hold." Private cars and other individually owned vehicles, on the other hand, proved almost impossible to regulate. For reasons to be explored, dispersed private ownership made automobiles nearly invulnerable to restrictive regulation.

Finally, traditions and legal decisions about the street made it very difficult to restrict street use in any way, and made it almost impossible to restrict access to a given street by private vehicles. This ultimately meant that the city was required by the pressure of business and public opinion to provide street space for whatever number of vehicles of whatever sort might appear.

All this, added to the fact that early twentieth-century planners harbored considerable hopes for the automobile as the long-sought agent of urban dispersion, meant that Chicago's policy toward the automobile would be largely positive. That policy was worked out in the context of a struggle, just as "traction policy" was. The struggle here, however, was not one between reformers and corruptionists or between city and companies. It was rather an ongoing battle of lobbies in the council chambers, accompanied outside by a deadly competition for public space—a battle for the streets.

Section I:
The Alternatives

> Are we quite rational in . . . fiercely contending that public utilities that pay should be controlled by corporations while the nonpaying utility is reserved for the taxpayer? Is it not, after all, largely a matter of expedience?
>
> *Lyman Cooley to the Western*
> *Society of Engineers, 1906*[4]

Street improvements and traffic control were both complementary and alternative means of dealing with street congestion. The city's responsibilities in both these areas were evolving rapidly between 1900 and 1917. In the process, a heightened awareness of the need to control downtown traffic emerged, only to be eclipsed in the popular crusade for street improvements spearheaded after 1910 by the Chicago Plan Commission. Both movements—that for traffic control and that for better and more "modern" streets—must be understood in the context of their complementary evolution.

The City's Responsibility for Streets

The city had long had responsibility for paving. In the mid-nineteenth century, while Chicago was still testing its legal right to grant street railway franchises, citizens were writing the *Tribune* to serve notice that they held the local government responsible for keeping the streets smooth and dust-free. In the 1850s, the upkeep and improvement of streets was achieved by three days of obligatory labor each year on the part of the city's adult males, commutable at 50¢ per day.[5] By 1900, this system had vanished, but the new proceedures were little better suited to contemporary reality.

Two agencies, in addition to the city's nine park boards, shared accountability for streets at the turn of the century. The Bureau of Streets contracted for work done from regular property tax assessments; the Board of Local Improvements, created in 1897 to take major street projects out of politics, controlled all special assessment work. Repairs, however, were done only when contractors saw fit to live up their guarantees. In 1906 the *Tribune* felt that the condition of the city's streets was a fine argument against municipal ownership of street railways.[6]

The failure was in part financial; each road building body still depended ultimately upon real estate taxes for its funds, and such taxes were not easily raised. This system had its basis in the idea that the chief benefit of a street was the access it gave to the property alongside it. In a large industrial city, where more and more vehicles made long trips through the metropolis, this system was taxing only some of the beneficiaries of street work. In addition, as autos proliferated, damaged old pavements, and gave rise to demands for new street surfacing, access to abutting property became a relatively less important justification for street improvements. Some way had to be found to tax the user of the street for its improvement.

Chicago's solution was a vehicle tax—one of the first in the nation. Passed in February 1908, the "wheel tax" fees wee substantial: they ranged from $5 per horse for animal-power vehicles to $20 for an automobile. The tax was met with lawsuits and passive resistance, but enough was collected from the sale of the round metal tags to permit a modicum of street work in the first year.[7]

Most of the 264 miles of street work financed by this tax in 1908 was repair rather than new construction, and it was distributed fairly evenly throughout the city, Before long, however, regionalism reasserted itself and by 1913, 85 percent of the collections in each ward were spent within that ward—the result of action by an increasingly sectionalist city council.[8] Streets had taken a step toward the status of mass transit: the user paid at least part of the cost.

But the movements of individual private vehicles could not be traced to determine the use to be made of the taxes paid upon them. Thus, in the building a maintenance of streets, the old idea that a road's chief benefit was to those located near it lived on in the principle of special assessments for special improvements, as well as in the uses of the vehicle tax.

Traffic Regulation: The Tradition of Right of Access

Law and tradition were thus unclear as to how the cost of streets was to be regarded—whether as a burden of real estate ownership or as a price to be paid by actual users. The question of rights in the street—the aspect of tradition directly affecting traffic control—was no less muddled. By 1907, however, court decisions had made it clear that, for *practical* purposes (no matter what a city ordinance might say) no class of privately owned vehicles had priority in the streets.

The first charter of the Chicago City Railway had given horsecars the right-of-way on their own tracks, and traffic ordinances subsequently affirmed that other vehicles must yield to streetcars at the sounding of the trolley's gong. In practice, however, any vehicle could remain on the tracks until a reasonable possibility of moving off arose, and the possibility, wrote Bion J. Arnold, "in the judgment of the average teamster, . . . decreases in ratio to the weight of the load."[9]

Further, during the first decade of the twentieth century, court rulings in effect defined the entire width of the street as public space and adopted what would later be known as the doctrine of "last clear chance" in assessing blame for accidents involving transit vehicles. "The public have the right to use and travel upon the entire street including that portion in which streetcar tracks are laid, and are in no sense to be treated as trespassers . . . ," the Illinois Supreme Court determined in 1906.[10]

At first glance, such rulings may have appeared to contemporaries as "progressive." They seemed to give "the public" access to the street on an equal footing with the "traction monopolies." In practice, however, they meant that the majority of citizens (transit riders) could be delayed by the actions of anyone who wished to use the streets for whatever purpose. As autos proliferated, this problem became critical.

In essence, while the streetcar had the legal right-of-way, its operator was required to behave as though other vehicles had priority. Bicyclists might ride on the tracks, children might play on them, or teams might turn across them: whoever had the last clear chance to prevent a collision was responsible for the results.[11] The only action open to the track-bound streetcar was to stop. By 1915, Chicago streetcars on the most heavily traveled lines were stopped for over a quarter of the time they were on the street; 44 percent of the time they lost was due, not to passenger stops, but to traffic congestion.[12] The routing of major lines through the CBD spread the effects of these delays throughout the city.[13]

By and large, transit spokesmen looked to the city to save them from the street competition they could not win. The police did act to keep the tracks clear. Serious efforts began in 1906, and for a time they were met with success in the CBD. Major abuses by teamsters contributed to the congestion; these could be controlled by policing. By 1912, however, the limits of a law enforcement approach were beginning to become clear, as courts demanded proof that the obstruction for which a teamster was arrested was needless.[14] The physical nature of the street, the constraints of the teamsters' business, and the appropriation of parts of the street by other users made the use of the car tracks by wagons legally defensible in many parts of the city.

The city could not effectively set aside one part of the street—the car tracks—for the use of mass transit vehicles, nor could it effectively give transit vehicles the right-of-way over other traffic. A third alternative, the designation of certain streets for mass transit or freight traffic only, was also ruled out. The city believed it could not exclude automobiles from any street. Arnold took this position, and occasional suggestions that private traffic be banned from the Loop during rush hours attracted no following.[15]

The single exception to this rule was the boulevard. As the special province of quasi-independent governmental units (park districts), boulevards were placed off limits to traffic thought to be incompatible with the function of parks as islands of bucolic peace in the noisome city. Wagons and streetcars were thus excluded from the boulevards, which became increasingly the preserve of the automobile. The important role of boulevards, and the manner in which they were transformed by the private car, is dealt with below. Here the point is that law and tradition made it easier to build special streets for automobiles than to regulate traffic in favor of mass transportation.[16]

These alternatives were laid before the city in 1904 in a study of paving materials commissioned by a businessmen's association, the Commercial Club. Thus the very ancient concern of businessmen and property owners with the fashion in which the city discharged its established responsibility for street pavement led to the first thorough discussion of the problem of traffic congestion.

Chicago had long been committed to granite block pavement in the CBD. Granite was durable, and dray horses found it easy footing, but the material had serious drawbacks from the merchant's point of view. Wagons with narrow metal tires chipped the edges off the blocks, and the resulting cracks became traps for the dust which blew through downtown streets with every gust of wind. The shattered blocks magnified the clatter of the thousands of wagons traversing the Loop, and automobiles were badly shaken when their drivers tried to speed across stone pavement as they did over smooth-surfaced boulevards. "When the brisk wind blows," wrote one British observer in 1902, "you cannot see across the road for the gritty dust . . . roads are rough cobbles, skull-sized in the main streets. . . . When it rains the streets are masses of greasy filth."[17]

Loop merchants wanted asphalt or macadam pavement which could be oiled to keep down the dust, but these pavements were not as durable as granite streets, nor could teams of horses drawing overloaded wagons keep their footing on smooth pavements during the winter months. In 1904, to throw light on the pavement question, the Commercial Club commissioned a study of the alternatives by engineer John Alvord.[18] Alvord produced a detailed study of the paving alternatives, but he also made three points which spoke to the use and character of the streets themselves.

First, he noted, the mixed use of business streets by pleasure, transit, and freight vehicles was an integral part of the paving problem. Because each used different modes of propulsion, no surface could accommodate them all.[19] Various uses required various surfaces. The coming of balloon-tired vehicles with internal combustion engines in all fields of transportation eventually mitigated this problem, but for 1904 Alvord's conclusion meant that different kinds of vehicles would have to be segregated within the same street on different pavements or that special purpose streets would have to be designated or constructed.

Alvord's second point concerning traffic underscored Bion Arnold's earlier judgment that better equipment would not, in itself, bring better mass transit. On all major streets, at all times of day, Alvord found, most private vehicles traveled not on the city-paved portion of thoroughfares but on the trolley tracks.[20] Some means had to be found to control the behavior of teamsters and pleasure drivers, or mass transit had to be removed from the crowded streets into subways. Yet law and tradition regarding the streets, together with the transit policy embodied in the 1907 settlement, made both alternatives impossible.

Finally Alvord stressed that the key problem of street improvement finance was public relations. The general public cared little for street improvements outside its own neighborhoods. Alvord warned that the major improvements he advocated would have to be "sold" to taxpayers. When, three and half years later, the city got the "wheel tax" Alvord had advocated, his assessment was vindicated: the substantial majority of vehicle owners simply refused to pay the tax.[21] Soon thereafter, the CBD busines community began a massive propaganda effort for the remaking of Chicago's streets which culminated in the drive to sell the *Plan of Chicago* to the public.

• • •

Street building and traffic control were thus the two major alternatives considered in the struggle to deal with turn-of-the century downtown congestion. Both were very largely dependent upon public relations—the "selling" to the public of new taxes for modern CBD streets or new forms of traffic regulation. Law and tradition meant that neither could be forced upon an unwilling public.

Urged on—and often led—by downtown business, Chicago made a major effort to revise public attitudes toward both citywide street building and traffic regulation. By 1917 it had largely succeeded in gaining acceptance of the idea of general funding for the rebuilding of the city's street system. Chicago's record on traffic regulation during this period was mixed. Changing technology and cooperation among businessmen helped resolve some of the problems created by freight vehicles in the CBD. The city's burgeoning auto traffic, however, proved largely uncontrollable.

Section II:
Traffic Control

The Teamster and the Trolley

> The other fellow says, "get out of the tracks." Well, we might say, "get the streetcars out of the tracks."
>
> *William Cavenaugh (team owners representative), 1916*[22]

During the first two decades of the twentieth century teamsters and trolley motormen engaged in a deadly competition for downtown street space. The inflexible nature of heavy, horse-drawn freight wagons and track-bound streetcars, the increasing use of smooth-surfaced pavements throughout the city, and the complex nature of both the CBD and the teaming business itself combined to create this first round in the twentieth-century battle for the streets. The most notable result of this contest—which neither party really won—was Chicago's first sustained effort at traffic control.

Between 1906 and 1917, smooth-surface materials (creosote, asphalt, and macadam) became the standard paving used by the Board of Local Improvements and the city's Bureau of Streets. CBD merchants as well as auto interests helped to bring about this change. As as a result, CBD pavements—oiled to calm the dust and sloped to promote cleaning—became the special bane of the teamster. In outlying areas, where city pavement often degenerated into a mire, streetcar tracks were sometimes the only road. Indeed, a demand for a streetcar extension at the city's far south end could be interpreted as a request for a free road rather than for trolley service per se.[23]

The CBD witnessed the most serious battle for street space between teamster and trolley. There teams and trolleys between them made up 44 percent of Loop traffic in 1906. There too the movement for smooth streets had made the most headway.[24] But it was the complex nature of the CBD and the character of the teamsters' business there which made a solution to the problem of Loop most urgent.

In the 1900–1917 period, the two-square-mile area which included the Loop was a true "central business district," not merely a commerical, administrative, and banking center. A produce market flanked the Loop on the north, warehouses and rail yards stood to its south, and light manufacturing establishments dominated the area to the west. Between 1900 and 1910, the transport of goods took up more of this area's street space than did any other use.[25]

In 1909, vehicles carrying goods predominated on State Street and on all non-boulevard streets intersecting it in the Loop. Teams worked their way north from the rail terminals at the Loop's south edge and spread east and west at every intersection, carrying coal for business and commercial establishments, transferring freight between railway terminals, and bringing sundry merchandise and food to stores, hotels, and restaurants. The number of teams in Chicago had tripled between 1891 and 1905, and many of those operating in the CBD were owned by specialized "teaming" concerns which constituted a major local business interest.[26]

But these were not the only teamsters in the Loop. The South Water Street produce market, hugging the main branch of the Chicago River, routinely accommodated 3,000 parked vehicles each day. In 1907, 19,856 vehicles per day had to negotiate this jumble of carts and wagons. The market brought a different kind of teamster into the Loop—the part-time driver not responsible to a major firm for his behavior. This led the Chicago Association of Commerce to lash out against "young and reckless Greeks and Italians who have no reguard for pedestrians, street car movements or the rights and difficulties of heavy traffic."[27]

Passenger vehicles, bound on errands of business or pleasure, combined with pedestrians to add to the miseries of the operators of heavy vehicles in the north and east sections of the Loop. Their importance in the CBC traffic jam was not fully appreciated early in the century: most of the passengers in the CBD streets traveled in streetcars, and most of the goods were hauled in heavy wagons.[28]

The crucial confrontation, then, was between trolley and teamster. With time this conflict was mitigated by technology. Outlying streets were slowly paved, horses were shod with rubber, the slant of some streets was lessened and they were salted in winter. Most important, the motorized truck, touted since the beginning of the century as a solution for the traffic problem, ultimately made freight-hauling vehicles more compatiable with the automobile than with the streetcar. But in 1915, "auto-trucks" made up under 14 percent of the city's freight-hauling vehicles and constituted just 7 percent of the CBD's traffic.[29]

There was one source of conflict between mass transit and other kinds of traffic which technology could not resolve. This was the increasing use of curb space by vehicles of every sort.

The problems created by the loading and unloading of wagons in the CBD were as simple as they were insoluble. Merchandise, fuel, and supplies had to be delivered to the front of most Loop buildings; the quantities demanded had long since become too large to be accommodated in wagons which could negotiate the narrow alleys, and the time had not yet arrived when CBD property owners could see the necessity of devoting some of their own valuable ground space to loading docks. Food products, and coal in the winter, had to be delivered daily, and (because they were delivered to wholesalers in the predawn hours) they usually arrived in the Loop during and just after the morning rush hour. Consequently, wagons constituted a significant obstacle to other traffic as they stood alongside the curbs to unload. This situation was particularly bad in the South Water Market, but it prevailed throughout the Loop and in other commercial districts.[30]

The segregation of delivery from through-teams and other traffic was one plausible solution. A technological remedy for some of the congestion presented itself at the end of the nineteenth century in the form of a proposition for privately owned freight tunnels. Sixty miles of these subways were in operation below the Loop and the northern CBD by 1912, but they, alone, could not solve the problem. Tunnels were necessarily limited in their number and location; they could not accommodate bulky freight moving between railroad terminals, nor could they handle coal or produce.[31]

There was another obvious kind of segregation possibility—segregation by time. This system worked in New York City where the law required that deliveries to the congested district be made at night; Chicago team owners initially favored the idea. Railroads, however, refused to alter the timing of their coal deliveries, union teamsters demanded higher wages for night work, and store and building owners objected to the added expense involved in keeping a night staff on duty to receive the off-hour shipments.[32]

Teams could not be prevented from parking and thus forcing other traffic onto the car tracks, but the city scored its first significant victory in traffic regulation when it succeeded in moving teamsters off the car tracks in congested districts—at least long enough to allow a streetcar to pass. The technique adopted bespoke both past and future. It was archaic in that it employed traditional police methods to deal with traffic, but it presaged the growing importance of cooperation between the city and major business for the purpose of traffic control.

The "solution" was mounted police, especially trained to handle team traffic. By 1911, eighty-five police officers patrolled the Loop and the congested intersections two miles to the north and south, ordering teamsters out of the tracks and generally keeping the traffic in line. In the Loop itself, at each major crossing, police moved and held back traffic from each direction alternately with an effectiveness that amazed and delighted European visitors. In some respects, the operation had a military character. Many of the mounted police were former soldiers; Captain Charles Healey, their leader, was a retired military officer. Their major task appeared to be the imposition of an external order upon a chaotic situation.[33]

In fact, however traffic itself determined the rhythm of regulation. The officers did not so much direct traffic flow as create and clear small traffic jams alternately on each intersecting street. When traffic backed up on one street, the other was whistled to a stop, and the vehicles in the jammed street flowed across, with an officer orchestrating turns and trying to prevent unnecessary stops. The result was an uneven movement of traffic, with the east–west streets, which were narrower and backed up more easily, receiving the greater amount of time. North–south streets, which carried the greater streetcar traffic, including the turning west side streetcars, inevitably suffered.[34] They had the right-of-way, but the more numerous vehicular traffic determined patterns of movement.

The difficulty was that the police continued to see traffic from a police perspective: it was a series of potential problems—traffic jams—at specific, static locations, not a continuous flow. The BOSE knew that traffic came in waves which could, potentially, be timed and perhaps controlled, but nothing was made of this insight until the mid-1920s. For the time being, however, the police system worked and worked well. Traffic moved as it had not before, though the constant presence of the officers was essential. If they withdrew for even a few minutes, profound chaos resulted. A photograph, made when the police were purposefully removed for a short time to prove this very point, still exists and is frequently reproduced to illustrate, erroneously, "typical" street congestion in 1911 Chicago.[35]

The police system worked because the business interests wanted it to work. The Chicago Association of Commerce (CAC) backed the traffic force soon after its creation, and started a traffic committee of its own which was to play a major role in the ultimate resolution of many traffic control questions.[36]

Mounted police Captain Healey met frequently with team owners, shipping and receiving clerk's associations, the city's Law Department, the CAC, and the teamsters' union. There was some attempt at coordination and routing of shipments, but the major accomplishment of these meetings was the education of the team owners and drivers to the necessity of obeying police. Indeed, progress was made chiefly through the cooperation of team owners, who disciplined men who had disobeyed the officers.[37]

The traffic squad soon expanded its patrols, first to the Englewood commuter railroad station, eight miles south of the Loop, where police were able to do what they had done downtown despite, Healey said, "the nature of the business carried on . . . and the nationality of the people. . . ."[38] The CAC provided telephone communications for the squad. Healey was treated as a local celebrity, was sent to Europe to study traffic control there (most of which he found inferior to Chicago's), and was then sent around with a stereopticon to lecture to Chicago businessmen on the values of regulation.[39] But by 1912, Captain Healey had encountered one factor he could not conquer with slide shows and diplomacy—namely, the convenience of the businessmen themselves, and of their customers.

The Unmanageable Motorist

The crowd [at the auto show] was a study for a Psychiatrist. There were the people who own automobiles . . . the pleasure class. Next there were the people who would like to own automobiles . . . and mingling with these on terms of perfect equality were grimy faced mechanics. . . . There were girls in many of the cars, and occasionally a bright eye would wander as if to say, "wouldn't you like to be blowing along a country road on a Summer's day with me beside you?"

Chicago Tribune, *Feb. 4, 1906*

Speed is home life.

George Hooker, 1914[40]

Downtown team traffic could be controlled to some degree because so many teamsters worked for major firms whose cooperation could be enlisted. The well-being of the teaming business was bound up with the viability of the CBD. Automobile traffic was another matter.

Motorists were private owners or chauffeurs: no employer or business group could pressure them into considerate behavior on the street. Indeed Loop business feared to alienate them despite their relatively small numbrs in the years before 1917; angry motorists might take their cars and their business elsewhere. Thus police and magistrates sought fruitlessly for the first two decades of the twentieth century for some means by which motorists might be controlled. In the last analysis, the motorist simply had to be accommodated, for he successfully refused to be regulated in the interest of other forms of traffic.

Eventually the auto triumphed in part by force of sheer numbers. Between 7 A.M. and 7 P.M. on October 23, 1907, 1,421 automobiles crossed the Rush Street bridge into Chicago's Loop. In just eight hours on July 13, 1915, 10,158 automobiles made the same crossing.[41] In the intervening period, the role of the automobile had undergone a rapid change.

The early public response to the automobile has been much discussed, but it deserves a brief review here. Contemporary hopes and fears surrounding the private auto are reflected in and helped to create the city's early, unsuccessful attempts to regulate the motorist. The complex response of early planners aside, reaction to the automobile in the first years of this century was ambivalent.[42]

"I have never known a case where the viewpoints differed so radically as they do from the inside and the outside of an automobile," wrote Frank Munsey in 1903. "The man who knows it only from the outside despises it and damns it on general principles . . . , but once inside of a really first rate automobile a marvelous change of heart comes over him."[43] To an avid "autoist," the motorcar was the solution to the traffic problem: Joseph Pennell declared that "the motor . . . is perfectly controllable, occupies half the space [of the equivalent horse-drawn vehicle] and travels now through the congested streets at double the speed. . . ."[44] It solved the pollution problem, too, according to one British MP, who wrote that "if it smells, the driver is to blame; it is within the reach of the man of modest means, and as little likely to break down as any product of human ingenuity."[45]

Before the turn of the century, Ray Stannard Baker had set out to show the readers of *McClure's* that the automobile was cheaper to own and more reliable than a horse and buggy and, not long thereafter, the onetime president of the American Automobile Association announced the era of "automobiles for the millions" and perhaps ". . . even for the workingman."[46] "Why," Winthrop Scarritt asked, "should men be compelled to live within a stone's throw of a car line?" The editor of *Cosmopolitan* thought the automobile could solve the mass transit problem too: small, flexible fourteen passenger buses could take the worker to his door and make commuting a pleasure instead of a horror.[47] Walter Fisher studied the city-owned omnibuses of London and found in them another argument for the 1907 ordinances. "[Soon] we won't care a rap for their [the streetcar companies'] roads," he told the *Chicago Tribune*. "The autobus is being developed so rapidly that with good paving the new vehicles will in time beat the streetcar all hollow."[48]

Despite these early hopes that motorization could reshape the city, it was far from universally popular. For one thing, in the years before the war, the private car was clearly beyond the means of most Chicagoans.

Writing for the auto industry, Henry Norman asserted in 1906 that families which could not afford a horse could easily keep an automobile. But in doing so he arrived at a minimum purchase price for the auto which was twice the yearly wage of the men who paved the streets on which the car would be driven. Even a well-paid member of the building trades would have to work for half a year to purchase such a machine.[49]

It was little wonder that a motorist who strayed into the Chicago slum known as Little Hell might be pelted with stones, as the *Tribune* reported, "for no reason." When one frustrated policeman shot out the tires of a speeding limousine which refused to stop, one hundred citizens of relatively affluent Rogers Park rallied to his defense. "The automobilists . . . are the monied class," a West Park commissioner remarked in response to a rash of speeding incidents. "The streets belong to the people."[50]

On occasion, Chicago politicians also referred to automobile owners in terms reflecting an appeal to working-class prejudices. But the scant evidence that remains about early Chicago motorists—from 1907 onward—portrays them, not as a group of millionaires and their playboy sons, but as a class composed largely of professionals, businessmen, and managers, with a significant scattering of skilled workers, living largely but not exclusively in the substantial neighborhoods served by the rapid transit lines and boulevards (see Map 2-1).[51]

Early middle-class ownership of automobiles and the reluctance of Loop businessmen to alienate their wealthier limousine-owning clientele may have helped insulate private cars from restrictive regulation. But the technology of the auto also made it difficult to regulate. For one thing, the auto was simply different from other vehicles: operations which were dangerous or impossible

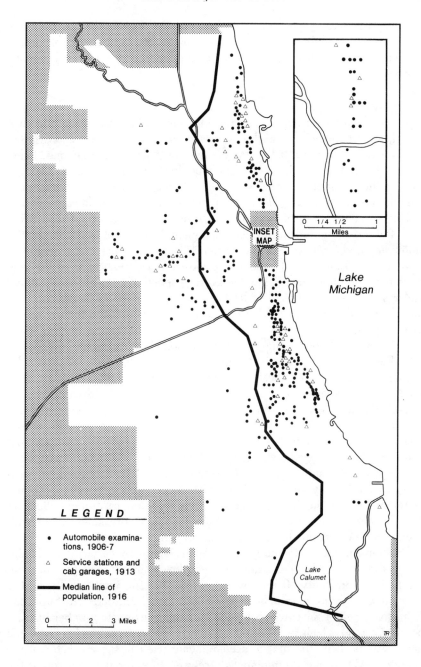

Map 2–1. Autos in Chicago, 1906–1913. Data from Chicago drivers' license examination record, 1906–1907; *Chicago Classified Telephone Directory,* 1913, *1916 Report.*

for horse-drawn carriages were comparatively easy even for the automobile of 1910. In addition, the motorcar's speed made acts of personal selfishness easier to "get away with" for the motorist than they had been for the driver of a carriage.

·   ·   ·

The most obvious characteristic of the automobile, and a major feature of early advertising for the machines, was its speed. Company demonstrators were pressured by potential customers to show just what the Thomas, Peerless, or Renault could do on the city's boulevards.[52] For its part, Chicago entered the twentieth century with a set of traffic laws based on immemorial conventions of highway traffic. In the past, the improvement of road surfaces had increased the speed of all vehicles together, so traffic law treated all alike.

But the automobile destroyed the homogeneity of traffic. At 20 mph the motorcar could come to a stop in the same distance as a horse-drawn vehicle traveling at 10 mph, yet Chicago, like most other cities, confined the new machines to 8 mph—just below the average speed of a streetcar in traffic.[53] The result was a great deal of lawbreaking, especially on boulevards and suburban roads, by citizens whose contact with officers of the law was otherwise very limited.

To be dealt with, the speeder had somehow to be identified and apprehended. Beginning in 1900 Chicago attempted to license all drivers by testing them in the mechanical and operational aspects of their vehicles. After 1908, vehicle tax tags provided another means of identifying drivers, but it is unlikely that more than two-thirds of the city's automobiles were registered at any time before World War I. And owners soon learned to conceal the tags in legal but barely visible niches on the automobile's frame.[54] The result was that police had to arrest traffic violators on the spot, and that was no easy task.

Drivers and automobile clubs protested low speed limits and, on occasion, objected to any sort of regulation. Police on foot had no way of stopping motorists, who could simply speed by. More than one frustrated policeman drew his gun and fired as he would at any fleeing lawbreaker—usually with results more disastrous to the officer than to the motorist.[55]

Even when a car was stopped, the owner was not necessarily chastened. When the driver was a chauffer, his employer simply paid the fine and escaped responsibility. If the owner was driving, he might challenge the policeman's right to "stop an honest citizen in a public street." The Suburban Evanston Traffic Court, which handled a disporportionate number of cases because a road which had become a favorite "speedway" ran through the town, was perhaps the first to accommodate itself to the character of the offenders. When one prominent lady refused to enter the courtroom, the judge held court in her car. Before long, he was taking evidence over the telephone.[56]

Chicago police did their best with the speeders. As the Department itself became motorized, and better organized, arrests for speeding soared: from 138 in 1909 to 4,962 in 1913, more than the number of arrests on any other charge except simple disorderly conduct. Nor were Chicago's courts as understanding as those of Evanston. In 1913, the first year for which such data are available, 4,536 of the 4,962 persons arrested for speeding were fined, while less than 46 percent of the total number of persons arrested for midemeanors in that year were given any sort of punishment or fine.[57]

But this upsurge in enforcement was only temporary. After 1913 the percentage of traffic violators fined began a steady decline. Much of the difficulty lay in the very severity of the city's traffic laws, which judges may have felt were inappropriate to the offenses involved. Granted discretion, judges tended to give low fines: the Lincoln Park commissioners complained in 1908 that the average speeding fine imposed for violations on it boulevards was $5—an amount which the commissioners did not consider an effective deterrent.[58]

But judges usually did not have legal discretion. A conviction for drunk driving, for example, required a fine of $200. Such a fine, imposed on a wealthy man, meant little, Traffic Court Judge Sabbath explained in 1916; to a laborer, it meant 400 days in jail. The judge's only alternatives were unreasonable severity or "not guilty" verdicts in cases in which the defendant was clearly guilty.[59]

"The man who is fined . . . leaves the court defiant," Sabbath explained. "He considers himself squared with the law." Sabbath preferred to lecture defendants on safety and acquit them. Other judges apparently agreed, since the conviction rate in traffic cases rapidly declined after 1913.[60]

Speeding autos, as will be seen, had an important effect in helping turn the park district's boulevards into primitive expressways. But it was the motorist ás shopper and commuter who administered the soundest defeat to restrictive traffic control. The parked car played havoc with prewar urban traffic.

Chicago's city council found that it had to deal with downtown parking in order to deal with CBD street traffic or streetcar crowding. Courts rejected council anti-crowding ordinances on the grounds that teams in the streetcar tracks prevented the streetcar companies from providing enough cars to alleviate crowding. But to remove the teams from the tracks the council had to find a way to clear the curb lanes of CBD streets. It was here that the aldermen discovered both the importance and the intractability of the parking problem.[61]

Six thousand automobiles came downtown on one summer day in 1915, across a single bridge. Over 2,000 of these cars parked for two and a half hours or more; in one hour the BOSE counted 1,029 standing vehicles in the 332 blocks it surveyed. Thirty percent of these were parked illegally under the lax regulations of the time.[62] Double parking was common. It blocked the doors of streetcars, forced wagons onto the tracks, and helped to render useless all the efforts of engineers and regulators.[63]

Some of these parked autos were cabs. Their owners had bought the right to stand adjacent to major Loop hotels from the hotel owners themselves. But many of the greatest nuisances were owned by businessmen. In the morning rush hour, chauffeurs pulled up in droves to discharge executives in front of their offices. In the evening, they returned and waited until the owners emerged, often idling at the curbside for up to twenty minutes during the height of the rush hour.[64]

This particular aspect of the parking problem points up the early relationship between the automobile and the relatively densely populated, high-class business and residential districts. By 1906 one developer had plans underway for two Lake Shore Drive apartment buildings with built-in garage facilities. The location of early automobile service and repair stations suggests the common routes of travel marked out by the automobile within the city. Automobile owners were concentrated along the main routes to the CBD in a way which hints at travel patterns similar to those created by the elevated railways.[65]

By 1912, for some residents of the fashionable south side Hyde Park–Kenwood area, at least, the automobile was already a means of rapid transit. They and the owners of dense new apartment buildings argued before the CLT that Lake Park Avenue was an automobile street and should be off limits to streetcars—precisely because Lake Park, through the boulevards, provided a fast and direct automobile route downtown.[66]

The automobile thus meant "suburban" home life without the loss of relatively easy access to the CBD. George Hooker thought a large number of the automobiles parked in the CBD were carrying people the commuter railroads served inadequately. The Commercial Club, seeking to assuage suburbanites' fear of road improvements, reminded them that "auto owners are people of means. . . . Their hearts return to the country as eagerly as the pigeon seeks its loft, and they bring their pocketbooks with them."[67]

Whenever they came from, the parkers defied the best efforts of the city council and the police. In an attempt to end cruising by chauffeurs, the council limited parking in the Loop to sixty minutes in 1911. The police issued hundreds of summonses, but the motorists were not fazed. Some refused to appear; others appealed decisions. By 1912, the city prosecutor was simply not acting on parking violations.[68] The reason for this inaction in unclear. Influence, and a feeling that parking violations were not important, may have played their parts, but the prosecutor faced a difficult task. City parking regulation was attempting to change an age-old custom relating to the use of the streets.

The precedent in law for parking limitation was clear enough. From the 1850s, the mayor had had the responsibility for seeing that obstructions were cleared from the city's thoroughfares.[69] Presumably, this provision was enforced with some success against horse-drawn vehicles: complaints about standing vehicles, other than wagons, figure prominently in discussions of traffic congestion only from 1910. Certainly horse-drawn carriages could not be left at the curb with as much assurance that they would remain. The major reason for the new emphasis on parking was probably the 19 percent increase in vehicular traffic in the CBD between 1907 and 1913, and the even faster increase in streetcar traffic.[70]

At the heart of the problem was the selective enforcement of laws governing the use of public spaces. Chicago had strict rules against street selling and unauthorized gatherings, but in 1914 the *Tribune* reported that fruit vendors and anarchist orators still milled about happily at the end of the Chicago Avenue streetcar line, a few yards from a major police station.[71]

The enforcement of street-clearing ordinances simply did not have a high priority with the police and courts. Thus the public tended to regard enforcement as discriminatory. A Chicago Motor Club spokesman told the CLT in 1916 that his organization favored an absolute ban on parking in some CBD streets, but that the first, vital step was blanket, uncompromising enforcement of existing ordinances. "I am as big an offender as anybody," he remarked in the presence of Captain Healey. ". . . They never summon me into court. If they did the judge would let me off. . . . I will not pay 50¢ [to park in a private lot] when there are fifty other people in my building who don't pay. . . ."[72] he suggested that a massive enforcement campaign would convince the public.

Certainly new laws did not work. On March 1, 1915, the council cut the permissible CBD parking time to a half hour. Illegal parking only increased.[73] Enforcement was the crucial problem. In 1916 Captain Healey could spare only two men from his traffic squad for the job of taking down the license numbers of parked cars, checking each car after a half-hour, and, after summonses had been served, appearing in court. There seemed to be no chance of hiring more officers; the city had severe budgetary problems from 1912, when its taxing power was cut by the state, and the council apparently could not see its way clear to hire officers to make arrests which the courts, once so vigorous in the prosecution of speeders, could not be made to uphold.[74]

This state of affairs was the more surprising when one considers the extent of disruption caused by all-day parkers. A parking time limit was really a minimal measure. Healey asured the council in 1916 that double-parking was commonplace; triple-parking was not infrequent. Some parking blockades in the late evening hours lasted so long that the streetcar company rerouted its cars to avoid them.[75]

The city had lost control of one-third to one-half of the space on many CBD streets. In 1915, vehicles forced onto the car tracks were taking up more space in the streetcar loading berths than the streetcars themselves. In 1914 both Healey and street railway president Leonard Busby spoke out for a total ban on Loop parking; Captain Healey, however, told the CLT, "You will never get it through the Council." It was, he said, a measure too damaging to business.[76]

Pressed into action by a political contest then underway between the council CLT, Mayor William Hale Thompson, and the state Public Utilities Commission, the city council took up the matter of a parking ban again in 1917. The arguments brought forth against this proposed ban, and the array of interests presenting them, revealed that the appearance of large numbers of motorists was forcing the city to seek a new definition of the public's right to use street space.

Opponents of the ban argued that all vehicles had the right of free access to the street, and that all should have the right to do business in and derive business from the street. Both contentions could be reduced to the proposition that the commercial usefulness of the Loop must be maintained and opportunities for its exploitation kept open.

The argument for freedom of access came from the Chicago Motor Club and from some aldermen from outlying wards. These men claimed that the usefulness of an automobile depended on the right of its owner to leave it near his destination. "The streets of the city belong to the public . . . for both vehicle and pedestrian purposes," Alderman Fisher observed. "I do not believe you can legally deprive the citizens of Chicago of the right to use the streets or their vehicles. . . ."[77]

Alderman Vanderbilt characterized congestion arising from the necessities of business as "legitimate congestion," a perspective similar to that of a Commercial Club representative who argued, some years earlier, that "pleasure vehicles" used for commercial purposes became thereby a desirable element in traffic. One alderman argued for cooperation with the Traffic Committee of the CAC as "the people most interested" in the problem.[78]

Bion J. Arnold, while he favored in principle the elimination of parking, shared the view that "to keep them [automobiles] off one place you have got to give them another place." Arnold appears to have felt that any measure which tended to exclude automobiles from the Loop would be unfair by legal standards, and perhaps by his own standards as well.[79]

The CBD business community, as represented by the Chicago Association of Commerce, stood aloof from the parking controversy. The CLT's chairman, Alderman Henry Capitain observed in the midst of the debate that the CAC did not "cooperate" in parking, although it was very active in the drive to facilitate the movement of Loop traffic by other means.[80] Business did press the city to provide off-street parking at public expense—a proposition which Capitain and his allies rejected but which would ultimately be part of a compromise on the parking question during the 1920s.[81]

The second objection to a parking ban reflected the business perspective that lay behind much of the early resistance to traffic control. Taxi owners, and the hotels which used their services, contended that cabstands were vital to their prosperity. Cabs could have been summoned by phone, as was done in New York, but most of Chicago's 1,291 licensed cabs were individually owned in 1917 and lacked garage facilities. Parking ban opponents argued successfully that cabstands must be preserved in the name of competition and opportunity for the small entrepreneur. Throughout the 1920s, small businessmen of all sorts remained the most vigorous opponents of parking restriction.[82]

Capitain and his allies argued largely from necessity: streetcars carried most of the CBD's human traffic, and one way or another, they had to be permitted to move through the Loop. Ultimately, without the organized support of major businesses and largely for political reasons, Chicago got its first ordinance forbidding the use of CBD streets for the storage of automobiles; the law applied, however, only during rush hours. The ordinance was a minor landmark; it represented the city's first attempt to segregate street uses by law—even though the segregation operated by time rather than by place.[83] For reasons to be discussed later, it proved largely unenforceable.

The struggle for a parking ban also revealed the lack of an underlying philosophy in Chicago's efforts at traffic control. In its mass transit policy, the city had a clear and well-ordered set of priorities. If mass transit's public definition was ambiguous, it was nonetheless worked out in a set of easily understood propositions, the product of years of intense debate. Traffic control policy reflected no such debate. Congestion problems were treated as they arose. At no point in the prewar period was Chicago required to identify its priorities for the use of the street. Thus the city had strict traffic rules and a progressive parking ordinance, but lacked the means of enforcing either.

The courts did try to accommodate traffic violators, but the city was not able to help by developing off-street parking spaces (in 1917 there was but one parking garage in the CBD). The city tried traffic lights in 1916 and experimented with pedestrian signals, but the new hardware only provided a way of implementing the same piecemeal policy. By 1922, the city council's traffic engineer could say, "We are afflicted with signalitis in Chicago . . . ," so vocal was the lobby for new traffic lights and timing devices—yet the signals were, until the later 1920s, always a response to traffic, rather than an attempt to direct it.[84]

This vagueness had both positive and negative effects. On the negative side, lack of interest in the fundamental causes of congestion—or unwillingness to face them—meant that suggestions which went to the heart of the matter could arouse no following. A subway commission in 1909 suggested staggered business hours, the BOSE recognized that traffic might be forced to adapt to rational patterns of movement, and, as already noted, the idea of excluding private vehicles from the Loop during rush hours was broached from time to time.[85]

But such suggestions had no impact. Private traffic was not even subjected to scientific study as was transit ridership. By 1916 engineers could tell the city where over 350,000 of its transit-riding citizens lived and worked. But through the mid-twenties auto traffic counts consisted of only a few scattered and incompatible sets of statistics.[86]

At the same time, the lack of a clear philosophy of street use made traffic control flexible. Judges and aldermen could adapt the law to suit the movement of traffic without having to deal with the confusing and politically tempting issues of sectionalism and corporate greed which attached themselves to mass transit issues. Most important, the business definition of the CBD and the tradition of free access to the streets made facilitation of auto traffic easier than restriction.

The great flexibility of the automobile itself gave it yet another advantage. Ironically, that most disorderly and unregulatable vehicle benefited from the movement for order and city planning to a much greater degree than did mass transit. This was so, once again, because all questions surrounding mass transit involved a wide range of political and social issues as well as the interests of the public and of corporations. The movement for a city plan, which amounted in many of its specific applications to little more than a movement for the separation and facilitation of forms of traffic and of city functions, had little political opposition, and much historical justification—as well as contemporary support. The city planning movement helped turn the automobile into a viable mass transit vehicle.

Section III:
Building Solutions to the Traffic Problem

Subways, Roads, and Boulevards

Liberality on road building will be repaid a hundred fold.
Daniel Burnham, 1909[87]

Main roads are for all of the people, all of the time, all of the way.
Commercial Club of Chicago, 1908[88]

The automobilist is killing or maiming many victims and kicking his dust and smoke in the face of the other fellow [but he] is the only one who can today get back and forth with ease over the great district of greater Chicago.
George Hooker, 1913[89]

Chicago built its traffic control policy by increments, adding new laws and procedures to meet the challenges raised by ever-denser, ever-faster traffic. The city also met the demands of traffic through planning—planning which was not directed specifically toward the automobile, but which ultimately served its interests better than it did those of any other form of transportation.

All the sorts of planning discussed here attempted to deal with traffic by separating it according to its destination and the type of vehicle used. However, because of the connotations placed upon mass transit from continuing political battles, the four types of planning involved really fall into two broad categories.

Subway planning was a category by itself. Saddled with all the political and regional issues connected with other mass transit planning and held back by the idea that subway users, as the sole beneficiaries of their construction, must somehow pay for any subway whose chief purpose was not the easing of CBD congestion, engineers and politicians were able to produce nothing more than plans during the decade which followed the 1907 settlement. The good roads movement, the movement for parks and boulevards, and the Chicago Plan shared a very different fate. Each was the subject of much favorable publicity; each was fairly well insulated from political controversy; and each, with time, came to be seen as the concern of the entire community—not simply of those who used improved streets and roads in their journey to work or to places of recreation.

Subway Planning

Subway planning in Chicago arose from two distinct motivations: the desire to move transit vehicles more efficiently, and the hope of reducing traffic congestion. Bion Arnold observed in 1902 that "the desire for a system of subways has arisen on account of the congested condition of the streets in the business district" and concluded that subways should be built "on a plan designed to reduce the congestion to a minimum and at the same time render the most service to the traveling public." Engineering documents as well as subway proposals by private capitalists reflected this set of priorities.[90]

Subways, then, were a part of the more general movement for traffic segregation through the building of new facilities (as opposed to regulation) which found its best expression locally in Burnham's *Plan of Chicago*. But because subways were transit facilities, they had connotations different from those of street improvements.

First, because subways were mass transit facilities, it was almost universally assumed through 1926 that transit riders alone must pay the cost of their construction. Some subway plans called for the companies to pay, others for the city to build subways to be rented to the companies: either way the cost would ultimately be carried by the riders.[91] Street improvements, as will be seen, were paid for largely by the property owners and renters of the city as a whole.

Second, since they were to be used by transit companies, subways became immersed in the perennial traction debates over regulation and guaranteed fair returns. Subway planning was entangled with, and made fruitless by, the traction impasse through 1938. Obviously, this was not the case with automobile-oriented planning.

Finally, the belief that subways should be built to relieve congestion, combined with the fact that all transit riders would have to pay their cost, made underground mass transit the subject of sectional jealousies which became particularly intense after 1910, and which helped to separate transit planning from the city planning movement.

Subways were treated as adjuncts of the CBD because downtown street congestion was generally perceived as the most serious aspect of the traffic problem. In fact, some inner-city streets outside the CBD were subject to traffic disruptions almost as severe as those of the downtown district. During the prewar period, however, no engineer employed by the city would advocate the construction of crosstown or diagonal subways before the problem of CBD street congestion was solved. This was not simply a result of the engineers' commitment to centralization: Bion Arnold appears to have felt that the first street car subway should be built one mile west of the center of the CBD, and that Loop merchants were a major obstacle to such a plan.[92]

The intensely sectionalistic politics of Wilson-era Chicago polarized the subway issue. Representatives of neighborhood business and real estate organizations argued that subways would be used to preserve the hegemony of the CBD at a time when it should have been dispersed by the effects of its own congestion. To men like Tomaz Deuther and Otto Schultz of the North West Side Commercial Association and the Greater Chicago Federation, the centralizing role of mass transit was not the inevitable byproduct of larger and more efficient units of production and distribution. Such men saw mass transit as an agent of active city policy, which they believed was directed by and in the interests of CBD merchants.[93]

City and company engineers on the BOSE joined Bion Arnold in declaring that new transportation lines should be built "in the general direction of existing traffic and not . . . to . . . suit particular theories."[94] But serving existing demands in practice meant relieving the CBD first. It was only there, for example, that Arnold favored the building of subways which might not be profitable in themselves during the first few years.[95]

If congestion alone justified the building of subways, men like Schultz and Deuther were right. A subway built solely to relieve congestion would preserve and expand existing centralization. Such a transit system would be a rigid, conservative force. But the BOSE was also right. Subways or rapid transit serving districts whose streets were not already overcrowded would surely be the subject of political bargaining, and would be unlikely to leave the city any freer to develop in response to the individual choices of its citizens than would a CBD subway. In the political context of pre–World War I

Chicago, this dilemma could be "resolved" only by the building of no subway at all.

But the absense of a subway did not mean, as some regionalists may have hoped, the withering away of the CBD. Rather it meant the gradual adjustment of surface mass transit and of the streets themselves to needs which politics had prevented transit *planning* from serving. Subways, as the BOSE observed in 1914, were the only means of cutting downtown congestion without "resorting to Napoleonic street widenings."[96] The absence of subways doubtless encouraged the movement for broader CBD thoroughfares.

Napoleonic street widenings were certainly not what the regionalists had in mind. Their objections to CBD rapid transit paralleled the disillusionment of planners around the nation with the ability of mass transit to decentralize a city; yet, as will be seen, Chicago's neighborhood business leaders continued to place great faith in surface (as opposed to off-street) transit as a means of dispersing the city's commercial activity.[97] For the present, however, the important point is that regionalist resistance to downtown subways (along with the CBD business community's determination that such subways be built) helped to divorce transit planning from the mainstream of the city planning movement in Chicago.

The Good Roads Movement

The development of modernized, specialized streets escaped the pitfalls surrounding the movement for subways. There was, of course, no political/economic issue attached to roadbuilding which could rival the "traction issue." As an accepted government function, thoroughfare building could not be subjected to the test of profit which was applied to subways. Instead, the value of good streets to the economic prosperity and health of the entire community was pressed upon the voting public. Notably, too, the program for better streets and city planning overcame the suspicions of outlying community groups which so bedeviled the drive for better rapid transit.

The good roads movement helped to popularize the idea that thoroughfares were a benefit to all and should be built from general funds. The general movement, and the role played in it by bicycle and automobile as well as agricultural interests, has been explored by other scholars.[98] The salient point here is that in Chicago, and in Illinois generally, the good roads movement paralleled and helped vindicate the drive for better city streets.

The major policy goal of the early good roads movement was to make rural roads a state responsibility. The movement sought to broaden the power of state (as opposed to county) government just as the power of city government would later have to be increased to accommodate automobile-oriented city planning.[99]

An Illinois State Highway Commission was created in 1906, but found itself faced with the widespread fear among the rural population that the

Commission had been brought into being to remake country roads for the automobile at the expense of the farmer.[100] Nonetheless, the state soon took an active interest in roadbuilding. State driver licensing began in 1907, and funds from this source were for some time devoted entirely to roadbuilding. By 1912, the Chicago Association of Commerce, the Illinois Federation of Labor, and the Illinois Farmers' Institute were agitating for state aid roads, and the public schools of the state were participating in a "Good Roads Day."[101]

The lobbyists' program was brought to fruition in 1916, by which time the building of roads by direct taxation had proven itself too slow. Citing the exigencies of national defense, Governor Edward Dunne called for a $129 million roadbuilding program, to be financed by bonds which would, as Dunne said, "let posterity pay its cost." By this time, Cook County had led the way—"wisely and patriotically," Dunne said—with a $2 million bond issue which won easy approval at referendum. The year before, also citing the needs of national defense, the federal government had begun its highway subsidy program, committing $5 million for the first year and increments of $5 million per year thereafter for five years.[102]

There were just 12.64 miles of county roads within Chicago in 1907. But it was these city streets (and other streets which were extensions of state and county roads) which, by the mid-1920s, would be pouring automobiles into Chicago at a rate great enough to require more revisions in the city's policy toward the private car.[103]

The Parks: Contributions and Complications

By 1917, the city's park boards and the Chicago Plan Commission had begun to prepare the way for these changes. The parks movement not only provided background for the planning movement; it also helped create the city's first special purpose automobile streets.

The relationship between the movements for parks and that for city planning has been well studied. Parks were thought to uplift the voter, mitigate the effects of tenements, and provide sources of neighborhood identity. At the turn of the century, Chicago, which had had a particularly impressive park system in the 1880s, was made especially conscious of the need for better parks by a flurry of concern over the decline in park acreage per person through the previous twenty years.[104]

This concern over parks was combined with a determination that they be kept clearly distinct from the city surrounding. Parks were to be havens of natural beauty and peace amid the noice and chaos of the city. Streetcars had always been ruled out of the parks and off the boulevards, and every attempt was made to see that even the busiest boulevards kept a rural appearance.[105]

Efforts were also made to see that parks maintained a genteel air. As one park superintendent observed in launching an 1897 crusade against "bums" amid the shrubbery, "The parks ain't no lodging houses." Commissioners also tried to prevent extraneous traffic from disturbing the atmosphere of parks: all vehicles were required to stop at boulevard crossings and park entrances, and "speed tracks" were built to attract bicyclists and automobile enthusiasts to areas where they would not disturb the pedestrian majority.[106]

By the turn of the century, however, pleasure vehicles threatened to take over park thoroughfares entirely. Bicycles found the only long stretches of usable pavement in the parks. "Racing paths" did not draw them off in sufficient numbers. Automobiles constituted an even greater hazard. The *Chicago InterOcean*, which organized the nation's first automobile race, conducted it in the city parks.[107]

By 1899, the South Park Board was trying unsuccessfully to exclude automobiles from park roads altogether. Thus one of its engineers described roads in parks as a "regrettable necessity."[108] Park police were the first law enforcement officers to face manifestations of the new technology in large numbers. It was they who first resorted to shooting out the tires of speeders in 1906, but their campaigns were ultimately futile.

The private auto transformed the parks. In 1906, West Park Commissioner Eckhart answered spokesmen for an automobile club requesting a 15 mph speed limit by telling them that "the board will not take any action at any time that will accord any special privilege to any one set of men." Two years later, park commissioners were noting the destructive effect of automobiles on park pavements and were taking steps to improve the surface. In 1908, the West Park Commission began the installation of traffic warning signals at heavily traveled boulevard intersections; the South Park District had already put in pedestrian safety islands—another concession to the inevitability of speed. By 1911, the West Park commissioners had raised their boulevard speed limit to 15 mph.[109]

The city's boulevards were naturally attractive to motorists. Their pavements were designed for pleasure vehicles, and they were free of streetcars and other large, slow-moving vehicles. "Parkways"—grassy strips alongside the main boulevard roadways—cut many boulevards off from smaller side streets. Park district rules and a 1917 city ordinance required other vehicles to stop before crossing or entering boulevards.[110] Thus, they easily became "speedways" for motorists, precursors of the "limited ways" of the 1920s and the "superhighway" plans of the 1930s. Park roads were able to respond so well to the demands of the automobile, however, only because park finances and boulevard building had become subjects of intense civic concern.

Boulevards, like parks, were advocated for a variety of reasons at the turn of the century. One plan of 1897 envisioned a south shore boulevard as a means of preserving the lakefront against nonpark uses.[111] Arnold and others saw the boulevarding of Washington and LaSalle Streets in the CBD as a means of segregating traffic. The *Tribune* sought a boulevard link across the

CBD as a civic symbol, as well as a means of alleviating Loop congestion. Some State Street interests wanted a commercial boulevard on their thoroughfare, and Edward Dunne wanted to make a boulevard of working-class Halsted Street.[112] Whatever their purposes, however, boulevards benefited—with the parks of which they were parts and extensions—from the increased concern over recreational space.

In 1905, Chicago business interests led the fight in Springfield which produced a law allowing Chicago park boards to issue bonds that would not become part of the city's debt. By 1906, park plans for the metropolitan area which would total $25 million were before the public. The South Park Board alone spent $4 million in the four years before 1907 and, in that year, won approval of an additional $3 million bond issue at the same referendum which approved the traction ordinances.[113]

The park boards thus made a twofold contribution to the city's adaptation to the Motor Age. First, they acquiesced in the redefinition of the boulevards from pleasure drives to automobile preserves. Boulevard traffic counts are even more irregular for the prewar period than CBD counts, but a check at Jackson and Sacramento, four miles west of the lakefront, showed 765 automobiles in 1911 and 3,551 for a comparable period in 1914: an increase 2.5 times as great as the rise in city auto registrations.[114]

Giving rise as it did to bonding for city improvements which was not subject to the city's debt limit, the park movement also made a breakthrough in urban finance. But the park districts themselves, subject as they were to frequent charges in inefficiency and extravagance, could not have brought about this change by themselves. As apparantly was the case in cities around the nation, it was a united effort of the business community—much like that which had helped to put across the 1907 traction ordinances—which opened the way for park expansion.[115]

The awareness of the necessity of practical city planning—as opposed to beautification—began in Chicago, not with parks, but with Arnold's 1902 report on public transportation. The scientific study of Chicago traffic had begun with this report, which showed the importance of lines of travel and argued that those lines could be at least partially modified. By 1905, when aldermen could still scoff at city planning as something for those who wanted to "loaf around in a palm garden," order and planning were major themes of the movement for transit reform.[116]

Arnold's work, and the traffic studies which followed his report, showed the importance of team traffic in Loop congestion, and this in turn pointed to the need to consolidate railroad terminals, segregate traffic, and increase the size of the CBD. Each of these problems became targets of the essentially automobile-oriented Chicago Plan. While the traction struggle failed in its major goals, the transit planning movement which arose with it helped give credibility to the street planning movement which ultimately replaced it.

The Plan of Chicago and the Chicago Plan Commission

[We proceed] upon the assumption that the average man in official position is a man devoted to the faithful performance of his duty. . . . In Chicago we place implicit trust in our city officials.

Charles H. Walker (chairman of the Chicago Plan Commission), 1913[117]

It was Franklin McVeagh, the same man who later dismissed municipal ownership of traction on the ground that the city council lacked the wit to run an inkstand, who first brought the need for a city plan to the attention of the Commercial Club. McVeagh was far from the first Chicagoan to make such an observation; but Daniel Burnham's *Plan of Chicago*, commissioned by the Commercial Club in 1906, became a local watershed. The plan fused the movement for a "city beautiful" with the ongoing drive for an efficient, accessible metropolis which until then had been focused on transit planning and the struggle with the traction companies. It served as the single most important agent in diffusing the idea that good streets were essential to a livable city and created an invaluable tool for the accommodation and channeling of the flood of private cars which was to come.[118]

Burnham's *Plan of Chicago*, which first saw print in 1909, was above all a call for order in urban growth. Burnham and Bennett wrote that the desire for a plan had arisen from a general dissatisfaction with "Chaos."[119] The heart of the plan s program for order was street building. Beautiful streets, Burnham argued, could make life pleasant and "felicitous": broad, well-located streets could renew the city's circulation and reinvigorate its people and its commerce simultaneously.[120]

But streets were to do more than edify, inspire, and promote easy movement. They were to take over transit's role as a weapon against slums. The planners described in some detail the section of the city centering around Chicago and Halsted Streets—the jumble of factories, warehouses, railyards, and slums popularly known as Little Hell. Electrification, they wrote, could cut railroad pollution but, as to the slum itself, the remedy was "the same as has been resorted to the world over: first, the cutting of broad thoroughfares through the unwholesome district"; second, the enforcement of sanitary regulations.[121]

Burnham's plan, then, was more than a program of street building and landscaping, though these were the features which received the most attention from the men who oversaw its implementation. The *Plan of Chicago* drew together and helped to give quasi-official status to the attitudes which made it easier for Chicago to adapt to the automobile. It also provided a beautiful and concrete visual image of a Chicago rebuilt around broad, well-landscaped streets.

The *Plan of Chicago*—the book presented to the public in 1909—deemphasized mass transit, much as city policy had deemphasized it, by segregating it from other considerations. This was done subtly and, perhaps, unconsciously. Burnham advocated the extension of street railways into the suburbs along the rights-of-way of his proposed interurban highways, and he devoted four pages of discussion to a comprehensive subway system which reflected the contemporary concern with spreading the CBD while preserving the centralization of the existing system. Indeed Burnham's entire plan was a transit plan, for—while transit vehicles are absent from most of the drawings—Burnham evidently expected them to be present on or under the real streets, benefiting from the widened thoroughfares and the more direct diagonal routings.[122]

Transit, however, was incidental, a utility to be omitted from the broad vistas of idealized roadways. The only extended treatment of mass transit in the plan is grouped with its treatment of railroads in a chapter devoted chiefly to the control of a necessary nuisance—the presence of steam trains in the downtown district.[123]

Had the Burnham plan been carried out in most of its details, mass transit would have benefited immensely. Through World War I, however, the plan was not implemented except as it affected the lakefront, several CBD streets, and a single west side diagonal thoroughfare. As implemented, the Chicago Plan before 1917 was a program of urbanizing parks and parkways, and turning them to the service of the CBD. The plan's major impact was made through the propaganda generated in its behalf, propaganda which emphasized facilities for making the private motorcar a more viable means of urban transit.

Publicity, as Alvord had point out, was an essential element in any movement for publicly financed street improvements. Boulevards were especially unpopular with the general public. According to the managing director of the commission set up to implement the plan, "When you mention boulevards in Chicago you present a horned beast with a forked tongue." An attempt to widen Halsted Street undertaken prior to the Chicago Plan had met strong opposition from small property owners and ward politicians and was ultimately withdrawn despite the support of the city's most influential business leaders as well as that of Mayor Edward Dunne.[124]

With the publication of the plan, therefore, the Commercial Club appealed to city government for the appointment of a body whose function it would be to study and interpret the plan and to work for its acceptance by the public. That commission was given the status of a public body by a 1910 referendum approving the plan as a guide for the city's development.[125]

Thereafter the Chicago Plan Commission (CPC) worked for the adoption of various features of the plan and strove to prevent or alter projects which conflicted with it. But the most important work of the CPC was in popularizing the plan, a job it did so thoroughly that it is probably fair to say that the features the Commission chose to emphasize became for most Chicagoans the whole of the scheme. Through its executive committee, the CPC minimized the role of mass transit in the plan as it became known to most Chicagoans.

The CPC, as originally constituted in 1909, was composed of 238 members and included small business and labor spokesmen, along with CBD merchants and public officials. From the beginning, however, the Commission was run by its executive committee consisting, after 1910, of Chairman Charles H. Wacker, Vice-Chairman Francis Bennett (an architect and co-author of the plan), one alderman from each ward, and a dozen of the city's most prominent businessmen.[126] The records of the early meetings of the entire Commission make it apparent that that body was intended merely to approve and lend credibility to the decisions of the executive committee. The executive committee determined what communications came before the Commission, and the Commission was expected to give committee decisions unanimous acceptance and to engage in minimal debate.[127]

The executive committee early determined to avoid points of political controversy, and this decision structured the way the CPC presented the plan to the public. Wacker therefore ruled out of consideration all matters pertaining directly to housing conditions and mass transportation as affairs over which other agencies had jurisdiction.[128] The Commission did devote great energy to two remaining matters—the improvement of streets and park facilities, and the popularization of the ideas of the plan.

From the time of the appointment of Chicago Association of Commerce Manager Walter Moody as managing director in 1911, Commission members engaged in hundreds of yearly meetings with neighborhood groups and civic betterment associations. The CPC issued two summaries of the plan which gained worldwide exposure and acceptance and mailed copies of an outline of the plan to every Chicago citizen who owned property or paid over $25 monthly rent.[129]

Perhaps the Commission's most important achievement was the assignment of *Wacker's Manual of the Plan of Chicago* as required study material in all the eighth-grade classes of the city's public schools after 1912. The contents of this manual, which was taught to grade school students by rote memorization, as well as the contents of Moody's speeches and other widely distributed plan literature convey something of the vision of a good urban life which the Commission sought to present to the city's people.[130]

As in the plan itself, broad streets and public spaces were presented by CPC spokesmen as a panacea for urban ills. The widely held idea that city life led to physical degeneration was used repeatedly as an opening argument for the necessity of the plan.[131]

Eighth-grade children memorized the fact that the world was urbanizing, that the "first three things of which our cities are demanding the conservation [are] the health of the people, the streets by which they may conveniently go from place to place [and] the parks within which they may

conveniently find recreation." They memorized, too, the fact that the plan
was a tool of social control: "What do the scientists tell us are the results of the
narrow and pleasureless lives of our people? A: Outbreaks against law and
order."[132]

"In all our educational propaganda," Wacker told the secretary of war,
"[we emphasize] the health and happiness of the people, hygienic and social."
Wacker described the plan for the readers of the *Chicago Examiner* as a "man
building plan," and Commissioner of Health George Young complained to the
same readers about Chicago's "muddled districts" with "good and bad housing indiscriminately mixt," for which the only solution was ". . . to create a
basis for good housing by carrying out the major street projects of the Plan of
Chicago [for] . . . housing congestion is largely a result of our non-solution of
Chicago's transportation problem as it affects our streets."[133]

In short, the Chicago Plan took over the major social functions attributed by some nineteenth-century thinkers to mass transit, including the task
of "making Chicago one great centralized city."[134] In *Wacker's Manual* only
two specific transit improvements were mentioned: a Loop subway for suburban commuters, and the placement of streetcars in the middle of a widened
12th Street—neither of them ventures like to have had great appeal for the
students who memorized them. As in the original Burnham plan, mass transit
was treated as a utility and was lumped with steam railraods.[135]

Streets, on the other hand, were so important to the city plan that the
CPC taught they must be built without regard to the city's budget. Moody
drew a round of applause from an audience of government and business
leaders with the declaration that "we ask ourselves too closely 'will it pay?' I
sometimes think . . . [that] in other countries . . . they do not so much ask
themselves 'will it pay from a monetary standard?' but 'will it pay as an
investment, in the happiness and contentment of our people?'"[136] Of a city's
debt, when incurred for street widenings and park improvements, Moody
argued: "solvency as it pertains to the City of Chicago does not spell progress;
in our case it has spelled retrogression."[137]

Arguments like these, which could never have been made in behalf of
mass transit, appear to have done nothing to diminish the standing of the CPC
with the business and political elite of Chicago. By 1919 the Commission's
managing director could refer to it as an "adviser" to the city council, and from
the beginning of its existence it rejoiced in the unanimous and enthusiastic
support of the press.[138]

•   •   •

Between 1910 and 1917, the Chicago Plan Commission selected parts of
Burnham's *Plan of Chicago* and pressed for their immediate execution. The
main thrust of its efforts was the enlargement of the CBD and the completion
of its transformation into a purely commercial center (to the exclusion of
wholesale and freight transportation uses).

The Commission attempted four major endeavors, one on each side of the Central Business District. To the south, it sought removal of railroad yards beyond 12th Street, 1.5 miles from the center of the business district, with the hope that the Loop could then expand in that direction. East of the CBD, it pressed for a large lakefront park and recommended that railroad tracks which ran along the lake be placed in a depression, where they would be out of view. To the north, a major new boulevard was to be built across the river, extending the high-class commercial district in that direction. To the west, the CPC tried to prevent the installation of railroad terminals so that business could expand in that direction as well.

The details of the Commission's efforts on these four projects are of limited, local interest. However, in pressing for these four improvements, the CPC developed tools and tactics of great importance. In these first four battles, the Commission found or created the techniques by which it later sought to help the city adjust to the coming of large numbers of private cars.

The widening of 12th Street and its elevation across a half-mile stretch of railyards just south of the CBD were the first tasks undertaken by the Commission. Once this viaduct was completed, the CPC hoped to convince the railroads whose tracks blocked the southward expansion of the Loop to remove their stations to the new boulevard, freeing a half-square-mile of land for commercial development. [139]

The Burnham plan had contemplated a boulevard on 12th Street, but the Commission found, almost at once, that local property owners—many of them small merchants or cottage dwellers—looked upon boulevards as automobile speedways or pleasure promenades and were disinclined to pay for either. [140]

The Commission's first tactic was to alter the plan by substituting a widened street with trolley service in the middle for the projected boulevard. Its second innovation was the presentation of this new plan to property owners in frequent neighborhood meetings, where it was argued that the widened street would act as a stimulus to local commerce. Still encountering stiff opposition to assessments for construction which local citizens felt would take their property at court-determined prices and leave them with land of unknown value, the Commission sought changes in the assessment procedure which might allay some of the opposition. [141]

The CPC pressed the state to give the city the power to spread assessments over a wide area of the south and west sides, and to condemn more property than would actually be used for the improvement, so that part of the land taken could be sold to finance construction. [142] These two procedures, the spreading of assessments and "excess condemnation," became basic techniques for financing plan improvements which met with opposition from the affected taxpayers.

A final tactic was the exposure of resisters to public ridicule. *Wacker's Manual* and Moody's public addresses heaped scorn on the alleged parochialism of those who opposed the transformation of their streets. In a fairly typical blast, Moody described how "[i]n a public meeting . . . a man . . . he was not even a property owner, he was a long-haired proselyte of sensationalism . . . he howled 'how are you going to benefit Chicago by widening 12th street?' He had no more idea of the tremendous importance of the Plan of Chicago . . . than the man in the moon." The Commission's conception of the public interest, which centered around beautification and the enhancement of the CBD, was presented as transcending all other concerns.[143]

Twelfth Street was widened and elevated; the project was essentially complete by 1912. Neither the removal of the railyards nor the spread of the CBD toward the south commenced until the end of the 1970s, but the CPC had acquired both an important victory and an essential set of tactics.[144]

These tactics were applied with greater long-term success when the Commission sought to improve Michigan Avenue, the street at the east end of the CBD. The Commission called for a double-decked bridge across the Chicago River, where none had been before, to separate freight from passenger traffic, and the widening and boulevarding of the narrow street which continued Michigan Avenue north of the river.

All of the aforementioned tactics were used to promote this massive alteration of the city's "front yard." Especially notable was the extension of the "benfitted area" which paid its cost to include a broad section of slum, whose residents in reality had neither connection with nor use for the new boulevard. By 1918, the Michigan Avenue improvement was underway. It was ultimately one of the CPC's most outstanding successes: the new North Michigan Avenue became in the 1920s a second CBD, catering to more affluent consumers and relying for access to a large degree upon the automobile and the taxi.[145]

To the east and west of the Loop, the CPC faced a different set of challenges. Here it had to deal, not with tax assessments and property owners, but with major corporations—in both cases, railroads. Its tactics thus were different, and its success mixed.

East of the Loop, Chicago's lakefront in 1910 was occupied by a landfill through which stretched the tracks and yards of the Illinois Central Railroad. The Railroad held riparian rights along most of the lakeshore. Long negotiations ended in an arrangement whereby the Illinois Central assisted the city in filling in the area southeast of the CBD and for five miles south of 12th Street along the shore. In return, the railroad gained the right to run trains along a depressed right-of-way through the landfill.[146]

The importance of all this was twofold. The Commission had helped settle a long-standing controversy with a major railroad and was able to do so, in part, because of the prestige of the plan. And the city got more than a "free" park: the railroad's landfill later helped to provide the right-of-way for the city's first "proto-expressway," the South Shore Drive. The city's physical environment, the plan, and the Plan Commission's ability to bring together business and political leaders combined to produce this important piece of groundwork for the later expressway infrastructure.

On the Loop's west edge, the Commission was not so successful. There it sought to block the building of a union railway terminal by the Pennsylvania Railroad and its allied lines. This station, a half-mile from the center of the CBD at Adams and Canal Streets, was in consonance with the original Burnham plan. The CPC, however, strongly opposed the projected station because it wished to see all railroads removed to the area south of 12th Street, thus clearing "for business purposes a large part [of the central district]."[147]

The Commission, however, could not unite the downtown business community behind this proposition. State Street merchants in particular, though they supported the plan in general, were loathe to see a large percentage of city commuters delivered to 12th Street, 1.5 miles from the center of the traditional retail center. As a result of Loop business and railroad opposition to a 12th Street Union Station, the Pennsylvania Railroad had its way.[148]

Here, then, the CPC failed in its objective because it could not unite the business and political communities behind a long-range plan which had immediate costs for major CBD interests, and because it could not force a major, non-Chicago corporation to spend money in a way the corporation believed inefficient. The CPC never repeated the mistake of launching a major effort without a consensus in the business and political communities.

One further example of the Plan Commission's efforts deserves attention because it reveals both the importance of mass transit to the aspirations of neighborhood business and the reasons for the CPC's inability to deal with public transportation.

In the autumn of 1913, CPC Chairman Charles Wacker noted a feeling among residents of the city's northwest section that "our plan is all for the benefit of the downtown section." The Commission thereafter held hearings to determine what project might next be undertaken on behalf of the city's west division. What emerged from these hearings was the realization that the major interest of west side business and civic leaders lay not in boulevards or street widenings but in mass transportation.[149]

The CPC therefore undertook its only major effort in behalf of mass transit—a drive for streetcar subways to close the breaks in two west side transit lines occasioned by the Chicago River and by railyards. The proposal for streetcar subways on these streets, respectively 2 and 2.5 miles west of the city's north–south axis (State Street) and within a mile of the city's population axis, had originated with the BOSE as part of a 1913 report on subways. Unlike the BOSE, however, the Plan Commission was willing to recommend that these west side subways be built before downtown subways.[150]

In the sectional climate of Chicago politics between 1910 and 1930, such a proposition stood little chance of success. Any effort at subway building involved not only sectional politics but difficult and highly politicized negotiations with the streetcar companies as well. Committee on Local Transportation Chairman Capitain told the CPC quite frankly that the building of west side subways before that of a downtown subway was politically impossible. Despite a serious effort, the Plan Commission eventually had to concede defeat.[151]

• • •

The Chicago Plan Commission, then, succeeded in putting across major downtown street improvements at a time when CBD subways were politically out of the question. When it sought to bring about mass transit improvements, however, it was soundly defeated. In attempting to relocate the planned Union railroad station, it met opposition from railroads and downtown business. In seeking streetcar subways for the neighborhoods, it was thwarted by sectional politics. More than corporate power and local jealousies were involved. In the decade following 1910, transit was still of more immediate importance to businessmen, and perhaps to most citizens, than were improved streets, and *because* it was so important, mass transit could not be dealt with effectively.

Nonetheless, the Plan Commission made an extremely important contribution to Chicago's local transportation history during this period. The improvements it sponsored—Michigan Avenue and the south lakefront landfill in particular—laid the groundwork for the transforming role the automobile was about to assume. But the most significant work of the CPC before 1917 was not the facilities built on its initiative, but the changes in public attitudes it helped to bring about.

The plan helped to transform the public image of the automobile street from a promenade for the rich to a necessity of modern urban life. Burnham, himself a daily automobile commuter, recognized the potential of the private car as an agent of dispersion. His plan is a landmark in Chicago transportation history because, in itself and in its interpretation by the CPC, it marks the transfer of the center of attention from transit planning to street planning. Burnham's plan really said nothing which had not been said before. It simply, quietly, shifted the focus of the drive for a more efficient and livable city away from the context of political and corporate villains, profits, and regulatory battles.[152]

The world of the Burnham plan, therefore, was a far different world from that of the Arnold plan, or of any transit program. It was a world in which questions of cost, business efficiency, political corruption, and local jealousies were to be put aside, replaced by enthusiasm for a program of publicly financed improvements that purportedly benefited all alike. It was also the world of the automobile.

The very unreality of this vision points up an essential difference between mass transit and the automobile—one transcending technology. A major part of the difference between Burnham and Arnold, between the Chicago Plan Commission and the Committee on Local Transportation, between traffic regulations and transit regulation, was this: that every issue—political, social, and economic—associated with mass transit had to be faced at once by whoever tried to deal with or plan for it. Street and city planning (as that came to be understood) could be discussed in terms of an idealized vision. Simply, streets never became politicized in the way mass transit had been.

With time, the technology of the automobile, by making possible individual ownership of the "transit vehicle," would combine with spreading affluence to allow most citizens simply to forget the whole complex of issues associated with urban transportation. But in 1917, streetcars and subways were more significant in policymakers' thoughts and in people's daily lives and, on the whole, more efficient than boulevards, parkways, and automobiles. Mass transit was the pressing planning and transportation issue of the 1907–1917 period, and it became increasingly evident during this time that, under regulated private ownership, transit could fulfill the aspirations neither of planners nor of straphangers.

Chapter 3:

The Failure of Regulation: Mass Transportation, 1907–1915

Section I:
Service and Profit

> The street railway franchise ordinances of Chicago . . . and the Cleveland settlement ordinances of 1910 . . . represent the high water mark thus far attained in . . . municipal franchise granting in America.
>
> *Delos F. Wilcox, 1911*[1]

Chicago adapted itself slowly, often only half-consciously, to the needs of motorists. By the early 1920s, the city's traffic policy—so pragmatic as to seem hardly a policy at all—had been much assailed for its backwardness.[2] Yet the city's treatment of the automobile was, in the long run, both flexible and, from the point of view of the motorist, progressive.

Chicago's 1907 traction policy, by contrast, was extremely well thought out—the product of seven years of organized research and intense political discussion. Between 1907 and 1915 Cleveland, Detroit, San Francisco, Toledo, Seattle, Toronto, and Des Moines, among other cities, reached settlements with their transportation companies which included features first set down in the Chicago ordinances and which were to some degree modeled on the Chicago settlement.[3]

Delos F. Wilcox, the nation's foremost independent student of the politics and economics of traction, was far from being alone in his opinion that the Chicago ordinances were an examplar of a fair and progressive traction settlement. When, in 1911, Bion Arnold drew up a list of provisions a city might wish to seek in a settlement with its transit companies, he could think of only one thing—subways—which was not provided for in the Chicago ordinances.[4]

The Chicago settlement then is of broad significance. The settlement represents some of the best early twentieth-century thinking on the problem of the relationship between cities and their transit systems, and it is to a great degree typical of the sorts of arrangements made in American cities outside the Northeast. Further, it embodies two important assumptions built into most of the Progressive Era attempts to deal with big business, locally and nationally: that public services, whenever possible, should be provided under the spur of the profit motive; and that government regulation could best assure good service and honest dealings.[5]

Thus the fate of the Chicago settlement carried implications for transit reform on the national level. Three East Coast cities—New York, Boston, and Philadelphia—found a different sort of solution. In these three cities rapid transit underwent a change in definition. By 1917 all three cities had taken steps which would ultimately involve them (indeed, if they had not already involved them) in financial repsonsibility for rapid transit. Other cities had arrangements with their transit companies which provided greater security for the investor or more thorough regulatory power for the city; however, as Wilcox noted, no city but Cleveland matched Chicago's combination of provisions protecting all parties.[6]

Could regulation, overseen by impartial engineers, make good public service and reasonable profits compatible in urban mass transportation? Between 1907 and 1914, after the most offensive abuses of the nineteenth century had been eliminated and before the profitability of mass transit had been seriously undermined by extraneous factors such as wartime inflation and the competition of the automobile, the Chicago experience provides a testing ground for regulated private ownership.

The ordinances can be examined in four general areas. First, how well and at what cost to the future were the specific service provisions of the settlement fulfilled? Second, did the provisions of the ordinances succeed in freeing mass transit from political manipulation while assuring effective local regulation? Third, could engineers help plan the city's transportation future and mediate between city and companies without themselves falling prey to political controversies? Fourth, how did the settlement itself affect the city's efforts to gain a comprehensive, unified, modern transit system—in other words, what was the impact of this settlement on planning and technological innovation?

The 1907 ordinances set five basic service goals: new equipment,

reasonable speed and comfort, through-routes, extensions, and the creation
of a single, low fare for service throughout the city.

•   •   •

The Chicago settlement produced new equipment in abundance: by
1917 all of the surface lines' cars were of twentieth-century design. Five
million eight hundred thousand dollars was spent on new rolling stock alone
during the first three years under the ordinances, and the new cars did make a
difference. The BOSE estimated in 1911 that faster equipment had made it
possible for a worker under the new regime to live conveniently over half a
mile further from his place of work. In the first year under the ordinances, the
number of passengers per seat also declined: the dream of a seat for all riders
must have seemed to be approaching realization.[7]

But in the long run, new cars did not produce greater speed and
comfort. Traffic congestion, increased ridership, and the companies' necessary commitment to maximum profits offset the benefits brought by larger and
faster cars, and ultimately made the situation worse than it had been in 1907,
when the ratio of downtown passengers to seats was 1.4 to 1. By 1911 there
were 1.6 passengers per seat in the Loop, 2.12 per seat in the greater CBD.
The proportion of seated passengers on streetcars leaving the Loop declined
through 1912, and the same situation prevailed on the elevated railroads.[8]

Unmanageable downtown traffic and an increase of nearly 20 percent in
streetcar ridership between 1907 and 1914 contributed much to the new
levels of crowding.[9] Obviously neither factor can be traced to the 1907
settlement. But the companies' search for profit helped make crowding
worse.

In 1912 the BOSE found that each day the surface companies were
sending roughly 20 percent more cars than were needed into the CBD. In
1911 the four competing elevated railways were routing a 35.2 percent excess
of cars around the two-track Loop. The large number of cars slowed traffic and
meant that fewer cars were actually available at crucial points in the CBD
during the short peaks of the rush hour.[10] Each company, in order to obtain
the maximum proportion of the downtown business, was sending its cars
further into the congested Loop than was necessary. Although this was
inimical to good transportation, one transit manager explained that it was still
a legitimate business practice and the duty of managers to traction
stockholders.[11]

•   •   •

Through-routes—streetcar routes which enabled the passenger to ride
across the city through the CBD without changing cars—were another de-
sideratum of the 1907 era. In theory, they would cut the number of cars in the
Loop (since cars would pass through the CBD without doubling back) and
unite the divided sections of the city. Here too, however, the companies'
legitimate concern with profits reduced the effectiveness of the settlement.

Companies objected to through-routes because of the long "dead" mileage sometimes involved and because they could not agree among themselves upon a fair method of dividing the receipts collected on cars that ran over the tracks of more than one company. Surface companies consequently operated through-route cars at infrequent intervals and—it was alleged—marked them with poorly designed signs placed in hard-to-see corners of the cars. The elevated companies simply refused to through-route cars before 1911. The city alternately bargained with and threatened the companies, to no avail.[12]

• • •

The 1907 ordinances also promised extensions—a minimum number of miles to be built each year as directed by the city council. Because of the capitalization provisions of the ordinances, between 1907 and 1910 the companies constructed one hundred miles of track *beyond* what was required by the agreement. After 1912, however, the north side company resisted potentially unprofitable extensions. The company president observed that there was no reason why "we should build a line to lose all the money we can make on other lines, simply to carry out the ordinance." Beginning in 1914, the companies were either unable or unwilling to fulfill the quota for annual extensions and again used the extensions as a bargaining counter in negotiations with the city.[13]

Extensions into undeveloped territory were unproductive for the system as a whole in any case.[14] Yet it is significant that the promise of annual extensions, like that of more seats for the straphangers, proved impossible to fulfill under the 1907 compromise largely because of the status of transit as a private business.

The fifth service promise of the ordinances, citywide service at a single low fare, could be obtained under unsubsidized private ownership only at considerable cost to the viability of the system and to the prospects for future improvements. Before addressing this point, however, it is necessary to look back at the reasons the city was able to gain the service improvements it *did* under the ordinances. The key was a quirk in the profit-limiting feature of the settlement.

For the surface companies, all returns above 5 percent on the companies' valuations were to be divided according to the 55/45 formula. Thus the greater the companies' "book value," the greater the proportion of their earnings they would be able to retain. Further, the ordinances called on the BOSE to add amounts to the valuations according to formulae which did *not* bear any direct relationship to the amount of securities issued to cover specific improvements. Thus, by raising their valuations in specific ways, the lines were able to increase the amount of cash they could allocate according to their own internal needs.

By the terms of the 1907 agreement, the cost of track, cars, and other equipment classified as "rehabilitation" was added to the companies' purchase price without deductions for the equipment replaced. Ten percent was added to the price of these items as construction profit and five percent as brokerage charges, and the companies were empowered to issue bonds and claim a 5 percent return for the total amount. Since the ordinances specified the reconstruction items in terms of *minimum* quantities, the companies were able to add as much as they wished to the properties under these attractive financial terms. In addition, during the rehabilitation period, all ordinary renewal work whose cost, when added to operating expenses, exceeded 70 percent of the year's gross was added directly to the companies' capital accounts.[15] Under these procedures, the companies added $86,320,854.16 to their original $53,596,728.13 capitalization between 1908 and the end of 1914. Of this amount, $3,540,688 was renewal money added to the valuation under the rehabilitation period provision and the remainder was new equipment and purchased properties. By 1912, Arnold estimated that $30,000,000 of the companies' capital accounts was water, the amount having tripled since the original valuation.[16]

With this system of incentives, the 1907 contract worked well enough, although the long-term effects of these procedures on both the city and the companies were very bad.[17]

The attempt to achieve the city's fifth service objective—a single fare throughout the city—further revealed the limitations of regulation. The city could regulate service only within its boundaries, and it had to give at least the appearance of treating all sections alike. In addition, to gain fare concessions from the companies, it had to allow still more overcapitalization. All these factors combined to assure that the city could achieve its objective of "one city, one fare" only at the cost of making the system as a whole less rational and, it might be argued, less equitable.

After 1907, three small south side companies and some outlying lines on the far northwest side remained outside the agreement. Most of these lines were providing a low level of service in undeveloped sections of the city, and some served territories both inside and outside the city limits. The separate operation of these companies violated the principle of "one city, one fare," so the city set out almost at once to arrange for the absorbtion of those lines which ran within its boundaries by the two large companies operating under the 1907 agreement.[18]

The purpose of all this, in the case of the small south side companies, was to extend the 5¢ fare to the city limits (fourteen miles south of the Loop) and, as Walter Fisher observed, to bring about a subsidy for unprofitable service by a "going concern" (the Chicago City Railway). This aim was achieved, but only after over $2 million in "intangible values" was added to the capitalization of one smaller company.[19]

When this line, the Calumet and South Chicago, began exchanging transfers with the Chicago City Railway, a passenger could in theory travel over twenty-five miles for a nickel. In a few locations the new arrangement did tie together unnaturally divided sections of the city. At the same time, however, it carried to a new extreme the inevitable subsidy of service and real estate development in new territory by patrons of more heavily traveled routes.

On the north and west side, the "one city, one fare" policy was implemented in an even stranger fashion. Here the city pressed for the absorption of the in-city lines of the weak Chicago Consolidated Traction Company by the large Chicago Railways system, with the goal of extending the 5¢ fare to the northern and western city limits. Chicago Railways was, for a variety of reasons, in a good position to resist the city's pressure, but the city eventually offered an especially attractive overcapitalization "prize" for the assumption of this unwanted burden. By 1910, an agreement had been reached.[20]

There was a heavy planning price for this unification of service within the city. By severing the CCT's suburban lines (wires were actually cut at the Evanston border), Chicago lost all direct surface transit connections with suburbs immediately north and west of the city. One could now travel fourteen miles from the south end of the city to the Loop for a single fare, but a trip to western and northern suburbs six to nine miles from the CBD required two fares and a change of cars. This was even more senseless because in some places the areas on either side of the city limits were already being built up. Protests, especially from the western suburbs, were bitter and protracted, but the petitions and lawsuits were of no avail.[21]

Thus the city went to considerable lengths to get single fare service within its borders, but was unable (and perhaps unwilling) to maintain direct connections with contiguous built-up areas outside the city limits. At the same time the city encountered great difficulties in gaining completely unified service within its own borders. There even "universal transfers" (valid from any car to any other car going in the same general direction) proved difficult to achieve, despite the promises of 1907.

The companies had insisted on the exclusion of the three-square-mile CBD from the transfer provisions of the 1907 ordinances and agreed to its inclusion in 1914 only with reluctance. One reason for this was the prevalence of transfer cheating, which—in the atmosphere of hostility toward traction companies which pervaded Chicago and other cities—assumed the status of an organized "racket." Drugstores, newsboys, and, in one instance, an entire force of salesmen backed by their employer conspired to defraud the companies of their fares.[22]

The transit companies, for their part, applied the strictest rules they could to transfer use, rules not always compatible with the most rational use of the system. Eventually the Chicago system did allow wide transfer use. A bargain was struck in 1914, in consequence of which the transfer to paid-fare ratio in Chicago came to be higher than that which had led a New York court to order an end to free transfers on the surface lines of that city in 1907. As in the matter of extensions, however, the city's power over transfer privileges was not real regulatory power; the city had to bargain for the broadened transfer privileges.[23]

Thus, the 1907 settlement failed to achieve some of its most important service objectives and attained others only at the cost of still-greater overcapitalization; overcapitalization which in later years made unification of the surface and elevated companies on the one hand and public ownership on the other more difficult to achieve. In all cases, the fact that public transit was operated as a business exacerbated the city's difficulties.

But the settlement had another set of regulatory goals: it was to give the city the power of day-to-day regulation as well as the power to insulate this regulatory function from partisan politics through the agency of the BOSE. Here too, the settlement failed.

Section II: Politics and Regulation

> When an alderman is in the car, why they [the passengers] are likely to take your head off . . . they think that if you are doing your duty the companies would give the kind of service they are promised.
>
> *Alderman Anton Cermak, 1912*[24]

In Chicago, politics and regulation were inseparable. The city's transit policy had to yield both profits for investors and votes for politicians. Further, the power of city regulation was subject not only to the review of the courts and the passive resistance of the companies, but to the fiat of the state legislature, from which the city derived whatever powers it had. Scholars have shown that federal railroad regulation was indeed the servant not of the public interest but of interest groups and that local transit regulation's inflexibility on the point of fares helped cripple the industry during and after World War I. But the point to be discerned in the Chicago experience is that the regulation of mass transit, by its very nature, placed city politicians in an extremely difficult position. Here it was the lack of real power and responsibility which made regulation, too frequently, destructive or ineffective.[25]

Between 1907 and 1914, Chicago's power to regulate mass transit was abused, frustrated, and finally eliminated altogether. Before the power of regulation was withdrawn by the state, however, the city council had demonstrated that, BOSE or no BOSE, regulation would be political. Public expectations, political advantage and rhetoric, and the need to bargain with the companies for important if unprofitable innovations set the tone and determined the content of transit regulation during this period.

In the case of the elevated companies, with which the city had no contract ordinances, meaningless regulation continued to be imposed in an attempt to gain worthwhile concessions such as through-routing. The city, for example, forbade vending machines on elevated platforms and harassed the companies in a variety of other ways.[26] With the surface and elevated companies alike, the city engaged in a number of minor regulatory crusades, such as the battle to provide lower car steps for hobble-skirted women and the campaign for "fresh air" (windowless) cars in winter (the public assiduously avoided the latter).[27]

When regulation which attempted to make mass transit adapt to passing fashions succeeded, it simply reduced the resources devoted to transportation. And the surface companies counted the fines paid for violations as "operating expenses"—a procedure which, under the complicated provisions of 1907, meant that the city itself paid a large share of the fines.[28] In the regulation of transportation service as such, the city met with still more serious obstacles.

Service regulation required, in addition to the legal power of regulation, two things: information on which to base the regulations, and the will and ability to draw up specific, legally enforceable rules. Throughout the period, both were lacking.

Until 1914 the city council depended for its information on the complaints of aldermen and citizens, and upon the BOSE. In the ordinary course of events, complaints were referred to the council Committee on Local Transportation; the Board of Supervising Engineers made on-the-spot checks of the factor causing the complaint and reported its findings, which were then presented to company representatives for adjustment.[29] This procedure entailed serious problems.

The first problem was the assumption by the BOSE of a business definition of adequate service. It considered infrequent service on lightly traveled lines to be justified. Crowding thereby became the only reason for which the Board would recommend the addition of runs to a line or the operation of all trips to the end of a car line. This policy amounted to the engineers' partial nullification of the city's occasional attempts to extend service ahead of substantial population. The transit system as a whole thus suffered the expense of constructing the extensions and maintaining unprofitable older lines, but the service provided was not necessarily sufficient to meet the developers' needs.[30]

A second difficulty was the fact that the city lacked the means to keep a constant check on service. The companies resisted giving schedules to the council and, on occasion, made changes for the sake of a CLT or BOSE inspection which were subsequently rescinded.[31]

The most serious problem, however, was political, for aldermen were unable to arrive at an acceptable standard of service on heavily traveled lines. Before 1914, the city had no standard which could survive a court test.[32] The events of 1914 brought the council to a hurried effort to arrive at an enforceable standard. They also presaged the end of all city control over local transportation service.

In 1913 the Illinois state legislature passed a bill creating a state commission, to be appointed by the governor, for the regulation of all public utilities. The precedent for such commissions had been set by Wisconsin in 1907, and by 1914 state regulation had assumed the proportions of a national movement, opposed by many utility interests but supported by some in the transit industry.[33] In Illinois, however, it was Samuel Insull, the utility magnate who controlled, among other things, the northeastern Illinois electric power industry and all of the Chicago elevated railroads, who provided much of the impetus for the creation of a state commission.

Like Yerkes before him, Insull seems to have hoped that such a commission would put the utility industry on a more stable basis by placing the regulatory power at the state level where local politicians—and consumers—would have less impact. Illinois Governor Edward Dunne—the same man who, as mayor of Chicago, had fought hard for municipal ownership—sought the inclusion of a separate commission for Chicago. In this Dunne failed, but citing the need of the state as a whole for a utility regulatory commission, he reluctantly signed the bill.[34]

The Public Utilities Commission created by this 1913 law removed all city control over the day-to-day operation of mass transit and in so doing precipitated a crisis which showed that the city could not have effectively regulated mass transit even if it had retained the legal power to do so.

When the PUC was created, the city rushed to create an "opposite number"—the Chicago Department of Public Service. Early in 1914, this Department and the city council Committee on Local Transportation began the search for an enforceable service standard on which to base a legally viable ordinance with which to test the authority of the PUC. The aldermen and other local politicians knew that if the PUC could act successfully on service complaints, it would change the focus of transportation politics by providing a viable alternative to local regulation.[35]

This search for a service standard exposed the fundamental limitation of the elected city council as a regulatory body. Neither the BOSE nor the Department of Public Service (DPS) could formulate a service standard except in terms of the number of standing passengers to be allowed in given periods of the day. BOSE engineers in particular believed that the elimination of standing passengers would reduce profits unreasonably and thus violate the 1907 agreements.[36] To some extent, this amounted to an admission that the straphangers really did pay the dividends.

The city thus had to *limit* a condition (standing) which, in the popular belief, should not have existed at all and which the propaganda for the 1907 ordinances had promised to eliminate. Aldermen steadfastly refused to "vote for straphanging" by endorsing a service ordinance that recognized its inevitability. Ultimately, the council could not bring itself to accept a standard which defined acceptable rush hour service by limiting the amount of crowding instead of attempting to eliminate it. [37]

By 1914, then, it was apparent that the city council could not make enforceable service regulations. More than partisan politics was at fault for the city's incapacity. Aldermen were held accountable for service by the public, and the public appears to have had a conception of good service which was in reality inconsistent with the profitable operation of the transit companies at a 5¢ fare.

The position of the aldermen was uncomfortable in the extreme. They could neither fulfill nor escape the responsibility of assuring an impossible standard of service. In their difficulty, they lashed out, not only at the companies, but also at the Board of Supervising Engineers, which was the bearer of the bad tidings that the public's standard of service was not a realistic one. The conflict between public expectations and the requirements of over-capitalized, profit-seeking transit companies drove the aldermen into a confrontation with the engineers who had been expected in 1907 to depoliticize the traction question.

Section III:
The Politics of
Engineering

Public service commissions feed out of his hand. Public service corporations rise up and call him blessed. . . . There is something fundamentally fair and equitable in his mental makeup that leads to a profound confidence in his trustworthiness and ability. . . . [Bion J.] Arnold would have been an ornament to the United States Supreme Bench, but he would never have filed a minority opinion. . . . He believes that eleven men are more likely to be right than one.

Scientific American, *1911* [38]

You can't do big things unless you make them political.
Bion J. Arnold, 1913 [39]

Much has been written concerning the role of engineers in the Progressive Era. Bruce Sinclair has detailed in particular the efforts of Morris L. Cooke to bring mechanical engineers to the side of city governments in their struggles with local utilities. Cooke's effort was unsuccessful, however, and most engineers who dealt with utilities, including those associated with the academic community, appear to have favored either the corporate position or simply noninvolvement with questions of social policy. Indeed other scholars point to an elitist, antidemocratic element in "engineering progressivism."[40]

Engineers involved in the regulation of Chicago transit between 1907 and 1917 shared a seemingly inherent bias in favor of the corporate position on some issues. Bion J. Arnold, the dominant transit engineer of the period in Chicago, certainly entertained a not entirely unjustified antipathy for local politicians. He favored complete regulatory control of transit by engineers, not elected officials.[41] Yet disdain for democracy and a yearning after elite rule do not appear to have been important factors in the failure of Chicago engineers successfully to fill the role of neutral regulators. The engineers faced the unenviable task of reconciling public expectations with corporate needs in a field where the two may simply not have been compatible. Further, in the political context of pre–World War I Chicago, any meaningful engineering judgment about mass transit was inherently political.

Most of the Chicago transit engineers active in public life *were* transit businessmen. Arnold, the owner of a small interurban railway and an incorporator of a major Chicago elevated line, was no exception. The first "city" engineer on the BOSE was Charles V. Weston, who resigned shortly after the Board went into operation to become president of the South Side Elevated Railway (he was replaced by his brother George). Arnold and the Westons agreed on the need to include "intangibles" such as franchises in the valuations of street railways—a standard company position.[42] Clearly, then, the BOSE was never simply an ally of the city.

Yet Arnold was certainly not an ally of the companies either. Between 1902 and 1914 he was invited by fourteen other city governments to study transit systems around the country; city governments provided a major part of his engineering firm's business. As he had in Chicago, he often told these cities that they had the right to expect more of their transit companies than they had been receiving.[43]

According to Arnold, city (not state) regulatory power was essential to transportation planning. Local government must have ". . . complete control of its public utilities, in such a way that private privileges will not be allowed to seriously interfere with the growth and prosperity of the community." Such regulation, he believed, should be exercised by a board of engineers similar to the BOSE, but with far greater power. Its chairman should be ". . . relieved from changing political sentiment through a reasonable tenure of office," and elected officials should not be able to override the decisions of such a commission.[44]

In the years around 1907, most of the parties to the Chicago "traction wars" appeared to share this faith in the ability of engineers to coordinate the interests of city and companies. Only the Chicago Federation of Labor and Alderman Dever voiced loud dissent.[45] The hope that engineers could act as neutral arbiters of transportation questions was, however, quickly disappointed.

The neutral image of the BOSE was first undermined by the role of its members in the valuation commissions active between 1906 and 1914. The 1906 surface lines valuation, on which the 1907 ordinances were grounded, exposed the difficulty of the engineer's position. In it, Arnold acceded to a valuation totaling over $53 million. The figure eventually agreed on by the city ($50 million) contained at least $11 million of water by Arnold's own later admission. Later valuations of smaller lines discussed above were likewise political compromises endorsed by engineers.[46]

If the engineers on any of these commissions had stood firm for the city's long-term interests, they might have lost their usefulness altogether. This was the fate of a 1911 commission appointed to value the elevated lines. The city representative, J. J. Reynolds, who stood by what he claimed was an honest figure, was simply rejected by the companies for any comparable future commissions, and the negotiations in question ended.[47]

The engineers on the other commissions performed a service for the city by endorsing and "legitimizing" political compromises. Still, they made this contribution to city policy at the cost of long-term damage to their credibility as public servants.

After 1907, engineers who accepted commissions to draw up subway plans found themselves similarly trapped. Arnold's 1902 report, submitted at a time when the city lacked the power to act, was greeted with universal joy and wonder because it showed that *something* could be done. Later studies, however, took place in the context of heated political disputes over *what* should be done and inevitably provided ammunition for one side or another.

CBD traffic and subway studies by city engineers John Erickson and R. C. St. John in 1909 and by Arnold in 1911 made clear the impossibility of politically neutral transit planning. In each case the engineers—taking account of the city's financial limitations, the general desire that subways be publicly owned, and their own belief that half the purpose of subways was the relief of surface congestion—advised the city to build small downtown subways instead of attempting to start a large comprehensive system. Spokesmen for outlying interests, who saw downtown subways as part of a plot to perpetuate the dominant position of the Loop, attacked the engineers as well as the politicians who favored their plans. By 1918, Arnold in particular had become a convenient symbol for the alleged domination of city transit policy by CBD plutocrats.[48]

The best single illustration of the impossible position in which the engineer/planner found himself came in 1913 and 1914. During those years,

Carter Harrison II, who had been elected to yet another mayoral term in 1911, promoted the idea of a citywide subway, to be built by a private company under contract with the city. To gain a stamp of legitimacy for his program, Harrison introduced a new device into the city's transit history: he appointed a Harbor and Subway Commission made up of handpicked members who could be counted upon to produce a plan consonant with his wishes.

The resulting plan was impossibly expensive; no financiers ever came forward who were even willing to use it as a basis for negotiations. But Harrison pressed for a referendum on the plan. He hoped to use it as a threat in negotiations with the existing companies and as a means of consolidating his own political position.

Not surprisingly, Harrison's political rivals on the council Committee on Local Transportation turned to the BOSE for an evaluation of a CBD streetcar subway, a judgment they knew would be embarrassing to Harrison.[49]

The BOSE took the bait. At the end of October 1913 it filed a report advocating downtown subways built in cooperation with the existing streetcar companies. This report, backed by Arnold's personal prestige, drew the support of the downtown business community and helped defeat Harrison's comprehensive subway proposition at the polls in April 1914.[50]

The BOSE had made the only possible answer to the question asked it: if the city wanted subways at all, it could only afford to build small ones to be used by the existing companies. But the "truth" as the BOSE saw it had serious political repercussions for the mayor and served the interests of the companies by revealing the fact that the city could do nothing without them.[51]

Harrison never forgave the BOSE. Political attacks on the board originated in 1909 when some aldermen challenged the board's accounting procedures and pointed out Arnold's role in the valuation proceedings in Chicago and other cities. With the subway fiasco, however, the war between the board and the politicians began in earnest. A professional audit of the BOSE's records was made, the integrity of its accounting procedure was challenged, and a second, independent audit was commissioned. Apparently the latter was of no political value, as its results were never released.[52]

•   •   •

The subway question produced the most dramatic confrontations between the city and its engineers. But the inherent conflict between the needs of politicians and the task assigned the engineers is clearest in the day-to-day disputes between BOSE engineers and members of the CLT.

Aldermen were under constant pressure from their constituents to produce results. The BOSE, on the other hand, when asked for an engineering judgment on the possibility of a crowding standard, for example, had to reply in complex language which left the aldermen unsure of what they could or could not do. "I have a lot of regard for expert testimony," CLT Chairman Eugene Block observed in 1915, "but I have found that expert testimony applied in everyday business . . . is simply paper talk. . . . We cannot get our experts to stand pat. They come in here with reservations and equivocations."[53] Arnold found the aldermen equally frustrating: "When you get the information, you do not know how to use it and you belittle it," he responded. "You stand up," he told the chairman, "and you foam at the mouth, and you don't say anything." In the last analysis, Block's chief complaint against the BOSE was that it performed the very function for which it was created. "Why go after me?" Arnold asked the alderman after a particularly bitter exchange. "Because," Block replied, "you are standing between the city and the companies."[54]

It was from the BOSE that the aldermen learned what the financial and service provisions of the 1907 agreement meant—not a seat for every passenger, but, at best, a limit on the degree of discomfort passengers would be made to suffer. From the BOSE they heard, too, that unprofitable extensions meant service on relatively long headways and that "publicity of accounts" did not give the council the right to accurate ridership data or access to the companies' books. This was information of a sort which, no matter how impartial the judgment behind it, was bound to irritate the politicians.[55]

Attacks on the Board continued for the next two years, but no fundamental changes were made, aside from the replacement of George Weston as city representative by E. W. Bemis, a longtime municipal ownership advocate. In the spring of 1915, the BOSE declared that it would no longer serve as a research organization for the CLT and, after 1916, its annual reports, which had regularly contained traffic and engineering studies, were confined to the recording of data as required by the 1907 ordinances. The city found a more amenable investigator in R. F. Kelker, Jr., a former BOSE assistant. Kelker worked with the city first as an engineer for the Department of Public Service, then as the CLT's own engineer.[56]

Between 1915 and 1924, Arnold slowly withdrew from public life. His career coincided roughly with the rise and fall of the idea that regulated private ownership could provide the realizations of the Planners and the Straphanger's Dreams—an idea in which Arnold seems to have believed firmly. He was, in one sense, a would-be technocrat. A man with what one admiring contemporary called "a profound conviction as to his own inerrancy," Arnold urged the council to place regulation "into somebody's hands that is technical[ly qualified] to understand the problem from day to day." The public, in Arnold's view, *was* capable of understanding and passing upon questions of transit policy "if they could study the question thoroughly, but they will not do it."[57] Had Arnold obtained real power, he might have been a technocrat. In reality, Arnold had influence, not power, and he often used it creatively.

It was Arnold, after all, who had argued against the immediate self-interest of the companies and was responsible for making through-routing and universal transfers viable goals of city policy. Later he willingly argued against the interests of CBD merchants and in favor of early transit improvements outside the CBD. Arnold drew attention to the wave patterns in urban traffic and advocated underground parking garages and, in a study of Los Angeles, urban highways with rapid transit in their median strips. In other cities he advocated tax-supported subways and extensions, paid for by those whose real estate values would profit.[58]

Certainly Arnold conformed to the basic pattern which David Noble found among "progressive engineers" in general: he subscribed to the "corporate liberal" belief in "combining social reform with corporate requirements." Indeed this belief contributed much to his downfall. He proudly indentified himself with the 1907 settlement, recommended similar arrangements to other cities, and consistantly tried to assume a mediating position between city and companies. The problem was that, where the needs of the city's politicians and the transit companies really were incompatible, Arnold's position became untenable.[59] But if Arnold, who did much that was good for the city's straphangers, also contributed to the transit impasse, it was not simply because he favored technocracy. Rather it was because he added the weight of his credentials to the common sense of the time, affirming, quite reasonably in the context of what he could know, that mass transit could serve planners and public as a regulated private enterprise.

Section IV:
Beyond 1907: Policy versus Planning

> Here is a riddle for the children: "when is a subway not a subway? When it is comprehensive."
>
> Chicago Record-Herald, *1914*[60]

The 1907 settlements failed in many of their major goals. They could not make profit compatible with the best of service, assure the city of the power of regulation, or remove transportation questions from the political arena. But the effects of the settlement and of the ideas which produced it reached further than the immediate goals of the ordinances themselves. Between 1907 and 1914 the city persistently tried and failed to achieve unification of the surface and elevated companies and to begin the construction of subways. In each case, the policies set out in 1907 helped assure that the long-range goals of a unified and comprehensive transit system could not be achieved.

Unification

Unification of service among the surface companies, and between the surface and elevated lines, was the obvious final step in the rationalization of local transportation begun by Yerkes and advocated from the turn of the century by Arnold and the CLT. Unification of the surface lines would mean that cars could be run to conform to travel patterns, not to the needs of competing companies. On the elevated lines it would eliminate the practice whereby each company ran each of its trains around the Loop instead of routing it through to another line. Elevated through-routing would allow passengers to travel directly to their destinations and would make the expansion of the crowded CBD easier by opening up the fringes of the Loop to direct access from all quarters of the city.

Most important, if the surface *and* elevated companies could be unified together, routing and transfer policies could be established which would direct long-haul traffic to the elevated lines. This would end the existing anomalous situation in which slow-moving streetcars were crowded with long-haul passengers who did not include the elevated in their journey because to do so would require a second fare.[61]

Between 1907 and 1910, New York, Boston, and Philadelphia, the only other American cities with major off-street rapid transit system, all made progress toward unification of this sort.[62] In Chicago, such progress was very limited. In this failure, too, the ideas of 1907 played a major role.

Throughout the period, important financial interests supported unification: it was in the interest of some of the companies and their backers. By 1911, all of the elevated companies had been drawn together in a trust agreement financed and dominated by Samuel Insull, who also controlled the area's supplier of electric power, the Commonwealth Edison Company. Beginning in 1905, the city's surface transit companies had shifted to Edison power, and by 1911 they were almost entirely dependant upon Insull's company for electricity with which to run their cars.[63]

Insull did not, however, control all the city's transit lines. Rather they were separately controlled by men who had common interests and were in frequent contact with one another as well as with Insull. The relationship was less one of interlocking control than of occasional coordination.

Henry Blair, a banker, was the dominant figure in the north and west side Chicago Railways Company, and in two of the elevated companies as well. He was also a member of the board of Commonwealth Edison, and was in almost daily contact with Insull. Ira Cobe, a prominent Edison board member, played a major role in the affairs of the south side's Chicago City Railway. Yet Insull's involvement with local transit was apparently reluctant. He treated the unification of the elevated and power companies with the surface lines as essentially Blair's project.[64]

The unification of Chicago's transit companies, however, was not sought primarily for transportation reasons. As one trade publication observed, perhaps the chief motive for such a consolidation was ". . . the arrangement by which Commonwealth Edison now furnishes power for the surface railways. . . ." For Insull, the expansion of electric transit meant the extension of the off-peak market for his primary product: electricity. The failure of the elevated companies, which appeared imminent in 1911, would have damaged that market.[65]

The unification of all the transit companies, however, required the consent of the city and its assistance in gaining the repeal of nineteenth-century state laws which forbade such consolidations. That, in turn, meant placing the rapid transit companies, until now regulated by the state, in relationship with the city analogous to that assumed by the surface lines in 1907.

Throughout the 1911–1914 period, the elevated and surface companies attempted to gain concessions from the city in return for such an agreement. They asked for guarenteed returns and the acceptance by the city of valuations reflecting their outstanding securities, not the worth of existing properties.[66]

When the city council rejected this approach and stood firm for low valuations, the companies walked out of the negotiations. To the city, a valuation of the sort demanded by the elevated companies ($40 million above their "real" worth of $53 million) was politically impossible. It would almost certainly raise fares, and would provide excellent ammunition to the enemies of aldermen who voted for it. But the companies had to demand high valuations: their consolidation in 1911 had involved Insull in the marketing of new securities which had to be covered. A low valuation would undermine the companies' credit, and perhaps Insull's as well.[67]

By holding out for lower valuations the city missed the best opportunity it would have to unify surface and rapid transit service. In the event, as the financial positions of all the companies deteriorated, financiers demanded more, not fewer, concessions in return for unification, and the two types of service were not brought under one management until municipalization took place in 1947.[68] Yet in 1913, the year following the elevated companies' walkout, the city achieved a significant victory—its last for a quarter of a century.

In that year the streetcar companies came to the city council seeking unification of their own service. Fearing the effects of the newly created state Public Utilities Commission and fully convinced at last that unified operation would raise, not diminish, their profits, the companies accepted an ordinance calling for a citywide 5¢ fare and for transfers valid throughout the city, including the CBD. In return the city accepted a small degree of additional overcapitalization and made other, more minor concessions.[69]

Chicago achieved unification of its surface lines almost in spite of its own efforts; its policy tradition stood in the way of this progress. In attempting to build subways, the city was not nearly so fortunate. In this case, politically potent outlying interests, whose vision of the proper role of mass transit in developing a city differed from that of the traditional policymaking community, emerged to challenge the control over transit policy which CBD business had established in 1907. Their efforts, combined with the inevitable politicization of all transit issues, blocked subway construction throughout the pre–World War I period.

Subways: Neighborhoods against "The City"

Chicago's failure to unify its transit system was clearly a loss to the community as a whole. The city's inability to produce downtown subways, on the other hand, may be viewed with some ambivalence. As contemporary critics point out, and as many Chicagoans knew well by 1914, heavy rail rapid transit generally benefits downtown merchants and the urban and suburban middle class above all others.[70] If built from city funds or from transit fares, such subways would function as a sort of subsidy for those who used the CBD, paid by those who did not.

Yet, as will be seen, there were already important transfers of revenues and benefits going on within the system. Further, a 1916 study revealed that 48 percent of workers whose journeys could be traced were employed within the five-square-mile CBD, and so many trolleys were still routed through the Loop that the effects of congestion there were spread to the surrounding districts.[71] Loop subways would have served affluent shoppers and workers from the city's north and south lakefront, but they would also have made life easier for thousands of clerks, stenographers, and factory workers, and would have speeded transit service in the inner-city neighborhoods as well.

By failing to underwrite subways, Chicago left intact the tradition of self-supporting mass transit for the entire city and, given the fact of centralization, imposed significant hardships on riders of every class. Conditions like those in Chicago led to the use of public funds to support subway construction in three East Coast cities before 1917. Boston, New York, and Philadelphia, each in different ways, lent public credit or spent public money for the building of subways leased to private companies.[72] Chicago could not follow suit because both political needs and neighborhood interests intervened.

But the neighborhood opposition was not mere obstructionism. It reflected a different vision of the proper role of mass transit and an awareness of the special problems associated with Chicago transit as it existed. The degree of congestion at the peak of Chicago's rush hours was greater than that to be observed in other cities. In the three-hour morning and three-hour evening rush, most cities' transit systems, including Chicago's, moved about half their daily traffic. But in 1909, when the average American street railway was carrying 16 percent of its daily passengers during its busiest hour, 25 percent of Chicago's streetcar riders and 33 percent of its rapid transit traffic traveled during the busiest hour of the day. Synchronous scheduling of business hours and the concentration of a wide variety of activities in the small Chicago CBD contributed to this extraordinary peaking of traffic.[73] Of course, neighborhood interests argued that instead of building subways to accommodate this concentration of business, the city should improve outlying streetcar service and thus allow commerce to disperse.

The city's traditional political policies, combined with neighborhood interests' vigorous and persistent opposition to any subway, assured that no subway would be built.

•   •   •

The first set of obstacles to subway construction—what may be called political obstacles—led to a series of fruitless plans as well as to confrontations between city and companies and among politicians between 1909 and 1914. Since these added nothing new to the traction debate and produced no significant results, they can be dealt with in summary fashion.

Briefly, the city found itself financially unable to build any but the smallest CBD subway from public funds: court decisions making "Mueller certificates" part of the city debt combined with the city's inability to raise the debt limit for transit purposes meant that any major subway project would be possible only in cooperation with the existing companies. But if company money were to be used, the city would have to make politically unpalatable concessions, as became clear in negotiations between 1912 and 1914.

At the same time, the surface companies opposed the construction of a publicly owned streetcar subway because it might be used by the city as a bargaining tool when their franchises expired in 1927. The upshot of the impasse thus created was Harrison's 1914 plan, whose fate has already been discussed. The plan, which called for the city to grant a franchise to yet another traction company—and which was transparently impractical in any case—suffered a resounding defeat in all but the most loyal machine wards. Areas supporting the 1907 settlement and those opposing it, all emphatically rejected in 1914 plan.[74]

In part, then, both subways and unification were blocked by the same factors—the quest for political advantage and the fear of further overcapitalization. But neighborhood groups also played an important part in the defeat of subway plans, not only in 1914, but through the mid-1920s as well. A study of that opposition reveals that in Chicago at this time it was neighborhood interests who pressed for decentralization. Their tool of choice for this purpose was not the automobile but mass transportation.

Engineering studies of the subway question between 1902 and 1914 treated centralization and the dominance of the CBD as natural and inevitable. Subways might determine whether the journey to the Loop was difficult or easy. They might even help determine whether the Loop expanded slowly or rapidly. But the journey for most riders, it was thought, *would* be to the traditional CBD. Housing and even industry might decentralize. Commerce would not.[75]

After 1910, however, powerful coalitions of neighborhood-based interests argued not only that mass transit could shape the city but that it had already been used for that purpose. Membership in this coalitions shifted. Neighborhood business, including the large Goldblatt's chain of department stores, played an active role. So did the Cook County Real Estate Board—an organization of outlying realtors. Socialist alderman William Rodriguez, backed by many of the small businessmen of his west side ward, also argued that downtown business had used mass transit to shape the city at the expense of the people. From time to time the Chicago Federation of Labor took up a similar cry.[76] First and last, however, the spokesman for this point of view was the short, balding organizer, Tomaz Deuther, whose Northwest Commercial Association was based in the heart of Chicago's crowded Polonia district at the corner of Division and Ashland.[77]

In the simplest sense, Deuther and his allies represented the interests of the ethnic and neighborhood business communities which had grown up around the city's major streetcar intersections. Fearing the CBD subways would bring about a shift of customers to Loop business houses, they argued from the unspoken premise that the city must be an economic artifact equally open to all. Just as many turn-of-the-century Loop merchants had seen Yerkes era traction as a bar to the development of their businesses, just as they may have seen political corruptionists and irresponsible traction barons as threats to their hegemony in the formation of city policy, so Deuther and his allies saw transit planning as the enemy of small and neighborhood business—as the force which beat down neighborhood and ethnic entrepreneurs for the sake of the artificially preserved dominance of the Loop.[78]

With leaflets, billboards, and newspaper advertisements, Deuther, the NWCO, and the citywide Greater Chicago Federation drove home three major points: (1) concentration was artificial, not natural; (2) technology did not hold the answer to the city's transportation problems; and (3) only public ownerhsip would make a just solution. The movement based on these issues, one alderman observed in 1915, was the most important single factor thwarting the city's attempts to build a modern transit system along the lines advocated by CBD business and establishment engineers. An alderman active in transit policymaking during the 1920s voiced the same opinion.[79]

Congestion, Deuther argued, echoing Henry George, was simply "another name for . . . profitable business." Deuther seized upon a comment in a 1909 report by city engineer John Erickson. Erickson had warned that without subways the Loop would soon reach its capacity for traffic and ". . . a new order of development must come into existence that will scatter people about." The CBD, Erickson warned, ". . . may even develop into scattered centers." While Erickson's remark may have been intended simply to frighten downtown interests into active support for a subway, the opponents of centralized transportation took it as evidence that dispersion of commerce was the *natural* order of events—this without any reference to the automobile.[80]

Subways, the dispersionists argued, were primarily a means of preserving the CBD; they did nothing for the city as a whole. The entire "efficiency" of a subway lay in its ability to concentrate traffic. CBD subways would thus force more crosstown traffic into the CBD and lead to more skyscrapers, more layering of streets, and more congestion.[81]

Indeed, technology as a whole held no answers for the "traction" problem. "When we had horsecars it was said that the cable would relieve congestion and we would have no more straphangers," Deuther recalled. Cable cars were followed by electrics, then elevateds, but congestion and straphanging only increased. "Now you propose to build a subway. . . . The more you add to the present system the worse you make it." Monorails were better than subways, Deuther thought: people should travel in the light. But the key question was *where* the transit system would run.[82]

The remedy was to simplify streetcar routing, connect major crosstown streets where they were interrupted by the Chicago River, and work to make every part of the city accessible to every other.[83] Superficially, Deuther's program was aimed at a goal similar to Burnham's. The differences lay in the westsider's emphasis on mass transit, his belief in the naturalness and inevitability of decentralization, and his conviction that the city's congestion and transportation problems were the results not of a lack of planning but of conscious policy, directed from the CBD. "If you could look down on Chicago from an airship," Deuther told the Western Society of Engineers, "outside the Loop district the city would present a spectacle of a lot of large dishpans with the elevated embankments of the railroads surrounding them, all disconnected, but *all* directly connected with the downtown district."[84]

On Chicago's flat terrain, this was unnatural development according to the thinking of the neighborhood interests. It had been brought about because the "natural" function of the streetcar—that of a neighborhood utility—had been perverted. Subway plans were aimed, they thought, at exacerbating the situation and making it permanent by modeling Chicago's development on that of eastern cities. "No fact, idea or experience of New York," said Deuther, "has any bearing on Chicago."[85]

Until the mid-1920s, Deuther argued that dispersion could best be accomplished by means of publicly owned, streetcar-based mass transit. Existing transit policy was a bargain between big business in the Loop and big business in transit, designed to foster the interests of both at the expense of the rest of the city. Public ownership would at least remove the economic incentive for artificially centralized transit. Then nature could take its course.[86]

The impact of this rhetoric on the public cannot be measured; all referenda on transit plans were laden with extraneous issues. The results of these votes do show that the northwest and southwest side voters cast disproportionately negative votes on traction referenda through the mid-1920s, no matter who the political backers of these plans were. With time, transit planning made more and more concessions to the idea that future transit should make crosstown travel easier instead of concentrating on the accommodation of existing CBD-oriented travel patterns.[87]

Neighborhood opposition to subways points up two things. First, no subway could be built because subway building created a double-bind situation for politicians. A short, affordable CBD subway was bound to be defeated by outlying voters, yet bargaining for a citywide system would inevitably force the city into concessions to private companies which would be as politically unacceptable as a downtown subway.

Unlike the Chicago Plan, which could be implemented piecemeal at the expense of local property owners and renters, transit improvements required the city to commit itself to comprehensive plans—plans which, as will be seen, always included fare increases or other concessions the voter would not accept. The costs of street improvements were hidden and divided. The costs of transit improvements were visible and fell exclusively on the riders.

The second point is of equal importance. The NWCO was not alone in its expectation that the streetcar could and should fill the role later assumed by the automobile. This expectation, combined with the assumption that mass transit should make profits, build suburbs, reshape the city, and, simultaneously, charge no more than a 5¢ fare, was impossible to fulfill. Technological limitations aside, the public would not pay, either in fares or in taxes, for the service it demanded. Mass transit was thus condemned to "failure" by the standards of its own heritage even before the coming of large numbers of automobiles.

Chapter 4:

"The
Poor Man's Motorcar"

Section I:
The Public and Its Transit
System, 1907–1917

> . . . an old lady came over who wanted to go to the hospital: her son was dying . . . she waited one hour for a car. She was afraid to stand there. She stood under the light. . . . And she came over and asked me why we do not do something. . . .
>
> *Alderman William Carr, 1913*[1]

> The surface cars are the poor man's motorcar.
>
> *Otto Schultz (Northwest Commercial Association), 1912*[2]

The policy embodied in the 1907 ordinances contained several internal contradictions which prevented it from achieving its ends. Equally important was the fact that the public's expectations of mass transit were contradictory and impossible to satisfy.

In the simplest terms, public transit was asked to provide a combination of automobile-like convenience, far-reaching social and economic benefits, and low fares which would have been impossible for any self-supporting system, whether municipally or privately owned. The effort to attain this impossible combination of goals contributed to the overcapitalization of the companies, and overcapitalization, in turn, helped prevent a comprehensive treatment of the city's transportation ills. Only if mass transit had become a genuine public service—a sheer impossibility in the climate of the time— might popular expectations have had even a hope of fulfillment. The point is

104

not that the expectations of low fares, fast, convenient service, heated cars, and seats on which to sit were unreasonable as such. Rather these goals conflicted with the political side of transit policy because, in reality, they required a subsidy.

•   •   •

Before entering into a discussion of the public's expectations of its transit system, it is necessary to gain some understanding of the composition of the transit-riding public. As might be expected, this question—Who were the straphangers, and how did they use mass transit?—yields a complex and crucially important answer.

At first glance it would appear that mass transit in 1915 was, as it had been in 1906, disproportionately a concern of the CBD and of the middle class. As already noted, most of the riders of the elevated lines and one-third of the city's streetcar passengers traveled to and from the central business district. Further, the residence survey made by the 1916 Traction and Subway Commission revealed that 24.4 percent of the 350,007 workers traced lived within one mile of their jobs; a total of 34 percent lived within two miles. If one eliminates the 115,085 Loop workers counted in this survey (most of whom lived over five miles from their places of work), an even more striking picture appears.[3]

Of thirty large non-CBD work groups studied, twenty reported that 30 percent or more of their members lived within one mile of their jobs. For four groups, the figure was 50 percent. In one, centered around the steel mills of "South Chicago," 79.2 percent of the workers surveyed lived within a mile of their places of work. When combined with the fact that the average streetcar ride in Chicago was reported by the same commission to have been 4.16 miles, these striking statistics make it appear that mass transit in 1916 held relatively little importance for many lower-income members of the working class.[4]

This impression is, however, not correct. Exclusive concentration on the place of residence of the chief wage earner in families where those wage earners were employed by major concerns (the usual source of information in such studies) could lead to highly misleading conclusions—a fact of which the 1916 commissioners were aware.[5] Indeed, there is no better place to begin than with the commissioners' own statement. Bion Arnold, William Barclay Parsons, and Henry Brinkerhuff, who, among them, had engaged in transportation studies for many of America's major cities, observed that

> different occupations may be, and probably are housed under the same roof in the modern city. . . . There is likelihood that this residence will be located conveniently for the father, or, perhaps it would be better expressed, the breadwinner of the family. The children of this family, however, growing up under the modern educational system, are more likely not to follow in the footsteps of the parents as to occupation.

Assume the father a foundryman, living somewhere on the west side, centrally located as to three or four groups of factories largely composed of foundry workers. The sons are more likely to have taken up some more recent occupation, such as automobile, electrical, gas engine or lathe work. Assuming a daughter working, her occupation is likely to be clerical or light manufacturing. This would give a probable direction of travel for the three, the father south bound in the morning into the district of heavy foundry work, the son south or west bound into the electrical and gas engine districts, and the daughter east bound to the loop or the west side manufacturing, or perhaps to the Central Manufacturing, Montgomery Ward, or Sears Roebuck districts.[6]

The commissioners' purpose in reciting all this hypothetical detail was to stress the need for a system giving access from each section of the city to every other section. For our purposes, it underlines the fact that straphangers, even within one family, were a diverse group with diverse needs, and that mass transit (though it served the CBD best and could not, of itself, disperse population) was of great importance in making a variety of jobs accessible to workers who could not or did not wish to change their places of residence.

The commission's data confirm this impression. Workers employed in plants built before the unification of transit service were much more likely to live near their jobs than were workers employed in newer plants. Better transit had given access to new jobs. Further, some of the city's most heavily used transit lines did not enter the CBD at all. Two in particular, north–south lines west of the CBD, showed very heavy loadings with relatively short lengths of haul. In addition, several major lines which did enter the Loop reported very heavy passenger traffic outside the downtown district. Evidently the transit system was being used by a wide variety of people for diverse purposes. When one adds to this the city's high ratio of transfers to fares paid, it becomes evident that mass transit was being used flexibly, not simply for the journey to work in the CBD.[7]

A father employed for long hours at low wages might indeed wish to reside close to hs work. Studies of Detroit, Cleveland, and Pittsburgh confirm this fact, though (as will be seen) geography, demography, and economic and policy history combined to make each city different. Habit, the cost of new housing, and racial and ethnic prejudice worked against transit's decentralizing effect on residence. But transit *did* help determine what jobs were available, especially to younger residents of the inner city. Jobs, in turn, could help determine what sorts of housing they could afford in the future. In short, suburban extensions probably meant little to the working poor, but good service within the built-up sections of the city could mean a great deal.[8]

But the picture is still more complex. As will shortly be seen, transit riding during this period was very heavy on weekends. More sophisticated data collected after 1924 reveal that this pattern continued into the 1930s. Further, as late as 1933 outlying business districts, whose total retail sales were equal to two-thirds those of the CBD, were predominantly creatures of mass transit. In a twelve-hour weekday, 62 percent of the people passing through them used mass transit. Significantly, more than half of the patrons of these centers, which lay an average of 6.9 miles from the Loop, were women.[9]

The streetcar, then, was truely the automobile of the working-class family: it carried them to the parks on Sundays, to the shows on Saturday evenings, and to the important local shopping districts throughout the week.

It is this complex function of mass transit which will concern us now. The evidence suggests that, at least by the decade following 1910, mass transit had begun to support a way of life in which can be seen the beginnings of what has sometimes been called the automobile culture.[10] Most important for what follows, the major complaints against mass transit reflected the belief that it *should* fill a role like that which would shortly be filled by the automobile.

The Neighborhood Utility

The ridership of Chicago transit constituted many publics whose expectations could not be met for three reasons, in addition to the handicaps imposed by the city's transit policy. First, various kinds of riders, in various ways, expected the streetcar to afford them privacy, convenience, and certain amenities which were not to be had from a self-supporting transit system. In different respects, the streetcar was expected to be both an extension of the home and an extension of the neighborhood. Second, the attempt to use mass transit as a tool for the dispersion of population (and for the development of privately held real estate) and to make it, in other ways, a means of social betterment achieved but limited success and interfered with the system's ability to serve the majority of riders. Finally, in some respects, Chicago transit was bad by any reasonable standard. The way riders used the system and the need of the companies to make competitive profits, both contributed to this situation.

The expectation that public transit would reflect the values of the home and neighborhood and provide the privacy, convenience, and amenities later found in the automobile is apparent both in the way the system was used and in the complaints riders made about it.

Chicago transit between 1907 and 1917 remained both a cause for and (predominantly) a reflection of the city's class and ethnic divisions. At times ethnic considerations could even determine routing, overriding the companies' natural concern for efficiency.

"The streetcar," wrote Delos Wilcox in 1911, "is a democratic vehicle. In it all classes and conditions of people ride together. . . ."[11] Chicago's Alderman Stanley Kunz knew otherwise. From 1911, the representatives of the city's crowded fifteenth and sixteenth wards had tried to induce the Chicago Railways Company to route the Division streetcar east through the light manufacturing and slum district and downtown via State Street. Always the company had some excuse for turning the car back at Wells Street, and third of a mile west of State.

"I know the reason for this better than anybody," Kunz told Harvey Fleming of the Chicago Surface Lines in 1917. "The trouble is on West Division you have a lot of Jews riding." Fleming answered, "I don't think you want to mix them." Kunz observed that ethnic and class mixing was inevitable: "On Milwaukee Avenue [where Division–Downtown streetcars were routed] they bring down a lot of Negroes. It should be the policy of the company to give the public just as convenient traffic as possible."[12]

Five years earlier, in discussing the same routing question with Tomaz Deuther, Chicago Railways President John A. Roach had put the situation more succinctly: "You can't mix silk stockings with picks and shovels." Ethnic and class prejudice persisted. The Division car line was not brought east to State Street until 1937, despite the congestion caused by the presence of Division cars on already-crowded Milwaukee Avenue.[13] The problem of Division Street was unique in the candor with which it was discussed, but similar situations may have existed elsewhere. One short line, so lightly traveled that the city engineer remarked that "they run the cars in pairs so they do not get lost," ran much of its length one-quarter mile or less from other lines. This line served factories, while neighboring lines served business streets. Another minor line, which survived until 1948, "jogged" a quarter-mile off course at one point in order to serve a small black enclave which lay just two blocks from another line. The reasons for such routings were rarely so simple or blatant as that for the Division Street line, but company officials did occasionally speak of lines which served "a nice class of people" who deserved better accommodations.[14]

Complaints about class mixing on the cars were extremely rare—a fact which may or may not reflect the success of the system in segregating its passengers.[15] The nature of elevated service helped to segregate it automatically. Though there were many inner city stations, by the decade after 1910 they were, for the most part, lightly used. By far the greatest portion of the el's ridership boarded each morning four miles or more from the Loop. As had been the case in 1906, would-be riders in the inner city found trains too crowded to board, though deadly crowding accidents became relatively rare. Express trains, the lack of inter-company transfers, and the fact that, after 1916, el fares were higher than those on the surface lines, may also have helped segregate ridership.[16]

Whatever the reason, as Alderman Cullerton remarked on behalf of the lower west side, "Elevateds are of absolutely no use to us and . . ." their noise is "a nuisance to the schools and the churches." One inner-city neighborhood vigorously and successfully resisted an elevated extension which, their representatives contended, would serve only the more affluent residents of outlying areas.[17]

Many of the city's wealthier citizens could escape whatever class mixing did occur on the el. Most steam commuter railroads had many in-city stops, and the 1916 Traction and Subway Commission found that one which ran along the southern lakeshore was "potentially competitive with and actually supplementary to" the elevated.[18] The effect of transit service in bringing "outsiders" into a community, however, was another matter.

One far south side community (Park Manor), threatened with the advent of a new interurban in 1908, sent a representative to the city council to complain that the new line would "bring out a class of fruit vendors and Italians and various other undesirables . . . who will try to live with us and their children mix with ours in the schools." Alderman Charles Merriam protested the extension of streetcar service into the area just east of the University of Chicago on the ground that it would be "simply . . . tapping the undesirable section of Hyde Park, bringing undesirable citizens, not regular Hyde Parkers." A line linking Hyde Park with a section of the city largely inhabited by blacks brought similar objections, and real estate spokesmen observed that streetcars created tenements because "there are some people who want to step from their door into a [street] car, and where a carline goes, there you will find cheap flats." Among those objecting to streetcar routings which brought working class-citizens through affluent neighborhoods was Colonel Nathan McChesney, later an active member of the Chicago Plan Commission through which he sought to maintain segregation by other means.[19]

What may be called class interests also clashed over the extension of a streetcars to the city's lakefront. The lake was in most places insulated from the rest of the city by a belt of parks, and streetcars were excluded from all streets which ran through parks. Settlement workers and aldermen with working-class constituencies continually pressed the matter after 1909, however, and several south side lines were routed across or under boulevards to terminals within easy walking distance of the lakefront.[20]

This appears to have been no mere idle crusade, for the use of the streetcar for recreation was a major part of its function before the advent of radio, neighborhood motion picture houses, and widespread automobile ownership. On some lines, the Chicago railways found Sunday riding heavier than weekday patronage. "On Sunday, the loads that are carried [to Humboldt Park] . . . you could not describe it," one alderman complained. "They hang on as though they were hung on by their teeth."[21]

The limitations of streetcar routings could define the world of the working class, but better service could not in itself break down sectional or ethnic barriers. The hopes of the 1907 era that unified service and "universal" transfers would open new homes and jobs to the lower working class were only partially fulfilled. Instead, transit riding continued to reflect the divisions marked out by the Chicago River.

In part, this was due to the inability or unwillingness of non-CBD workers to move when transit did begin to link sections of the city more effectively. As already noted, workers in establishments built before 1907 lived largely in the districts in which the plants were located. Those employed in newer plants were more evenly distributed throughout the city. Very tentatively, this suggests that workers, or their children, were more willing or more able to change jobs than to change residence. This was especially true of those struggling for home ownership. One businessman protested a routing change on the ground that many of his employees had built homes along the old line and would be hard pressed to find a new means of getting to work.[22]

Another conservative influence was that of ethnicity and class. Through-routing, in addition to keeping unnecessary traffic out of the CBD, had been expected to link Chicago's divided sections. But the 1916 Commission found that very few passengers used through-route cars for travel between the city's sections. Travel across the branches of the river in the west division was practically nil.[23]

Confusion caused by poorly marked through-route cars may have been partly to blame for this. So, too, was the residential pattern of the city, which had been set before 1907 and could not be transformed in a mere nine years. As a Chicago City Railway spokesman observed in 1912, "The people living on the south, contiguous to the lake, are the same class as those on the north side . . . and those people mingle freely with one another. The people living on the west side do not mingle with the people on the south side to any appreciable extent."[24]

The transit system itself thus reflected the city's class and ethnic divisions. Complaints about the character and quality of service sometimes mirrored these divisions as well. Companies found that cars which had been perfectly acceptable on one route elicited complaints when placed on another line. Some company officials felt that the north side had, as one company president put it, "a more fastidious public" than the southern division.[25]

Riders sometimes contended that it was the service, not the demands of patrons, which differed from section to section. Citizens of the city's west side were particularly inclined toward this interpretation. "You think that because we are laborers and foreigners that that is the way you treat us," complained one northwest side victim of employee discourtesy. When south siders complained of dirty cars, newly through-routed into their territory from the stockyards district, one west side alderman made the issue, as he saw it, explicit. ". . . It would seem that you favor a rich man's car and a poor man's car," John Toman charged. "We got that kind of service for years on the West side, and it's time you got some of the same medicine."[26] Many middle-class patrons did expect mass transit to provide a high level of convenience and amenities and, in some respects, to reflect the values of their own homes. Such dreams were doomed to disappointment.

Again, the citizens' complaints shed some light on their expectations. Such information must, of course, be regarded with a degree of wariness. Complaints were subject to manipulation by politicians and press, and did tend to increase or decrease in keeping with the level of political axe grinding. But frequently repeated complaints can be assumed to reflect popular expectations, even though they may not describe actual conditions on the transit system.[27]

In this category are a variety of demands for what may be called "autolike" service. Riders asked for door-to-door service; seats; warm, clean cars; short waiting times; easy availability of transit at night; and special consideration for women. Some of these expectations conflicted with others, and none was fully realized. The demand for door-to-door service, for example, conflicted with another important policy goal: reduced crowding.

By 1913, the companies no longer wanted to route all their cars all the way into the CBD. Their own studies and their experience with worsening CBD traffic had convinced them that turning back some cars would increase their operating efficiency.[28] Many passengers, however, demanded that all Loop-bound cars travel to the center of the CBD. This demand placed the interests of those who worked or shopped near the middle of the Loop against those of workers in the light industrial establishments at its fringe. On one corner just outside the Loop, it was reported, "there are hundreds and hundreds of people from those factories waiting there . . . until the people from downtown have gone home, and then perhaps they can barely hang on to the rear of the car."[29]

Company officials complained that businessmen and aldermen pressured them into running all cars to the end of each line, even though equipment was badly needed elsewhere. But the aldermen themselves were under pressure to see that all cars ran to the end of each line. "You could not go out among the people and get enough votes to carry anything if you turned those cars," one warned the chairman of the CLT.[30]

The demand for automobile-like service, close to the user's door, may also have inspired a number of extension requests, some for lines within two or three blocks of alternative service. Such demands did not come only from developing "suburban" districts. In the heart of Chicago's congested Polonia, a line only 1.75 miles long traveled almost entirely on residential streets, never more than two blocks from another line. In 1912 its operating expenses were over three times as great as its receipts, yet it lasted until 1936.[31]

Some lines survived through simple inertia. One vestigial trolley wound through a desperately poor section on the near north side. It carried thirty-five passengers a day in 1912. When Alderman Capitain volunteered that he had never heard of the line, Williston Fish of Chicago Railways spoke up: "I rode it ten years ago." "Why is it even operated?" Capitain asked. "There is not any good reason," came the reply. "How did it get there?" "It is there because the tracks are there," Fish explained.[32]

Such oddities sapped little of the companies' resources. More serious was the public's demand that cars not only run near their homes but stop near their doors. Through 1913, Chicago streetcars were required to stop on each side of every street intersection and in the middle of long blocks, whether or not there were passengers who wanted to board or alight. One major line made 120 stops along its eight-mile course. This practice was no mere archaic survivor of a time when horsecar speeds made frequent stops only a minor inconvenience: when the city and the companies backed an ordinance reducing the required stops to one per intersection, they provoked a minor outcry from patrons forced to walk an additional 50 or 60 feet on the frequently muddy streets.[33]

The elevated companies also found themselves under attack when they attempted to close down little-used stations; some patrons felt that a half-mile distance between stations forced them to walk too far. With stations a quarter-mile apart, elevated spokesmen reported, the speed of rapid transit in outlying territories was no greater than that of streetcars. But in 1916 the rapid transit lines were still operating all the stations they had ever built, even though nine of these stations were handling fewer than 500 passengers each day, and one was servicing only 12.[34]

The same tradition made patrons resentful of breaks in lines. The objection might be to the simple inconvenience of transferring, or to having to "walk through the slush." Breaks in lines could also cause encounters between different social groups. One transfer point was at the edge of a "dry" ward, contiguous to an equally dry suburb. "Unfortunately," its alderman complained, "all of the Evanston bums come down there and have a good time and make it very disagreeable for the people who have to transfer." But a "split line" could also produce situations like that found at the break in the Halsted Street through route at the south branch of the Chicago River. There one citizen "got off a car on Saturday night in that sleet and storm and . . . there were 100 men, women and children there without the slightest protection from the storm, and we waited there twenty-five minutes before a . . . car came down. . . ."[35]

Riders wanted through routes for safety as well as for convenience. One woman who found herself waiting half an hour in the dark at a "tough" intersection told the CLT that she would have to change her job if through service, which had been ended, were not restored. Another "broken line" caused riders to walk across a bridge which became a favorite spot for robbers. Almost all crime complained of by transit riders occurred at junction points where passengers had to wait.[36] Transferring was also especially difficult for women with small children, and, it was said, recent immigrants, transferring amid the confusion of the Loop, frequently lost their way.[37]

This wide variety of demands for through service reveals that the streetcar was being "asked" to adapt to the needs of a highly diverse population. As that population spread, the proportion of passengers taking transfers rose, and it became ever harder to provide the level of service demanded.[38]

Waiting was another frequent complaint, especially among residents of outlying areas. Even a ten-minute wait could provoke complaints. But, as the CLT's engineer observed, "it is not good business to run cars for a comparatively few."[39] In newer areas, where traffic was light, this principle could mean that extensions would yield an unattractive level of service.

On new lines, even the smallest of the companies' cars might not fill more than half of their seats when run on a twenty-minute headway. Thus companies might operate an extension with a bare minimum of service. This they did in the ward served by Alderman Charles Merriam. On 71st Street, said Merriam, ran "not a car at all, but a recollection of a car."[40] Service of this sort can hardly have been what middle-class residents had in mind when they demanded extensions.

Night workers also complained of long waits. Night service improved between 1902 and 1912, but neither riders nor companies could be satisfied with the results. In 1912, despite a thirty-minute night headway on their busiest lines, Chicago Railways representatives reported that all but one of the lines which had night service were losing propositions. But night service remained inconvenient, and riders could not even avoid long waits by planning to arrive at a stop when the streetcar was due: even the city council had difficulty in getting schedule information from the companies.[41]

Patrons could adjust to light service in a variety of ways. One outlying community demanded that the company provided benches for waiting commuters. Inner-city residents simply avoided lines with long headways and walked to those with more convenient service, thereby adding to the crowding on the latter lines.[42] For people with the means and little time, however, the constantly improving automobile must have provided the best alternative.

All these matters related directly to service. Some riders also demanded what may be termed "amenities." Public demands brought a city crusade to see that cars were kept heated to a minimum of 55 degrees. Smoking was permitted on streetcar platforms until 1918, and special smoking cars were provided on some elevated trains. Regular complaints of "dirty cars" also reflect the belief that the streetcar should "fit in" with the rider's home experience.[43]

Still another category of complaints was that concerned with the treatment of women. Some women apparently expected to be offered seats on crowded cars—a futile hope which crossed class lines and was usually voiced "for" women by males.[44] More often the objection was to "indecent crowding," a circumstance which disturbed people of every social group. One frustrated shopper described a December trip from the Loop by elevated, just before the evening rush hour: "I could not have gotten aboard myself but was shoved on with the crowd and could not get any further than the platform [of the car]. The jam was so great that . . . a box that I had in my arms was crushed right in. I could not put my arms up to protect it to save my life. . . . I could not turn around or go in or out. We stood on the open platform all the way to White City [eight miles]. . . ." In such situations, what one woman referred to as "unpleasant circumstances" were bound to arise.[45]

Varying groups of transit riders had varying wants and needs. To this long list must be added the impact on the system of the behavior of some of the riders themselves. Alderman Cermak reported a familiar example: streetcars jammed to capacity at the rear (entrance) end had vacant front platforms. Trying to move through such cars was "like taking part in a football game."[46] In summer, passengers crowded onto the open platforms of elevated cars, blocking those who sought to get in or out. If a company employee asked them to leave, one elevated spokesman explained, "they will say to him, 'you go to the devil.'" If the guard attempted to use force he would be "answerable to the law." A streetcar company spokesman observed that if no smoking rules were enforced, "half the passengers and the conductors would be in the police station."[47]

The extent of some Chicagoans' proprietary attitude toward transit facilities was remarkable. Where one set of elevated lines ran on the ground, six miles from the CBD, some citizens drove wagons astride them, to the peril of all concerned. When the surface lines rerouted some cars, patrons demanded to be taken along the old route and refused to get off until their wishes were complied with.[48]

For good or ill, then, organized interests and individual riders had considerable influence on the way mass transit operated. One of the ways in which interest groups had their greatest impact in the shaping of the system was by pressing for extensions into undeveloped territory.

Extensions: Costs and Benefits

One of the selling points of the 1907 ordinances was the alleged ability of transit to spread population and raise the value of outlying real estate. The extension provisions of the ordinance were amply fulfilled but, as one alderman observed, "it is done primarily at the expense of the fellow who lives in the city and has to pay for the ride" (see Map 4-1).[49]

The ordinance mandated a number of "cornfield extensions"—lines which, as on alderman put it, enabled residents to raise crops at night by electric light. On occasion, too, the companies still built for real estate speculation. One company, reluctantly building a line to a remote village at the city's southern tip, chose a particularly pastoral route. "Why . . . serve . . . trees and bushes?" Alderman Dever asked the company's representative. "Because we are building," the man replied, "for five or ten years hence."[50]

Some of the fens through which that line eventually ran remained bucolic for years thereafter. Trolley service could not, of itself, bring residential development. This does not mean that the extension of lines ahead of population was, in itself, bad policy. It was the way such improvements were paid for—ultimately from the fares of all riders—which made them counter-productive for the system as a whole.

A trolley line in a lightly populated area was not necessarily useless. On the city's far south side, clusters of industrial development and what amounted to small towns dotted the sometimes swampy landscape. It made sense to link these settlements together. In other parts of the city, streetcar extensions do appear to have helped to disperse population. Chicago's population center, which had remained the same distance from the CBD between 1890 and 1900, moved outward two miles between 1900 and 1916. Many factors led to this shift, but one study suggests that access to the central city was a major determinant of the *direction* of population movement during this period: settlement did, to some extent, follow the trolley tracks through 1920.[51]

But who really benefited from population dispersion, and who paid its cost? As middle class residents dispersed, there was presumably more room left in the inner city for those remaining. But as upwardly mobile workers departed, inner-city housing fell into decay. Of course, the increase in the city's housing stock could help those who did not move at once to move later, but it is difficult to determine the degree to which these benefits "trickled down" to the poor.[52]

Had benefited property been assessed for such improvements, as had long been the case with street widenings, Chicago's policy would at least have been consistent. Arnold, as noted, advocated this method of financing, and, indeed, such assessments would have made more sense for transit than for street improvements.[53] Property owners could pass along the cost of street improvements in the form of increased rents, but only residents with auto-mobiles would benefit directly from such projects. Any tenant with 5¢ could ride a trolley.

As it was, riders on the most heavily traveled lines subsidized real estate promotions which had no obvious benefit for them. Chicago's extension policy helped link together remote sections of the city and facilitated dispersions, with its attendant benefits and evils, but it did so by sapping the strength of the transit system as a whole and taxing transit patrons in general to support an improvement in life style in which not all could share.

Crowding

The propaganda for the 1907 ordinances had promised "seats for all," and through the next decade some Chicagoans tried to collect on that prom-ise. El riders and patrons of outlying lines complained of the absence of seats. Inner city straphangers more often lacked even standing room. Crowding remained a persistent, and apparently justified, complaint.[54] Rush hour crowding was often more than an inconvenience. One rapid transit spokes-man reguarded 110 passengers on a 48-seat car as "the comfortable maximum load." A comparable situation prevailed on the surface lines.[55]

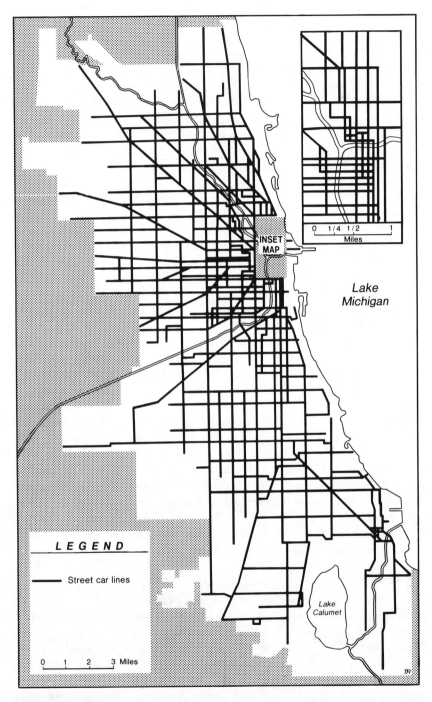

Map 4–1. Street car system, 1926.

In the winter of 1914–15, the BOSE conducted a series of experiments to determine what constituted a comfortable standing load for surface cars. The allocation of four square feet per passenger, it was discovered, left the average standee reasonably comfortable. By loading more and more volunteers onto a variety of cars, the engineers found that the maximum acceptable level of discomfort was reached when each passenger had three square feet of standing room. Three square feet per passenger in the average Chicago streetcar meant a total of 85 passengers. A 70-passenger load meant four square feet of standing space per person[56] With such figures in mind, the meaning of crowding complaints can be placed in some perspective.

In January 1915, ten major lines were carrying loads exceeding an average of 85 persons per car during the heaviest half-hour of the evening rush period. All of these lines were on the city's densely populated west or northwest sides. Four miles west of State and Madison, on Kedzie Avenue, an average of 100 passengers per car was counted in the busiest half-hour, with many of the passengers presumably transferring from the lines that ran out of the CBD.

Some of the worst crowding was found on the lines just north and south of the Loop. On Chicago Avenue, one mile north of the city's center, an average of 104 passengers clung to each car as it passed the maximum load point, bound home from a light industrial district. On the southwest side Blue Island line which ran into the Loop, nearly a third of the passengers on each car were picked up in the area of sweatshops and light industry west of the Chicago River. By the time the cars had moved one mile from the CBD with their loads of laborers and seamstresses, each passenger had just over one square foot in which to stand: 114 were jammed aboard the average 40-seat car during the heaviest half-hour. These, and seven other major lines, exemplified the crowding problem at its worst.[57]

The reasons for such crowding were complex. Street congestion and the routing of too many cars through the CBD were only parts of the problem. Through-routing, the turning back of some streetcars and elevateds outside the Loop, the lengthening of el platforms, all were tried before 1916 with limited success. Both surface and elevated systems needed more space to operate downtown—subways or more effective traffic control.[58]

Some causes of congestion were beyond the control of city or companies. The timing of business hours in the CBD remained uniform, thus concentrating the rush hours in peaks. On the surface lines, snow and extreme cold exacerbated the effects of traffic and poor routing. Passengers contributed their share to the problem by habitually "jumping" the first car they saw, no matter how crowded, even when a second was in view.[59] But company policies also contributed to the problem.

Scheduling was the companies' chief contribution to passenger discomfort. Chicago Railways, the largest surface company, failed to develop a specialized scheduling department. Division superintendents made out the schedules, and company representatives were apparently just awakening to the seasonal variations in transit traffic in 1912. Further, the company had no effective means of schedule supervision. Motormen and conductors, therefore, tended to treat each car as an isolated unit, to be rushed to the end of the line as quickly as possible so that the crew could gain more resting time.[60]

By 1913 the elevated companies had a form of schedule supervision, and the south side surface company, Chicago City Railways, had developed separate weekend schedules and off-street loops at major traffic generators. When surface operations were unified after 1913, the Chicago Surface Lines brought in "college men"—professional transit engineers—to reorganize the system. Management was not professionalized at this point, however: the CSL's new president was a banker, and its superintendent of operations after 1914 was a former sailor who had worked his way up from the position of conductor.[61]

In any case, professionalization was not the solution of the riders' problems. Chicago Railways had to "scour the country" to find a scheduling expert in 1912, but the new man's first innovation was to cut the number of cars on the street and speed up schedules with the aim of increasing profits. Since he apparently lacked good information on the timing and location of the demand for service, crowding increased.[62] The conflict between service and business efficiency in this period was real, as may be gathered from an exchange between Alderman Anton Cermak and William Livingston of the Chicago Railways Company.

"At 5:14 in the A.M. there was 130 people [in a car] at 22nd and Blue Island, and 30 or 40 more had left the car in the lumber district," Cermak told Livingston in 1912. "Now would you call that congested?" "No . . .," answered Livingston. For the transit man, he explained, the loading of a line was measured not by the number of people riding at a particular place and time, but in terms of business along the whole line for the entire day.[63]

The point is that transit managers had to try to conform the operation of their lines to the industry's conception of efficiency. Some lines, like west side crosstowns, had particularly short zones of heavy riding. Cutting congestion on such lines meant adding cars and crews which would travel much of the line with light loads. Chicago Railways Company was committed from 1912 to a policy of keeping earnings per car-hour on each line roughly equal to those of the company as a whole ($2.54 per hour or about 60 paying passengers per car-hour in 1912). When a route fell below this guideline, the company cut service. For relatively lightly traveled lines this policy could mean cuts in service even during periods of substantial traffic.[64]

Even more important to managers and investors than the earnings of individual lines were the companies' overall operating ratios. For street railways, operating ratios were a standard measure of efficincy unrelated to capitalization. "Operating expenses," which in Chicago included renewals of track and equipment but not depreciation, were simply divided by the company's gross receipts. The lower the resulting number, the better the company in question was assumed to be doing.

The operating ratio for all U.S. traction companies was 58.7 percent in 1912 and 63.8 percent in 1917. Companies reporting profits averaged ratios of 56.8 percent in 1912 and 60.9 percent in 1917. According to Bion Arnold, good service could require a ratio of as high as 65 percent, depending on the system. Between 1912 and 1917, the Chicago surface companies averaged operating ratios of 64.7 percent. The elevated companies also showed relatively poor operating ratios compared with those of similar firms elsewhere in the country (51.4 percent in 1916 as opposed to a national average of 46.2 percent in 1917).[65]

The streetcar companies worked hard to reduce their operating ratios to the national average, and they achieved their greatest success during the years when crowding complaints were most frequent.[66] The city's geography and political traditions made comfortable service difficult to combine with favorable operating ratios.

One-third of the surface lines' passengers traveled to and from the CBD, as did most of the riders of the elevated. For them, politics and regional suspicions made subways impossible. Outside the CBD, where most streetcar passengers came and went, trolleys not running crosstown service carried more than half of their riders for distances of two miles or less, with perhaps two-thirds taking transfers in 1916. Such short-haul business and multiple transfers meant specialized service—allocation of equipment to short, heavily used sections of longer lines. At the fringes of population, the demand for extensions sapped the resources of the companies and so lowered the level of service throughout the city. The results were predictable and, not surprisingly, inner-city straphangers suffered most.[67]

"How often has one seen women and children faint in intolerably crowded streetcars?" one community newspaper asked in the summer of 1918. This was not part of a political campaign: the Polish-American *Dziennik Zwiakowy* was describing the everyday experience of inner city workers:

How often are passengers riding on the platform step exposed to serious injury by losing their footing or by being brushed by parked automobiles? Everyone has undoubtedly been inconvenienced . . . by riding past his or her destination simply because it was impossible to squeeze out of a crowded car. The system as it is operated now is responsible for people arriving late to work. It causes them inconvenience on rainy days because the streetcars are so closely packed that it is impossible for them to take on more passengers. All these facts are unfortunately well known . . . there is no need to prove them.[68]

A variety of demands—from businessmen, politicians, and reformers as well as from riders—helped to shape Chicago's transit system and, at the same time, to produce the situation described by the *Dziennik*. The point is not to judge which of these demands were "legitimate" and which were not. The fact is that the demands were made and that, in a profit-seeking system, they could not be met to the satisfaction of all. Among the reasons for this failure, one stands out above all the others: profits, compensation to the city, specialized service, dispersion, and the enhancement of the CBD, and more—all were expected to be paid for by the fixed 5¢ transit fare.

Section II: Overcapitalization and the 5¢ Fare

> . . . if you want to keep adequate service at the lowest rate [of fare] you do not want to amortize.
>
> *Bion J. Arnold, 1913*[69]

In the 1907—1917 era, neither transit riders nor the public at large would pay the full cost of the service they demanded. This was so because the best advice available told Chicagoans that mass transportation could help the city evade the social problems created by congestion and poor income distribution at no cost to the city or the straphangers beyond the 5¢ transit fare.

The 5¢ fare, its inviolability a legacy of the "traction wars," was, with the inability of the city to subsidize mass transit, one of the most important causes of the failure of the 1907 settlement to achieve the results expected of it. Low fares, high levels of service and extensions, and guarenteed profits combined to assure that the companies would not amortize their capital accounts. Overcapitalization did not shrink: it grew. Given a privately owned transit system, competing for capital with other corporations, overcapitalization was of major importance.

Because the 1907 ordinance guaranteed a 5 percent return on the surface companies' valuations, overcapitalization reduced the amount of funds available for service improvements. More important in the long run, it added to the "purchase price" of the companies. Thus overcapitalization made the unficiation of service and comprehenisve modernization of and planning for the system more difficult, since none of these could take place without a merger of existing companies or municipal ownership. Overcapitalization, then, was both a contributing cause of the city's transportation problems and a symptom of the failure of the attempt to combine fixed fares with unsubsidized private ownership.

Overcapitalization: Its Degree and Its Causes

Between 1907 and 1917, the capitalization of mass transit companies in the United States increased by 43 percent; that of Chicago's street railway companies jumped nearly 280 percent. Bion J. Arnold admitted that the BOSE valuation contained, by 1912, about $30 million of "water." "I have done my best with that," he lamented, "but it grows." Elevated securities in 1916 represented an amount over 18 percent in excess of the value placed on the properties by the Traction and Subway Commission.[70]

According to the rough measure provided by capitalization per mile of track, the Chicago transit companies reported securities nearly 25 percent greater than the average of companies grossing over $1 million in 1917: $208,000 per mile in Chicago against $161,900 per mile nationally. Chicago rapid transit is particularly difficult to compare with that of other cities because in 1917 Chicago alone had a major rapid transit system with no subway mileage. But the Chicago Surface Lines showed a capitalization per mile of operated track of $101,701 in 1907, $117,490 in 1912, and $152,501 in 1917, as opposed to $104,147, $106,928, and $112,303—the averages for all other street railways.[71]

One of the reasons overcapitalization seemed acceptable was the fact that transit ridership was expected to continue to grow at a rate faster than that of the increase in population. By analyzing the rates of growth of European and American cities during the late nineteenth century, Bion Arnold arrived at an estimated future increase in ridership and receipts for Chicago which had the appearance of conservatism and scientific accuracy. In 1916, the Chicago Traction and Subway Commission still felt justified in basing its financial plan on calculations of this sort. Indeed, these predictions were not extravagant; through 1926 they proved rather accurate.[72]

Amortization of the securities of transit companies thus held little urgency: there was no reason to believe that such bonds could not be refunded when they matured. In the period of relatively mild inflation before 1915, costs and wages hardly entered into the equation. The economic future of transit, whether privately or publicly owned, could be assumed to rest not on the market for its service (which was assumed certain) but on the character of the service required and the terms on which it would be rendered.[73]

Short-term, twenty-year franchises also contributed to overcapitalization. Companies could not hope to amortize their capital accounts during such a period, but, they could reasonably assume, the city could not evict them at the end of the franchise period without either buying up their securities or finding other investors who would do so. Combined with the guaranteed 5 percent return already discussed, these facts made overcapitalization an acceptable, if not actually attractive course.[74]

Industry economists and valuation experts added theoretical justification for this approach. They argued that the valuation of a utility should reflect not the depreciated value of its equipment, nor even the money actually invested in it, but rather its capacity to earn revenue. This view—that utilities should be valued as "going concerns" and not according to the resale value or original cost of their equipment—was eventually adopted by federal courts. The doctrine had far-reaching implications, but the point for the present is that it added another justification for overcapitalization.[75]

If companies did not fear overcapitalization, the informed citizen could also treat it lightly. The 1907 ordinances supposedly set fares for all time: no matter how high the valuation rose, the fare was to remain 5¢. The ordinances also set aside maintenance and depreciation "allowances" from each year's gross receipts for the purpose of keeping up the properties. New equipment, it was assumed, could be financed through new securities which, because of the assured market for transportation, need not be retired.[76]

All of these justifications were based on illusions, as became increasingly clear after 1914. Even before wartime inflation reduced transit profits, before courts began using capitalization as a basis for setting fares, before transit's ridership began perceptibly to erode, it had become apparent to anyone reading investment manuals that overcapitalization could undermine even a utility company's credit. Indeed Chicago street railway bonds never sold without a discount after 1909.[77] Chicago's transit policy did as much to prevent amortization as did the delusions just discussed.

The Role of the 5¢ Fare

Between 1907 and 1917, rapid improvements in service and equipment, combined with the 5¢ fare, made significant amortization impossible. To discuss the failure of the street railways to amortize, one has first to understand what happened to the system in these years. Between 1907 and 1917, Chicago's surface lines acquired or built 461 miles of track and took delivery of 2,283 new cars. The added track and equipment alone constituted a streetcar system roughly equal in size to that of Boston or of New York City excluding Brooklyn.[78]

This surface system was, from 1914, operated on a "one city, one fare" basis. So, independently, were the elevated lines. This fare structure, combined with rapid, unsubsidized expansion, helped lead to further overcapitalization. One school of transit reformers, led most notably by Hazen Pingree in Detroit and Tom Johnson in Cleveland, had sought to make transit service more accessible to the poor and the working class by cutting the fare. After 1910 both cities had some service at 3¢ fares. Municipal ownership advocates, from Chicago's Mayor Edward Dunne to Delos F. Wilcox, pointed to the lower fares in British and some American cities as evidence of the greater efficiency of publicly owned systems and of the high profits presumably being hidden by large privately owned American lines.[79]

The 2¢ or 3¢ paid by Chicagoans above these lower fares was, it could be argued, the price of overcapitalization, or at least the toll extracted for the legitimate profits of investors. If true, this meant that improper financial practices—or the simple fact of private ownership—were responsible for a major part of the cost of transit service to its users.

This argument does not hold true for the Chicago surface system. If one takes the valuation of the Surface Lines in 1916 (the companies' most prosperous prewar year) and deducts the approximate amount of "water" Arnold confessed was in the original valuations, along with all brokerage and construction profit, the 5 percent guaranteed return on the remaining amount for 1916 is found to be $1,999,659 less than what the companies actually received. But this amount, divided among the cash fares paid in that year, amounts to just over one quarter of a cent. This was the amount added to the fares by overcapitalization.

If all profit left to the companies under this valuation is divided among the cash fares paid—to provide a result something like what would have been had under a service-at-cost plan—the average fare could be reduced to 4.38¢: still much in excess of the nominal fares charged in Cleveland and on some Detroit lines. Finally, if the valuation is depreciated according to the most radical plan advanced during the period, and capital is allowed no return whatever above the 5 percent, the fare could be reduced to 4.05¢.[80]

This last figure would have left the companies with a rate of return few investors would have found attractive. If some other cities could operate under 3¢ fares, why could Chicago not cut its fares below 4¢, even under this manifestly idealized plan? The answer lies in the geography of the city and in the demands made on mass transit by Chicago's fare and extension policies.

• • •

There was a built-in inefficiency in the Chicago system, rooted in the local reform tradition which insisted that transit subsidize dispersion of population through universal transfers and frequent extensions. Transit was also to pay for the enhancement of the CBD through subways built from the "traction fund," or to contribute to municipal ownership. In other words, universal transfers and other service reforms exaggerated the common tendency of fixed-fare systems to cheapen long haul travel at the expense of the short distance rider, and the traction fund set aside an average of more than $2 million each year which served no immediate transportation purpose.

The traction fund, which in 1907 appeared to be a preparation for public ownership, could in reality serve no other purpose than the building of CBD subways. The companies' valuations quickly raced ahead of the accumulating money—the fund could hardly make a beginning at reducing the agreed purchase price. Subways, as already noted, were politically out of the question. The fund simply removed this money from the picture until 1938 when, combined with federal funds, it was used to begin a CBD subway—hardly a comprehensive solution to the city's transportation needs.[81]

Much more important was the policy of making each section of the city accessible to every other section by means of a single fare with universal transfers. This policy did not wipe out slums or tie together all the disparate sections of the city. It did, however, aid dispersion and make new jobs more accessible. The policy in itself was not a bad one; the way in which it was financed—from the fares of all straphangers and at the expense of company credit—is open to question.

By the mid-1920s, 75 percent of Chicago's streetcar passengers did not enter the CBD. A 1925 study of Los Angeles, where 55 percent of streetcar riders passed through the CBD, pointed to the Chicago experience as evidence that crosstown service could make a real difference in the pattern of a city's development. In the California city, crosstown travelers had to rely on the automobile.[82]

Yet to attribute the increasing separation of home from work to mass transit policy or street improvements alone would be extremely dangerous. Detroit, which had little crosstown service in 1915, showed the same tendency of workers in new plants to live at a distance from their jobs as was found in Chicago. Indeed in the Michigan city, where the rapid growth of the auto industry dispersed jobs and helped even out population despite inadequate streetcar extensions, only 28.2 percent of industrial workers surveyed in 1915 lived within one mile of their jobs. A detailed comparison with Chicago is not possible, since the 1916 Commission surveyed workers of all sorts, but if CBD workers (some of them in light industry) are removed from the Chicago count, it appears that about 38 percent of non-CBD workers lived within a mile of their places of work. In Cleveland, on the other hand, where crosstown service was poor, 48 percent of factory workers surveyed in 1919 lived within walking distance of their jobs. The 1917 figure for Pittsburgh, where service was poor and the terrain hindered movement, was 44.6 percent.[83]

Clearly geography and the timing and pattern of industrial development could have at least as much to do with the relationship of job to residence as did transportation policy. But transit policy could accommodate or discourage movement across a city, and Chicago's transportation policy did have its impact. Throughout the 1920s the relatively neighborhood-oriented Surface Lines continued to carry four-fifths of the city's transit traffic. As the construction of belt-line railroads and the emergence of the motor truck helped to disperse jobs, the streetcar system which was in part a product of the city's regulatory policies could help make new jobs accessible to those without automobiles and probably contributed something to the thinning of population in the city's white, ethnic inner city.[84]

Indeed, geography and policy combined to make the operation of Chicago's transit system different from that in most other cities. In 1916, the Chicago Surface Lines carried its average passenger 4.16 miles—farther than any comparable American streetcar company and nearly as far as most U.S. rapid transit lines. The ratio of transfers to fares paid in Chicago was nearly 75 percent—almost certainly the highest in the nation. The high proportion of transfers issued reflects the necessity of long, angular rides, and the fact that many passengers transferred more than once.[85]

Viewed from another perspective, the number of passengers carried per mile of track was substantially lower in Chicago than in Boston, New York, or

Philadelphia. Passengers per car-mile were also fewer in Chicago than in other cities by 1917—a situation which had not prevailed seven years before. Many of the problems which affected transit in Chicago, such as extreme rush hour crowding and peaking of traffic, were common to other cities, but geography, policy, and demographic change had by 1917 combined to make Chicago's transit system less efficient from a business point of view than those of many other cities.[86]

Geography and demography, in particular, created much of the system's "inefficiency." The two major low fare cities, Detroit and Cleveland, had overall population densities roughly similar to Chicago's, but their populations were confined to areas of under 50 square miles. In 1910 Chicago covered 191 square miles. In addition, both Detroit and Cleveland had forms of zone fares. Given these factors, the inability of Chicago's system to achieve fares below 5¢ is easily understood.[87]

But the 5¢ fare can be looked at in another way. The average single, nontransfer ride on a Chicago streetcar was found by the 1916 Commission to be between 1.5 and 3.5 miles, depending on the line. For most major lines it averaged around 2.5 miles, though a few through-routes showed figures as high as 4.22 miles. It was the 41.5 percent of streetcar riders who transferred one or more times who brought the citywide average trip to 4.16 miles. In a sense, the short haul passengers subsidized these distance travelers because those who rode but one or two miles paid the same rate of fare as those who traveled 4.16, or even 25 miles. There is no meaningful way to measure this subsidy paid by short haul riders, but the average fare per ride (i.e., paid and transfer rides divided into gross receipts) was 3.01¢ in 1908 and 2.82¢ in 1916. The difference between these figures and 5¢ represents, in a very rough way, the subsidy paid by those who did not transfer to those who did.[88]

The above figures do not necessarily prove that a 3¢ fare would have been possible with the elimination of transfers. Even if lower fares had been achieved, wartime inflation would have soon forced them up again. This mathematical exercise does, however, point up an inconsistency in the transit policies of single-fare cities.

Chicago transit policy held that riders must pay the entire cost of mass transit service and improvements. This policy strongly influenced planning. In explaining some of the limitations of the plan produced by the 1916 Traction and Subway Commission—the most comprehensive transit plan made during this period—the Commission's chief engineer told the CLT plainly that "the plan worked out must be one which would give a self-supporting system. That is the understanding of the commission of what you have required."[89] There was, however, no question of making each rider pay in proportion to the amount and quality of service he received. The Commission recommended a 2¢ transfer charge between surface and elevated, but political resistance to this idea was strong, and the inclusion of a *possible* 2¢ surface–elevated transfer in the ordinance resulting from the 1916 study was one of the reasons for its defeat.[90] No one suggested graded charges for service within the surface system.

The poor, as the 1916 Commission's engineer observed, lived close to their work.[91] Lower fares for short trips would not have enabled them to move, but they might have allowed more people to ride to work. Even a mile, in the Chicago winter, constitutes a considerable walk. Universal transfers and the rapid expansion of the Chicago surface system after 1907 did make access from one part of the city to another much easier than it had been, an important measure for increasing the availability of jobs. With the flat 5¢ fare and other profit-limiting features of the ordinances, however, this improvement could only be had through increased overcapitalization, which handicapped the transit system in future years, and through a subsidy from short to long distance riders.

In sum, it is difficult to assess Chicago's fare and transfer rules as social policy. It is impossible to say with certainty who the short haul riders were or whether graded fares would have significantly discouraged dispersion. What is clear is that the effort to achieve social benefits from mass transit while insisting that it support itself at fixed fare meant disappointment for most straphangers' dreams and financial decline for the transit companies and placed serious handicaps on the city's transit planning.

An Assessment

The service given under the 1907 ordinances, then, was limited by a wide variety of factors, of which corporate rapacity in the form of overcapitalization was only one. Whatever the evil effects of overcapitalization, the policy which produced it also led to rapid modernization of the streetcar lines by encouraging the companies to spend money while the ordinances' "bonus" provisions were in effect during the first five years. The resulting system was flexible and, outside the CBD, notably fast. For the decade of the 1920s, the 1907 system gave Chicago an excessive amount of street railway securities, but also a fleet of transit equipment which, although increasingly outmoded, was nonetheless twice as large as any which could have been purchased with the same real investment in the postwar period.[92]

If the trend toward increasing overcapitalization could not have been prevented, neither could that toward ever-longer rides for a single fare. The depth of resistance to CBD subways and the disastrous public reaction to zone fares in other cities, together with the city's determination to bring all lines under the 1907 5¢ fare agreement, suggest that it was unlikely that Chicagoans could have been persuaded to pay fares graded according to the "amount" of transportation they received.[93] But the inefficiency introduced by the "one city, one fare" policy, together with the constant extension of lines through 1912, meant a thinning out of service.

The system, not its management alone, produced poor service and overcapitalization. Economic inefficiency was inherent in the Loop routing which companies, and large segments of the public, wanted. It was also implicit in the traditions which demanded two stops per block, a single citywide fare, and "cornfield" extensions. Fundamentally, it was inherent in the definition of mass transit as both a business and a public service. As a public utility its operation was judged by a standard of maximum public benefit—however poorly conceived—rather than one of simple adequacy and business efficiency such as might be applied to a gas or water utility. Yet transit was also expected to operate as a business, and profitable enough to pay between 5 and 9 percent of its annual gross to the city in the form of a special tax.[94] As a business, it had to compete not only with other businesses—bus and jitney companies—but with private automobiles, which impeded its use of the right of way which it and its patrons had built and maintained.

If ever there was a transit system which, because of the things expected of it, should have been operated as a public benefit under responsible public control, it was mass transit in Chicago. But in Chicago, as in most American cities, transit paid cash returns to local government as well as to stockholders. Between 1907 and 1917, through taxes and the traction fund, the surface lines alone transferred to the city and to other governments nearly $37 million of the straphangers' money. The cost of street paving, sweeping, and plowing must be added to this amount. The companies' share of the annual profits, excluding additional profits received through the effect of overcapitalization on the 5 percent guaranteed return, was, for the same period, $17 million.[95] This rough figure, $54 million, gives some idea of the cost of regulated private ownership to the transit rider in the first ten years of its existence.

The city, of course, would have had to borrow and pay interest on money for the purchase and improvement of the lines. But the over $178 million paid to the companies under the 5 percent clause between 1907 and 1931 (when profits ceased) might well have been sufficient to cover such interest and, given the inflated valuation, a reasonably generous allowance for graft. Wages might have been high under public ownership, but by the 1920s the private companies were paying transit wages higher than those in most other cities, including those with municipally owned lines.[96]

Such figures, of course, are entirely hypothetical. The city could not achieve municipal ownership. In 1907 the Illinois Supreme Court had declared the Mueller certificates to be part of the city's debt: city ownership would have been impossible without a major increase in the city's debt limit and some arrangement which would effectively mortgage the lines to the holders of the old traction securities.[97] But, with the creation of the state Public Utilities Commission in 1914, regulation by local government had also become impossible. The two traditional tools for transit planning by local government had been lost.

Public officials dared not admit these facts. Every major political figure between 1907 and 1928 argued publicly either for municipal ownership or for strict regulation. All soon learned, if they did not already know, that neither was possible. Transit politics thus took on an added air of unreality, because transit progress meant the abandonment of progressive reform as it had been known in Chicago—the renunciation of any meaningful claim on the part of the city to control over its transit system. This was so because, as will be seen, the private companies could not raise large amounts of new capital unless the city abandoned attempts at regulation and the transit issue was effectively depoliticized. Whatever degree of political democracy had existed for transportation issues had to eliminated, because political leadership was unable or unwilling to educate the public to these new realities.

It required a decade and a half—between 1914 and 1929—for the reality of the city's impotence to "sink in" and for political and economic interests to create a new transportation policy synthesis that could be successfully "sold" to the public. New political allignments, the slowly declining importance of transit to outlying business, and the effects of years of frustration helped make this policy change possible.

Those fifteen years were crucial for the redistribution of functions in local transportation. Between 1914 and 1929 the automobile rapidly took over the role of mass transit for much of Chicago's middle class. Politicians, local and downtown business, reformers, planners, and traction companies, all worked hard to accommodate the automobile and so contributed to its eventual domination of local transportation.

All interests were able to work together to help the city adjust to the automobile because it threatened, by the 1920s, to tie up all traffic and presented a very evident challange to the hegemony of the CBD and to the survival of politicians who did not find ways to deal creatively with it. Too, the auto was remarkably free of political connotations. Though it spawned some of the nation's largest corporate giants, the privately owned motor car remained a symbol of choice and freedom not, like mass transit, a symbol of coercion and of the power of big business.

In short, city government wanted to accommodate and channel the automobile, and business helped to see that it got the power to do so creatively. These facts, together with the continued inadequacy of mass transit, combined with ever-improving automotive technology to assure the eventual triumph of private over public transportation.

Chapter 5:

The Taming of the Automobile, 1918–1930

Section I:
The Automobile and Traffic, 1918–1924

> The ultimate test of a transportation system lies not in any techno-economic indices of efficiency, but in the extent to which it finds acceptance within the total value scheme of the community it serves.
>
> *A. S. Long and R. M. Sobelman, c. 1970*[1]

> The great American idea is "I am going to do what I want and the hell with the other fellow." Any skilled engineer can prepare the [traffic control] plan . . . but building the groundwork is a man-sized job.
>
> *Evan J. McIlraith, 1946*[2]

The rise of the private car to the dominant role in urban transportation cannot be explained by simple formulae. There was no straightforward, unhindered competition for public favor between mass and private transit; public policy, by defining them differently, accommodated one mode and stifled the other.

Nor was there a conscious effort on the part of any public body to promote auto use at the expense of mass transit. The city consistently sought to enhance the efficiency of both modes; but it lacked the tools with which to deal with mass transit, which therefore stagnated while the private car improved.

Chicago's policy toward the automobile was determined by many interests, of which CBD business was the most conspicuous. In the 1920s, as before, traffic and street-building policies changed in response to conditions. Both the rules and the enforcement procedures of traffic regulation were changed to reflect the improving technology and spreading ownership of motor cars. Both the style of street improvements and the way in which they were financed were adjusted in keeping with the auto's evolution from recreational to transit vehicle.

In effect, an understanding was reached between Chicago's motorists and its ruling community. Motorists accepted, for the most part, the modified form of traffic regulation and began to pay a larger share of the cost of the streets through which they drove. The city accepted the automobile as a means of mass transportation and attempted to plan for and accommodate it.

This compromise was far from perfect. The city never really managed to plan for the automobile: its street improvement and expressway policies, with few exceptions, aimed at relieving traffic jams rather than guiding the development of the auto's role in urban life. The private car could never completely fulfill the Straphanger's Dream: its cost excluded many, and the dispersion it facilitated brought problems of its own. Yet compared with its transit policy, Chicago's adjustment to the private car was a model of practicality. Costs were shared in a way which, if not entirely fair, at least made more sense than that devised for mass transit. Regulation made the auto easier to use. In short, the ways in which the auto's policy definition differed from that of mass transit made it possible for the city to deal constructively with the private car.

• • •

One of the reasons for the "success" of Chicago's traffic policy was the rapid spread of the automobile. By 1930 Chicago had one private car for every eight residents. But the auto in the 1920s was not the dominant mode of transportation in the central city. Changing traffic policy and the continued building of specialized streets for the automobile were less responses to Chicagoans' decision for auto ownership as such than they were consequences of the traffic crises created by the recreational as well as the commuter uses of the private car.

The Postwar Traffic Crisis

There was one auto for every sixty-one persons in Chicago in 1915. By 1920 the figure had reached one car per thirty residents; in 1925 it was one for each eleven. Automobile traffic presented American cities with serious problems before 1918, but it was in the five years following the end of World War I that "motorcars" became the predominant element in urban traffic jams. The tie-ups created by postwar traffic alarmed retailers, politicians, and auto industry spokesmen alike. Nationally, the industry's chief trade journal warned its readers that unregulated traffic could limit the industry's expansion. "If you don't help to regulate traffic, Mr. Dealer," *Motor Age* advised, "some day traffic will regulate your bank account."[3]

In Chicago, commercial leaders were equally concerned. The nationally known traffic engineer Miller McClintock observed in 1926 that downtown and neighborhood business had been clamoring for action against congestion since the beginning of the decade. Business, McClintock declared, was rapidly dispersing. Even neighborhood business centers, built up around the intersections of many major streetcar lines, would soon loose patronage to smaller, less congested centers further out.[4] Banks, too, were suffering. "The little local fellow," exclaimed an alderman who was also a director of an outlying bank, ". . . [is] getting pretty nearly all the savings accounts." One major newspaper lamented in 1924 that the death of the Loop was "probably near."[5]

Congestion: Its Causes and Dimensions

Congestion was indeed on the rise in downtown Chicago. The problem was not a new one, but it was rapidly getting worse. As before, travel patterns reflected a broader policy.

Chicago had led in the development of the modern elevator office building, and the city continued to encourage the building of "skyscrapers" after the Great War. A citywide zoing ordinance passed in 1923 was altered at the last moment to exempt the Loop from height limitations.[6]

In addition, builders found Chicago a cheaper town than most in which to build. Chicago real estate and contracting firms had organized to keep down the cost of construction by agreeing to hire only those workers who would accept the wage rates set by Federal Judge Kenesaw Mountain Landis in his landmark union-breaking decision of 1920. The result of these factors was a building boom for Chicago which got under way even before the general postwar depression had ended.[7]

For all the talk—and action—toward the expansion of the CBD which had accompanied the Chicago Plan, the city's major real estate developers concentrated their efforts on the further building-up of the Loop and of the new North Michigan Avenue business district. Thus the proliferation of tall buildings would have increased downtown congestion even if the automobile had not become so popular so fast. Growing efficiency in the use of CBD land, combined with the shift to a highly inefficient use of street space by automobiles, created a major traffic problem.

All these changes together produced what was perceived as a "traffic crisis," analogous to the "traction crisis" of the prewar years. Traffic jams were really only the newest manifestation of the population dispersion and concentration of certain kinds of business which had begun before the coming of the private car. Postwar traffic was nonetheless a *genuine* crisis. The city tried and failed to deal with this crisis between 1918 and 1924. Old methods of traffic control, it turned out, could not cope with large numbers of automobiles.

The Results of Congestion

The number of motor vehicles registered in Chicago increased more than threefold between 1918 and 1923—from 78,507 to 262,609. The number of vehicles entering the CBD over a single west side bridge in 1920 was equal to the number entering from the entire western section of the city in 1916. Reliable traffic counts for the city as a whole are not available before 1926, but it is reasonable to assume that automobile congestion increased to a comparable degree throughout the city. Deaths due to automobile accidents rose more slowly than the number of motor vehicles, but their increase was still alarming: from 279 auto-related deaths in 1918 to 587 in 1923. Streetcar service, more than ever before, was hobbled by traffic congestion. City engineers estimated in 1920 that streetcar riders collectively were losing 13,000 hours per day in auto-related traffic tie-ups.[8]

The automobile presented a threat to commerce and to public safety for two major reasons: (1) during the early 1920s, the city was unable to discipline the various parties competing for street space; and (2) street improvements, despite the efforts of the Chicago Plan Commission and the Board of Local Improvements, did not keep pace with the growth of auto traffic. Old-style, restrictive traffic regulation could not deal with the new volume of traffic for two reasons. First, neither the walking nor the motoring public accepted existing regulations as reasonable. Second, police lacked organization and received little backing from the courts.

The Combatants

The motorized vehicle did not win control of Chicago's streets until the 1920s. Through the middle of that decade, its chief competitor for street space was the pedestrian, who had long been accustomed to using the street as an extension of the sidewalk when the latter became too congested for easy passage. The pedestrian, said the police lieutenant in charge of Loop traffic control in 1922, was the major element in downtown traffic congestion.[9]

By early 1920s, the motorist and the pedestrian were engaged in an angry and often deadly struggle for mastery of the street. Pedestrians ducked around traffic policemen who tried to hold them back at street corners. They surged across automobile thoroughfares near beaches and mobbed Loop streets during rush hours, blocking all traffic.[10]

Pedestrians were exceedingly difficult to control. Twelve policemen could not keep the crowds at one Loop intersection from spilling into the streets. In New York and other densely populated cities, walkers were likewise too numerous to be arrested, ticketed, or otherwise regulated. Only in relatively uncrowded Los Angeles was some measure of control achieved.[11]

Businessmen and automobile clubs advocated anti-jaywalking ordinances as a solution "There is no more reason for the pedestrian to be permitted to run wild on that portion of the street set aside for vehicles," a Chicago Motor Club spokesman complained, "than there would be for vehicles to run wild on the sidewalks." Indeed pedestrian control was an issue of traffic segregation.[12]

One way to keep the pedestrian out of the street was to give him more sidewalk space. Proposals for double-decked sidewalks received some favorable attention during the early 1920s in Chicago and other cities. Tomaz Deuther claimed to have been the first to have broached the idea—as a hoax designed to show the logical consequences of centralization. In a 1912 letter to the *Chicago American* signed by "U. R. Pfuhl," Deuther advocated not only double-decked sidewalks but a pedestrian mall as well. The *American* was sufficiently impressed by the "Pfuhl" idea that it prepared an accompanying illustration.[13]

For the most part, however, policy ignored the pedestrian. The city's commissioner of public works actually suggested—in 1923—that CBD sidewalks be narrowed to provide more room for cars.[14] The pedestrian problem was handled by default in the general reform of Chicago's traffic laws during the later 1920s. This reform organized vehicular traffic and increased its speed and density to the extent that pedestrians had to accept the motorists' domination of the street if they were to remain safe. The problem of pedestrian use of CBD streets, however, was never satisfactorily resolved, and it remained a major part of the overall traffic problem.

Automobile drivers were just as difficult to regulate as pedestrians, and their resistance to restrictive traffic control won more organized support. Auto clubs, some engineers, and a part of the downtown business community opposed any traffic regulation which appeared to limit a driver's use of his vehicle. The individual motorist, for his part, often simply ignored traffic regulations he thought unreasonable. Courts gave the appearance of supporting his defiance.

The nation's automobile clubs were adamant against what they considered negative traffic regulation. They fought such laws across the country and worked against politicians who opposed their interests. In the view of the automobile industry, the only legitimate regulation was facilitative—regulation which would encourage the use of more and more automobiles.[15]

Accordingly, the clubs—along with a number of national publications which followed their lead—insisted that traffic regulation was a matter for engineers, not for police. When Chicago strayed from this positive, engineering model of traffic control, the automobile trade chastised it.[16] On the whole, however, the automobile interests had little of which to complain in Chicago traffic regulation. City engineers, some businessmen, and local auto lobbyists held off major changes in traffic regulation until the talent and skill were available to treat the traffic problem "scientifically."

The forces which kept Chicago from restricting the automobile during the early 1920s are evident in the unsuccessful movement to ban all parking in the Loop. While resisting this movement, automobile and commercial interests articulated a program for parking control which highlighted their belief that it was the city's responsibility to accommodate the widest possible use of the automobile.

The Parking Ban: Business and the "Traffic Expert"

The dominant idea expressed in Chicago by persons described as "traffic engineers" was that traffic should move as freely as possible. Before 1924, the city's most frequently consulted "traffic expert" was R. F. Kelker, Jr., a transit engineer employed by the city council's Committee on Local Transportation. Kelker opposed any regulation that savored of restriction, but he did support a parking ban.

Apparently Kelker wanted to stop Loop parking for the same reason he disliked traffic lights and stop signs: he was against anything which impeded the flow of vehicles through the streets. The engineer was also concerned with the effect of Loop parking on the majority of CBD commuters. "The streetcars are about 10% of the vehicles leaving the Loop," Kelker told a group of aldermen contemplating the parking question, "and they carry 74% of the total people. There is your answer." Other advocates of parking restriction took approximately the same view.[17]

•   •   •

During the early 1920s, the movement to eliminate Loop parking had the support of virtually everyone whose business gave him frequent contact with the Loop traffic problem. Aldermen with a special interest in transportation, particularly U. S. Schwartz and Henry A. Capitain, favored the ban. So did representatives of the cartage industry, two successive chiefs of police, the street railway unions, and, of course, the streetcar companies themselves.[18] Together, these interests constituted a coalition for the modernization of street use.

Throughout the 1918–1924 period, motor club and auto dealer representatives contended that parking was a part of the automobile owner's right to use the street for which he had helped to pay. The motorist, one angry spokesman told the CLT in 1920, was harassed by a dozen police agencies and now—with the parking ban—about to be hounded from the Loop.[19]

Many smaller Loop retailers, especially those catering to wealthy patrons, joined the motor clubs in opposing a ban on midday parking. Their businesses, they believed, required that motorists be allowed to park in front of the stores they patronized. Customers might complain if required to wait at a store while their chauffeurs "brought the car around," and businessmen, it was said, objected to walking two or three blocks from public parking spaces provided outside the Loop.[20]

Major Loop commercial interests, led by the Chicago Association of Commerce, opposed the smaller businessmen and advocated a ban on all Loop parking from 1920 onward. The CAC sought to make the automobile and the CBD compatible by the creation of off-street parking spaces to replace those eliminated by a parking ban, and tried to make the city responsible for the creation of these facilities.[21]

Loop merchants, automobile clubs, and the CAC, all approved of plans in which the city's responsibility for the street would be extended to the provision of off-street parking. The head of the Chicago Motor Club argued that the city was obliged to see to the building of garages and proposed that public funds be used to subsidize private garages where municipal land was not available. The Motor Club itself ran a parking facility on the unused lower level of the newly double-decked Michigan Avenue, and demanded police protection for cars in public lots which the Club did not patrol.[22]

The Chicago Association of Commerce endorsed a plan to build an underground garage in Grant Park, which lay between the Loop and the city's lakefront. When that plan failed, the Association endorsed a plan for a nonprofit, privately financed garage. Even the more progressive members of the CBD community did not propose that automobile facilities, like mass transit, return a profit to those who provided them.[23] Their efforts yielded results: by 1926, although it lagged in the construction of private garages, Chicago led the nation in the amount of public land devoted to the storage of automibles.[24]

Yet, despite the provision in 1922 of a private bus service connecting above-ground parking facilities in Grant Park with three Loop retail stores, Chicago had not found a means of reconciling the shopper's convenience with the necessities of traffic control.[25] Through the early 1920s, the hope that the traffic problem could be solved by getting streetcars off the surface and into subways (built at the straphanger's expense) helped divert attention from the need to restrict the auto's use of curb space. This fact, combined with continued resistance by auto clubs and many businessmen, allowed the parking problem to remain unresolved. A no-parking ordinance was passed by the city council in 1920, but Mayor William Hale Thompson vetoed it, citing a "storm of protest" from downtown business and evidence that a similar law in Los Angeles had "completely killed business [and] made that city look like Walla Walla, Washington." Subsequent attempts to revive the mid-day parking ban met with no success before 1927.[26]

Traffic Law Enforcement: The Crucial Flaw

A major reason for the movement to ban all parking in the Loop was the inability of police to enforce the existing law, which limited mid-day parking to thirty minutes. In the early 1920s, police were finding compliance with any traffic laws difficult to obtain. "We have plenty of law so far as . . . motor traffic is concerned," the CLT's chairman complained in 1924. "What we need is enforcement."[27]

In the early 1920s Chicago's position with reguard to traffic offenders was particularly hopeless. During the entire decade, enforcement was hand-icapped by the way in which police themselves were organized, the nature of early traffic law, and the attitudes toward traffic violations held by courts and by the drivers themselves.

Throughout the 1920s each of the city's nine park districts had its own police force. City police were responsible for traffic in the city streets, but the boulevards—four of which ran through the heart of the business district—were patrolled by three different park districts. City and park police had no formal means of communication with one another, and, through 1930, had to send messages by mail or by telephone. Worse yet, city motorcycle police were assigned to fixed posts and were not expected to operate beyond their rectangular "zones." Confined as they were, the police tended to try to surprise traffic offenders, rather than to cruise. The comic image of the cop hidden behind the billboard awaiting an unwary speeder apparently repre-sented a common fact of Chicago life.[28]

The task of the city's downtown police was made more difficult in the early 1920s as their numbers were temporarily reduced, apparently in order to cut expenses. But even with unified command and sufficient numbers, the traffic police would have been hobbled by the nature of the laws they were expected to enforce.[29]

Chicago traffic ordinances were all subject to court review. The Illinois Supreme Court, for example, ruled in 1922 that a park commission could not require all vehicles to stop before entering boulevards. Any governing agency, said the court, could be required to prove the reasonableness of any regulation. Later decisions affirmed that local governments could not exclude any motor vehicle from any street.[30] But even "valid" laws were difficult to apply.

Chicago has been credited with the invention of the "traffic ticket," but through 1926 a police officer still had to serve an offender personally if his citation was to be upheld in court. This regulation placed a special handicap on the enforcement of parking regulations, but it also made it necessary for police to pursue and overtake more serious offenders who might otherwise have been identified from their state or city license tags.[31]

And if police could catch a speeder or reckless driver, they might not be able to ensure his appearance in court. There was no state law requiring drivers to carry identification. For this reason, police had either to trust motorists to identify themselves correctly or to take them to the station and require them to post bond. The latter approach, of course, diminished the amount of time an officer could spend on his beat.[32]

More serious still was the fact that, because speeding and parking laws were so widely disobeyed, conformity to the law could seem to the motorist to be foolish or even dangerous. This was especially true in the case of the 20 mph speed limit, which prevailed on business streets and boulevards during the 1920s. One *Tribune* reporter who attempted to drive "legally" on the city's boulevards found that every other car on the road passed him by. Motor clubs and at least one judge held that the 20 mph limit was unreasonable. The attitude of courts toward traffic laws fostered disrespect for the entire concept of traffic regulation.[33]

The ratio of convictions to traffic arrests declined steadily throughout the 1920s. The West Park District was already complaining of the high proportion of dismissals in 1921. Between 1920 and 1925, the proportion of accused traffic offenders punished declined from 44 percent to 21 percent, while the average fine fell from $6.99 to $3.36. Throughout the early 1920s city prosecutors were fortunate if they could convict three out of a hundred persons in accidents attributed by the police to criminal carelessness.[34]

Traffic court was apparently an unpopular assignment with municipal judges, and the rapid turnover in judges assigned to the court may have contributed to their seeming reluctance to deal seriously with traffic offenders. For whatever reason, the standard of law enforcement was low: the *Journal of Commerce* found one judge worthy of praise for refusing to excuse drivers caught going more than 10 mph above the legal speed limit. Nationally as well as locally, by 1924 traffic regulation was, as a representative of the U.S. Chamber of Commerce remarked, ". . . proving a broken reed."[35]

At the heart of the traffic regulation problem was the motorist himself. Automobile drivers were perhaps no more selfish than pedestrians or straphangers, but drivers took up so much more space per person, and were so much more dangerous to those who got in their way, that their attitudes toward traffic regulation made a disproportionate contribution to the failure to traffic control.

During the 1920s drivers were still occasionally described in national periodicals as little better than motorized savages, but the motorist's chief contribution to the traffic problem may have been, not discourtesy or savagery, but an insufficient appreciation of the results of his actions. Just as riders, in the early days of the cable car, had tried "flipping" on and off the moving vehicles, with predictably injurious results, motorists frequently ignored traffic regulations in ways which a later generation of drivers might find startling. In the 1920s, the auto was still new to most of its drivers, and its deadly potential was not fully appreciated.[36]

The drivers who, according to the *Daily News*, raced past red signal lights on crowded Michigan Boulevard were expressing the same disdain for restrictive regulation as were jaywalking pedestrians. The difference was that *their* act of defiance could be much more dangerous for others on the street.

As late as 1931, the Citizens' Police Commission reported that "ordinary good citizens" would not obey traffic light when no officer was present.[37]

This disreguard for symbols of authority made it necessary for the streetcar company to install raised safety platforms at busy loading zones: motorists simply drove across painted "islands." Nor would many drivers stop before crossing boulevards or through-streets. "You can stand out any morning on these street corners that intersect the boulevards," one alderman noted in 1923, "and you will see a fellow coming down and he will just take a little look around and if there is nobody watching he will go onto the boulevard without stopping."[38]

There were other reasons why motorists were hard to control. Corrupt police enforcement lessened respect for law. In addition, traffic regulations were often confusing, especially in the early years of the decade. Besides the occasional conflicting city, park district, and state regulations, there were the idiosyncrasies of early traffic signals: some models used "red" to indicate "go," although it is not clear whether these were used in Chicago. Signals soon became uniform, but even in 1929, a CAC committee reported that a major traffic control problem was the inability of many drivers to understand traffic lights.[39]

Some driver behavior appeared to reflect no so much ignorence as genuine slowwittedness. "In the first three months of last year," R. F. Kelker told a convention of street railway engineers, "there were 17 accidents on the St. Paul [railroad] where people drove into the side of a moving train . . . between the seventh and the 18th car." "If we could get rid of the fellow who drives on Saturday nights and Sundays," Kelker concluded, "we would be much better off." Motorists' failure to yield the right-of-way, together with their speeding and driving on the wrong side of the street, were contributing causes in half of the 1,488 accidents studied in the 1926 traffic survey.[40]

Finally, class feeling probably continued to play a part in the motorist's resentment of traffic law enforcement. Throughout the 1920s, native-born white males had a higher rate of arrest for traffic offenses than they had for other misdemeanors. Between 1919 and 1925, auto-related misdemeanors made up 39.2 percent of misdemeanor arrests for white males of native birth as against 29.7 of the misdemeanor arrests for the population as a whole. It is likely, therefore, that for many white native-born motorists, their dealings with traffic police constituted their first contacts with law officers in the role of "arrestee." Traffic law enforcement was one aspect of the police officer's "order-keeping" function which often required him to make arrests. The role of the urban police was changing during this period, and the definition of traffic offenses—misdemeanors or problems of public order—was still unclear. In 1923, the objections of many drivers to being treated as "law breakers" put an end to Chicago's last major effort to deal with traffic offenders as petty criminals.[41]

With the support of his new police chief and the encouragement of the press, Mayor William Dever initiated a program of jailing speeders in the fall of 1923. Since drivers were apparently aware that the liklihood of conviction on speeding charges was small, Dever hoped to make the arrest itself an effective deterrent. The outcry against the new policy on the part of the driving public was so great, however, that the mayor was obliged to call a halt to the crusade within a year. Restrictive regulation simply would not work in the Motor Age.[42]

There were other reasons for the inability of the city to improve traffic conditions in the 1918–1924 period. Building contractors continued to block Loop streets with construction materials. The movement for night delivery of goods in the CBD made no more headway and the Illinois freight tunnels continued to be used at only 25 percent of capacity in 1920. Sidewalks were clogged with vendors during the Christmas holidays despite occasional city resolutions to eliminate them, and peddlers continued to use the streets in neighborhood business districts.[43]

The city's fundamental problem, however, was its inability to come to terms with the automobile. Chicago had either to build the auto new facilities or to find a way to adjust old streets to the automobile's use through regulations the motorist would accept. In the 1920s, thinking on street improvements, as on traffic control, was significantly altered by the fact of widespread auto ownership and the demands of heavy commuting traffic.

Section II: "The Magic Ribbons," 1918–1930

> There is practically no limit to the rapidity of service with automobile traffic. It is only a matter of providing road space to accommodate the increasing number of vehicles.
>
> *C. Z. Elkins (manager, Chicago Motor Club), 1924*[44]

> Streets are the magic ribbons along which the city outspreads itself when they are well paved.
>
> *Chicago Board of Local Improvements, 1931*[45]

Planning and the Automobile

Automobiles proliferated in the 1920s, and to some, the growing number of private cars meant the end of city life as generations had come to know it. The passing of the nineteenth-century city was, of course, viewed with delight. It meant that the social costs of the industrial society might be reduced by a means—the automobile—which itself promoted the growth of industry.[46] At the same time, however, the auto became an increasingly urban vehicle, used for travel to work and play within the city. The motor car was a traffic problem as well as a social blessing.

The urbanization of the automobile forced important changes on the city. As the decade wore on, the work of the Chicago Plan Commission came more and more to consist of efforts to accommodate the automobile. Simultaneously the city came to see streets in general and certain streets in particular as the special preserves of the automobile. Toward the end of the period, the city sought ways to make the motorist pay a larger share of the cost of the street improvements. Paying for urban highways ceased to be viewed as the exclusive responsibility of the real estate owner and the renter. With all of these changes, Chicago and its motorists progressed toward a recognition of the evolution taking place in the role of the automobile: a means of recreation and of rural transportation was becoming a basic mode of urban mass transit.

•   •   •

Through the early 1920s the auto industry itself viewed the private car as a rural and suburban vehicle. The average automobile owner, the National Automobile Chamber of Commerce found in 1923, was a resident, not of a city, but of a substantial-sized town. Furthermore, in 1924 the industry still viewed families with annual incomes under $1,500 as poor prospects for automobile ownership. One 1922 study reported that the "average" motorist held a $5,000 equity in real estate, had personal property amounting to $2,000, and maintained a bank account and life insurance. Investigations in the later 1920s correlated automobile ownership with the possession of other modern conveniences such as phonographs and radios.[47] Through the 1920s, however, ownership of private automobiles became more common and their use for urban transportation more pronounced.

As early as 1919, one handbook for auto salesmen advised that, if a saturation point for automobile ownership existed, it was "not the limits of present income but of use." Through the 1920s, while the price of mass transit and most other commodities rose, the cost of automobile ownership and operation decreased. Ownership also became easier, for the auto industry pioneered credit sales.[48]

Whatever his reason for purchasing a car, the auto owner quite naturally became an automobile commuter whenever road and traffic conditions allowed. As early as 1919 the NACC found that, while most automobiles were purchased for recreational purposes, 70 percent were also used for the journey to work or for business. As will be seen, urbanites probably did not use the automobile nearly so extensively for these purposes. Nonetheless, auto manufacturers were aware of the increasing importance of the urban market. Packard advised straphangers that its product was faster and more reliable than mass transit in downtown traffic, and motorcycle manufacturers promised to take commuters "home in a jiffy."[49] One auto dealers' guide suggested, as an opening pitch, "If you had a car this morning you could have saved ¾ of an hour of your time." With the coming of the closed car in 1923, the automobile became an all-weather vehicle, and its utility as a substitute for mass transit was increased.[50]

The results of this transformation were evident to thinking members of the transit industry. Transit service, one Chicago elevated spokesman complained in 1923, could now be "flouted" by "anyone who can scrape together $100 and put up a scrap container in his yard." In addition, the flood of autos crippled surface transit. In 1926 Chicago's streetcars, when traveling in the Loop, were the slowest in the country; outside the congested district, the same equipment was the nation's fastest.[51]

Heavy downtown traffic also neutralized the auto's most prized advantage—speed. In 1925 Chicago's elevated still boasted faster overall speeds than did autos on the major downtown boulevard. Flexibility, rather than point-to-point speed in traffic, was the automobile's real advantage. Still, it is worthy of note that reported 1929 auto times between the Loop and north suburban Evanston were but five minutes faster than the scheduled time for the same trip by rapid transit in 1914.[52]

The street congestion represented by the auto's slow running time presented the city with a serious problem. Motorists, unlike transit-bound nondrivers, could take their business—and their homes—elsewhere if the city did not accommodate them. American cities were frequently reminded of this fact by national observers and by spokesmen for the auto industry.[53] To keep the motorist's patronage, Chicago had to provide him with smoother, faster streets.

•　•　•

Planners nationally welcomed the automobile as an agent of dispersion, but Chicago *policy* continued to reflect the concerns of the city's "commercial-business elite." As in the southern cities examined by Blaine Brownell, business leaders in Chicago worked hard to make the CBD compatible with the automobile. Chicago leaders worked with equal vigor for mass transit improvements. Like many of Detroit's business leaders, they sought not just to save mass transit, but to rationalize and extend it. In Chicago too, one heard less of the virtues of the private car than one did in the south and more of the pressing need to accommodate it.[54]

During the 1920s, the Chicago Plan Commission continued to lead the crusade for street modernization. Only toward the end of that decade did the city's own street-planning bureaucracy begin to eclipse the CPC in this role. During this time, however, both the rationale used to justify street building and the kinds of streets advocated by the CPC and city agencies underwent a marked evolution.

As the decade wore on, the CPC made increasing use of auto traffic as a justification for street improvements. Other arguments—particularly the supposed social benefits of broad streets—continued to be made on behalf of plan improvements, but the necessity of accommodating the automobile assumed increasing importance. Thus the Commission warned "public spirited citizens" in 1924 that the automobile "demands" street widenings. Those who opposed such measures, it declared, were living in the age of the "horse and buggy." Street widenings were frequently proposed as the only means of "keeping up" with the ever-increasing, irreversible growth of automobile traffic.[55]

Street improvement plans underwent a complementary evolution. Wider traditional streets and double-decking (an idea with a long history before 1920) gradually gave way in the minds of planners to specialized urban highways for nonstop auto traffic.[56] Through about 1927, the CPC took the lead in directing this planning transition. Traffic management, as opposed to the implementation of Burnham's plan, was obviously the CPC's chief concern during this period. To this concern the Commission brought two basic emphasis, both founded in the Burnham plan and earlier planning traditions.

First, the CPC and others stressed the diversion of traffic from congested areas. Traffic was to be controlled not by restriction but by channeling. For example, the widening of LaSalle Street (an artery which ran through the western half of the CBD), would, it was promised, remove 50,000 vehicles per day from overcrowded Michigan Avenue. Similarly the Commission argued for the widening of a far northwest side residential street which would "give automobiles a through route . . . uninterrupted by streetcars."[57]

The Commission consistently struck a positive note. Street widening and traffic diversion were, CPC Office Manager Eugene S. Taylor stressed, both more important than and a precondition of traffic control.[58] With the supposed social benefits of street improvements in mind, the CPC urged that the city build rather than regulate.

The CPC's approach to the automobile was thus within the pattern found in the planning movement nationally. Chicago did not simply surrender to the automobile; its traffic engineers and planners tried to adapt the city and the auto to one another through channeling of traffic and accommodative regulation. Perhaps, as Blaine Brownell has suggested, planners who saw transit planning monopolized by transit engineers found in dealing with the private car an important means of making themselves a part of local government's planning process. Certainly Chicago's city planners were closed out of the transit planning process, by its politicization as well as by the position already held by transit engineers. As noted earlier, the CPC eliminated transit and housing issues from its consideration at the beginning of its existence, declaring that transit planning was the province of the BOSE. When it did attempt to see to the building of non-CBD subways, it was thwarted by politics.[59]

In any event, the 1920s saw widespread street improvements. The results of the city's efforts in the twelve years following 1918 are better illustrated than described. Map 5-1 displays the street widenings, double-deck bridges, and boulevards open to Chicago motorists in 1931.

The automobile had, by 1931, three good north–south crosstown routes; boulevards of varying quality leading directly north, south, as well as west from the CBD; and four important streets with widths of 100 feet or more leading through and around the CBD. Altogether, the city opened or widened 108.25 miles of street between 1918 and 1930. Little was added to the boulevard system, but major modifications were made on the lakeshore boulevards north and south of the CBD.[60]

The Board of Local Improvements alone spent $114,103,891 on street openings and widenings between 1915 and 1930, and an additional $174,623,912 on the paving or repaving of 2,384 miles of streets and alleys. A group of city engineers in 1937 estimated the cost of all street improvements between 1912 and 1937 at $450 million.[61]

Other cities greatly increased their spending for streets and highways between 1915 and 1930, but Chicago spent with a special urgency. One recent scholarly study of New York, Chicago, Philadelphia, Detroit, and Boston during this period has found that "highway" spending in Chicago was greater as a percentage of current capital spending than in any of the other four cities for most of the 1920s. According to Bruce Allen Hardy, Chicago's debt also increased more rapidly than that of the other cities studied.[62]

This expenditure changed the appearance of many streets and boulevards, and transformed a number into special facilities for the automobile. Yet the paving and street-widening programs of the 1915–1930 era did not solve the traffic problem. By the end of the 1920s, Chicago's planners had turned to a whole new kind of street—the limited access highway—in their search for a means of channeling the automobile.

The Cycle of Street Improvements

Wider streets and better pavements failed to solve Chicago's traffic problems for three related reasons. First, even with the great expenditure on street improvement projects, the amount of the new street surface produced was far too little and came too slowly to keep up with the increase in automobile use. Second, improvements in and coordination of the roads surrounding the city helped to generate new traffic to and from the city and added to the demands on city streets. Finally, street improvement in the city itself attracted traffic. By the end of the 1920s Chicago was clearly in a cycle of street improvement followed by renewed congestion and the demand for further improvements. The classic pattern earlier seen in transit improvements was apparent in street improvements as well: the better the facilities, the greater the congestion.

The money Chicago spent on street improvements was too little pri-

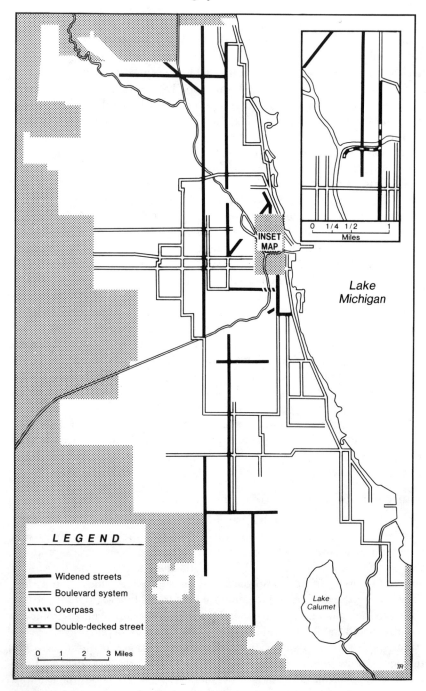

Map 5–1. Major street improvements, 1931. Data from BLI, *A Sixteen Year Record of Achievement.*

marily because the automobile—if it was to be used for mass transportation—required resurfacing, if not rebuilding, of most or all of Chicago's major thoroughfares. Preautomobile streets, which carried streetcar tracks on what may be described as a ridge at the center, were particularly unsuited to automobile traffic. The steep slope on old-style streets, which commonly produced a roller-coaster effect at intersections, caused motorists to seek the car tracks as the most comfortable roadbed, even in the 1920s. A great many "improved" streets, moreover, were paved with materials the motorist found unattractive or which had long since deteriorated under heavy traffic. [63]

The city could not keep abreast of the need for new sorts of paving. In 1928, 2,288 of Chicago's 3,563 miles of street were paved with asphalt or concrete pavements amenable to automobile traffic; 886 miles of street were unimproved. But in 1926, traffic engineer Miller McClintock observed that, while city records showed 2,450 miles of paved street, "these figures are distinctly misleading. A considerable amount of the earlier type of city pavement has become so roughened . . . that it can no longer be classified accurately as suitable surface for modern automotive traffic." "The most direct physical relief that could be brought to the Chicago traffic problem" McClintock reported, "would be an energetic and comprehensive paving program." [64]

Further, the important street widenings and rebuildings the city did accomplish were not completed until the mid- and late 1920s. Over half of the 112.75 miles of street widenings and openings done between 1915 and 1930 were completed after 1926. [65] The city moved slowly because street widenings, as conceived in the Chicago Plan, were great consumers of both time and money.

Street widenings were expensive because they almost always involved the condemnation of commercial real estate. The South Water Street improvement alone, which covered less than one-half mile of CBD riverfront, cost approximately $24 million. The 112 miles widened and opened by the city between 1915 and 1930 cost an average of over $1 million per mile. Costs were raised and action was delayed by property owners who sued for higher compensation for land taken in the widenings. Courts ruled over half the cost of the four largest projects to be "public benefits," paid for by the taxpayers of the city as a whole rather than by special assessment of the local landowners. On one occasion local property owners managed to prevent a street widening altogether. [66]

The Chicago Plan program was also held back by the same sectionalist spirit which made comprehensive transit planning so difficult. West side aldermen held up CBD improvements and, in one instance, wrangling among sectionalist aldermen prevented the presentation of a set of bond issues to the voters. [67] Because public money *was* available for street improvements, however, concessions could be made to neighborhood interests. The impact of sectionalism on Chicago Plan projects of the 1920s was thus not nearly as significant as the neighborhood opposition to subways in the previous decade.

Finally, the city's debt limit twice held back street improvements. However, on both occasions (in 1919 and 1927) Chicago's business community worked actively to have the limit raised by the state legislature.[68] This concerted and successful effort to facilitate public investment in street facilities was a sharp contrast to the strong opposition on the part of some of the same interests (in particular, the Chicago Association of Commerce) to the use of public money for mass transit.

While these various factors slowed street improvements in the city, a steady upgrading of roads in the area surrounding Chicago encouraged suburban motorists to drive to the city and enticed Chicago drivers into the countryside. Both phenomena increased the pressure of Chicago's too slowly improving streets.

The traffic which tied up CBD streets came, by a two-to-one ratio, from within the city itself. Nonetheless, as one engineer remarked, "the continued building of good roads and the consequent increase in the use of the automobile means a continued increase in the number of out of town cars using our streets, particularly those in the downtown district." In 1925 an average of 125,000 suburban automobiles crossed into the city on thirteen major roads each weekday.[69]

Improving state and county roads also generated traffic from within the city: Sunday produced the largest amount of traffic leaving and entering the city during the 1920s, as urban families fulfilled the prophecies of the automobile's champions and went to sample the sights and clean air of the countryside.[70] The county and state, like the city, built roads and worked to coordinate them in response to the ever-increasing traffic.

The State of Illinois approved a $60 million bond issue in 1918 for road construction. Cook County issued $7.8 million of bonds between 1914 and 1923 and another $15 million in 1926 as part of a $32 million roadbuilding program. In 1929 the county and city became eligible for some state money through a "state aid road" law passed in that year.[71]

The building of county and state roads in the Chicago region was largely uncoordinated before 1923. In that year, a Chicago Regional Planning Association was formed with the participation of many of the local governments of the Chicago area. In 1925 the CRPA began to seek agreements among counties surrounding Chicago on rules for road width and the coordination of roadbuilding to promote a smooth flow of traffic and to keep through-traffic out of Chicago. By 1929 the seven counties in the metropolitan Chicago region had adopted uniform rules for developers and had agreed on standard pavement widths. The Regional Planning Association took the same approach to traffic congestion as did the CPC: it widened roads; when this produced new congestion, it planned routes to divert traffic from crowded areas; and finally it turned to superhighways.[72]

Of course, better state and county roads made it possible for more and more automobiles to crowd in upon Chicago's streets. By 1929 the Chicago

Plan Commission was estimating that over 250,000 autos per day would soon be entering the city through twenty-one arteries. In the city, as outside, the best streets attracted the lion's share of the new traffic. Motorists flocked to the opening celebrations for major street improvements and before long rendered some of these new streets obsolete by filling them to overflowing with new traffic.[73]

The link between bigger and better streets and bigger and better traffic jams was not unfamiliar to engineers in the 1920s. One Los Angeles city planner observed in 1924 that "every positive solution which involves a physical improvement results in greater congestion, or, in other words, 'the cure is worse than the disease.'" In Chicago, however, this view—that there was a limit to the usefulness of street improvements—was expressed chiefly by men associated with mass transit or traffic control rather than by city planners *per se*.[74]

Faced with the inadequacy and cost of street widening, Chicago planners, engineers, and politicians looked to the development of a new kind of street—the superhighway. In doing so, they followed the mainstream of the American response to the automobile.

From Boulevard to Superhighway

Chicago's leaders did not invent the superhighway idea (it originated in Detroit), but they did not adopt it simply because it was being used elsewhere. Rather Chicago promoted and followed to its logical conclusions the idea of traffic segregation, which predated the automobile era. The policy route to the expressway led from the boulevard and through-street to grade separation and "limited ways" in a natural progresssion. Both the inadequacy of earlier solutions and the collapse of the city's traditional way of financing street improvements led the city toward the development of an entirely new kind of street—the urban superhighway.

Boulevards were Chicago's first through-streets and the progenitors of traffic segregation. Their development as limited access streets predates the automobile by decades: the requirement that vehicles on intersecting streets stop at boulevard crossings dated from 1863.[75] This rule was a tacit recognition of the fact that traffic and streets were divided into classes, one with prerogatives superior to the other.

The stop requirement was difficult to enforce. Nonetheless, by 1927 one Chicago paper, paraphrasing the president of the Lincoln Park Board, observed that "the great North Side breathing space had become a sort of glorified speedway." The speed limits enforced on boulevards were higher than those on commercial streets, the pavement on the whole was better, and the route was free of streetcars. As a result, during the mid-1920s, boulevards carried substantially more automobile traffic than nearby parallel business streets.[76]

Many congested commercial streets, however, were not near boulevards. It was natural, therefore, that the city would seek to obtain this unplanned benefit of the parks movement—the diversion of auto traffic from "car line" streets—even where no independent parkways were available.

One way of segregating traffic was the creation of "instant boulevards." Chicago invented this form of traffic segregation, which consisted simply in the designation of certain streets as through-traffic thoroughfares and the erection of "stop" signs at their intersections with ordinary streets. The first through-streets were "created" in 1922, and by 1925 Chicago had a substantial but rather poorly connected collection of through-streets and boulevards.[77]

Motor interests favored through-streets, but complained that they had been designated in a haphazard fashion, and without due regard for the traffic patterns of the city. "Chicago," declared the head of the Chicago Motor Club, "has run wild on the question of through streets. . . ." In 1924 the CPC acted to put order into the movement for through-streets, but the creation of such streets continued to be almost random throughout the 1920s.[78]

The greatest drawback to the through-street idea was that it threatened a rapid transformation of the character of residential streets. As one alderman observed, "The gentleman who wants through-streets wants to speed upon them."[79] Through-streets, if they succeeded in their purpose of diverting traffic, would mean a concentration of automobiles—and so of safety hazards—on what were for the most part residential streets.

Other means had to be found to divert the motorist from commercial streets and make speedy urban travel safe for drivers and pedestrians. Chicago had, within its planning tradition, two other ways of segregating traffic. One was the double-deck street; the other the limited access parkway. Both ideas were widely discussed and occasionally used in other cities.[80] Local needs and geography, however, appear to have shaped the timing of the evolution of the "limited way" in Chicago.

Two double-decked streets were among the most striking achievements of the Chicago Plan during the mid-1920s. Both were located along the Chicago River; both sought to divert traffic from the center of the Loop and to separate freight from passenger vehicles. Though neither was in full operation until after the 1930s, they provided good examples of what double-decking could do. They also illustrated its limits.[81]

Both projects were expensive—especially Wacker Drive, as already noted. Both placed one "deck" underground. This obviated the most serious property-owner objection to double-decking, since abutting buildings (almost all of them new in any case) did not "lose" their ground floors to the darkened tunnels which double-decking would have made of existing streets. But streets like Wacker Drive were possible due to geography and the concentration of business in the Loop. Its location seemed to "justify" the cost, and the River's presence allowed some light and air to reach the lower deck. Politicians continued to favor double-deckings, and the CPC was divided on the subject, but after the completion of Wacker Drive no further double-decked streets were put into service until the end of the 1970s, and none were ever built outside the CBD.[82]

Meanwhile a prototype for the expressway developed on one of the city's boulevards. Along the city's south lakefront, the park boulevard built on land filled in by the Illinois Central Railroad provided a kind of natural speedway. With the park and lake flanking it on the east and the railroad limiting access from the west, the "Outer Drive" could be reached from city streets only at quarter- or half-mile intervals. Full development of this drive did not take place until the 1930s, but by 1928 a grade separation in Lincoln Park, on what amounted to the north leg of the Outer Drive, had provided an example of a primitive expressway interchange.[83]

Burnham's lakefront pleasure drive thus began to develop into an auto highway. But the development of urban superhighways as a policy alternative did not arise from the Chicago Plan. Rather geography and finance combined to lead the city to this innovation.

Traffic congestion on the city's west and northwest sides provided a part of the impetus for the development of superhighway plans. The west side did not develop real limited access parkways, because it lacked a lakefront, and because west side boulevards all crossed major business streets. For this reason, only an elevated or depressed roadway—a new kind of street for Chicago—could provide a limited access thoroughfare for west side auto-mobile commuters.

An additional impetus to the movement for expressway building came from the fact that, by the time the west side traffic problem had emerged in its full dimensions, the city had effectively lost its traditional means of financing street improvements—the bond issue funded by special property tax assess-ments. Thus the expressway arose in Chicago both as a means of facilitating traffic and as a way of saving money, by forcing the motorist to pay a larger part of the cost of street improvements.

Expressway plans were under consideration in Chicago from at least 1926, and the Chicago Plan Commission developed the idea, complete with the inclusion of mass transit facilities (an idea which appears to have had its origins in Detroit). Most of the CPC ideas were put forth in a report of the city council's Subcommittee on Two Level Streets and Separated Grades, filed in May 1928.[84]

The Subcommittee, headed by Alderman John A. Massen, recom-mended a comprehensive system of expressways. The preliminary report suggested the use of railroad rights-of-way or adjacent land for the construc-tion of limited access roads to be used by automobiles, trucks, and intercity busses. In this and subsequent reports, Massen grounded his appeal for superhighways in the promise that they would help the city's business and industry keep pace in the competition with surrounding suburbs.[85]

The method of financing proposed for these "limited ways" marked a major departure from Chicago traditions. Between 1914 and 1927, Chicago

voters had authorized bond issues totaling $103,960,360 for street improve-
ment projects. Chicago's vehicle tax, though it was among the highest in the
nation, added no more than $35 million to this amount and was used for street
repairs, not improvements. When it is considered that all other street repairs
and improvements were made from real estate tax funds, these figures clearly
reflect a policy of subsidizing street improvements from property tax funds
and, so, from rents. Yet by 1932 Alderman Massen, writing on behalf of the
city council, was arguing that "streets and highways are a part of the organized
transportation system, and as such they must pay their own way."[86]

The initial proposals for elevated highways envisioned financing by a
regional taxing district, but by the early 1930s the city was seeking state
gasoline tax money for expressways linking Chicago with the surrounding
counties.[87] There were two reasons for this change. One was the fact that the
expressway altered the social and economic connotations of the street, and so
removed some of the rationale for public financing of street improvements.
Even more important, however, was the fact that, by the time the expressway
idea had come into vogue, Chicagoans had begun to vote down street improvement bond issues.

Chicago Plan street improvements had always been advanced with the
rationale that they would raise the value of land along their course and lead to
the development of higher-class business and residential uses. "Limited
ways," on the other hand, would *remove* traffic (and so access to business) and
would destroy parts of the communities through which they passed as well. In
short, they were facilities for automobile, bus, and truck rapid transit pure
and simple, with none of the connotations of social uplift so long associated
with boulevards and street widenings. Nor were expressways open to those
without automobiles. The early Massen reports hinted at the use of proposed
expressways by intercity busses, but the idea of placing mass transit on
expressways—a concept presaged in the Burnham plan—was not advocated
by any Chicago government body until the 1930s. Even then, the first
important plan to envision the use of expressways by local transit busses called
for the paving over of elevated tracks and the creation of expressways atop the
old rapid transit structures.[88]

Limited ways thus represented the final stage in the evolution of the
street from an extension of the local community, open to a variety of uses, to a
simple adjunct of motorized transportation. But this transformation was not
the only, or even the most important, reason why the city began to seek to
shift the burden of paying for street improvements to the motorist. As late as
the spring of 1927, Chicago voters had endorsed a $15 million plan bond issue
by a vote of three to one. Thereafter, however, voters turned down bond
issues for new projects and for the completion of street widenings already
underway.[89] What may appear at first glance to have been a general rejection
of publc responsibility for street improvements was, however, a much more
complex phenomenon. This tax revolt was, in fact, a rebellion against political
corruption and a response to the coming of the Great Depression.

The press and civic bodies had been generally favorable to public spending for street improvements between 1918 and 1927. Only during times of economic depression or when gross abuses were apparent, did newspapers, business organizations, and reformers turn against specific bond issues.[90] This confidence in city planning, however, was abused by local politicians.

Politicians worked hard to identify themselves with the achievements of the Chicago Plan. Through 1922, the CPC continued to praise politicians, thereby strengthening in the public mind the connection between the plan and politics.[91] The actual use of the money appropriated for the plan, however, can only have furthered public mistrust of local government.

Bond issues were sometimes presented in deceptive fashion. Major bond issues might be grouped together with minor ones, so that voters found it difficult to tell the difference between them. Sometimes different projects were made parts of the same referendum proposition: one could not be approved without the others. Bond issues were also used for relatively minor projects, such as traffic lights and, on one occasion, day-to-day operating expenses.[92]

More damaging than such "pork barreling" was the fact that the ultimate cost of most projects could not be correctly estimated in advance. The city might present a referendum proposition for a given improvement which would cover only preliminary work. The same improvement would then appear on ballots at subsequent elections when the city appealed for funds to begin construction. It might emerge again—perhaps several times—when courts ruled that the city must pay higher damages than had been anticipated, or that a larger portion of the improvement than expected had to be paid for by the city at large as opposed to owners of abutting property. Such "piecemeal financing," the Bureau of Public Efficiency warned, was "the forerunner of widespread dissatisfaction and mistrust of the manner in which such propositions are handled."[93]

Regionalist sentiment, which required that multiple projects be started so that each section of the city might see tangible results, was partly to blame for what the *Chicago Journal* called the "craze for bond issues." Property owners who sought to maximize the amounts paid them in damages and to minimize their assessments must share the blame for the fact that the city, in 1927, was littered with partially completed projects.[94]

But the greatest measure of responsibility must go to local politicians. "They have furnished a most fruitful source of extravagance," the CBPE reported in 1922, "of which the payment of extortionate fees to 'experts' is a conspicuous illustration." Toward the end of his final term, Mayor Thompson's administration was reported to have awarded a contract for a major plan project to a builder whose bid was $1 million in excess of the lowest submitted.[95]

Yet neither revulsion with the way bond issue funds were spent nor the cost of street improvements as such was enough to provoke a tax revolt. In 1925 Chicago's real estate tax rate was among the highest in the nation. This provoked complaints, but the CBPE endorsed $12 million of bond issues for downtown street improvements in the following year, noting that while the estimated cost of these projects had tripled, they were too important to be delayed.[96]

Chicago taxpayers finally rebelled against public spending as part of a protest against the assessment procedure and of a general repudiation of the Thompson administration. Chicago real estate tax assessment had been especially uneven throughout the 1920s. In 1927, assessments as finally fixed by the County Board of Review ranged from 74.1 percent of actual value in the most heavily assessed districts to 13.9 percent in the most favored area.[97]

There had been little public resistance to unequal tax assessments before 1927 because property owners knew little about the inequalities. Loop owners in particular may have feared that their assessments might be found too low if an investigation were held.[98] In spite of occasional tax-fixing scandals, therefore, there was little sign that a successful attack on the assessment system could be made.[99]

The break came in the summer of 1927. A special Cook County Commission—the result of agitation by the Board of Education and the Chicago Teachers' Union, which believed that Loop taxes were too low—filed a widely publicized preliminary report in July of that year. The report showed the Loop to be, not underassessed, but the most heavily taxed district in the city. Assessments everywhere were uneven. The small homeowner, as it turned out, was hardest hit after the Loop property holder. The result of this report was wide-ranging support for revaluation.[100]

Early in 1928 the Illinois State Tax Commission suspended all real estate tax collections in Chicago pending a new assessment. This, and the series of court decisions which followed, effectively removed the city's ability to collect taxes and finance its day-to-day activities for the rest of the decade. Thereafter Chicago became almost entirely dependent on downtown banks which bought the city's tax anticipation warrants. The banks, through a committee of civic notables, insisted on economy measures in return for their support.[101]

Revelations about assessment procedures combined with the coming of the Great Depression to give many Chicagoans a rationale for refusing to pay their property taxes. In 1938, over $47 million in taxes, dating from the years 1928–1930, remained unpaid.[102]

Tax delinquency cut the city's ability to spend and borrow. The determination of Chicago's creditors to force economy measures on the city government helped assure the end of massive, haphazard public works spending. It is important to note, however, that the financial community did not attempt to stop projects which were already underway, or even to eliminate all new ones, until the city's economy had begun to falter. As late as the autumn of 1928, the Chicago Association of Commerce joined the Chicago Plan Commission in attempting to restore public willingness to vote money for the completion of projects already underway.[103]

Some major civic organizations continued to support new expenditures through 1929. When specific bond issues *were* opposed by business and clean-government reformers, it was, until 1930, on the ground that the expenditures in question were ill planned, unnecessary, or to be supervised by the wrong agency—not on the basis of opposition to spending *per se* or to spending for roads.[104]

It was the abuse of the taxation system, not street improvements, against which the public appears to have been reacting. By 1926, 13 percent of the city's revenue was coming from special assessments—an amount much in excess of that gathered by the New York, Boston, Philadelphia, or Detroit from similar sources.[105] By the time of the vote on the April 1928 bond issues, the public had been exposed to eight months of publicity about the inequities of the tax system. Over half of the amount on the ballot was to be spent for additional payments to property owners, not for new work. Together, the old and new projects represented on the ballot exhausted the bonding power newly granted to the city. These very major expenditures were proposed by a city government which was engaged in a hotly contested primary election fight, the outcome of which was an overwhelming repudiation of the mayor who had endorsed the bond issues.[106]

All of these factors helped to bring about the April 1928 vote, which put an end to the great era of city spending for street improvements. Only after the tax controversy had begun to cool and a wide range of endorsements for bond issues was presented (in the spring of 1930) did public willingness to vote for further indebtedness revive. By 1931, however, bond issues were being advocated for straightforward work-relief programs.[107]

It was the Depression, then, which put an end to the city's traditional street policy. By 1931, the bankers whose loans enabled the city to carry on basic operations were demanding retrenchment as a condition of their credit.[108] Chicago's automobile policy, however, did not undergo any really fundamental alteration.

The city's street improvement policy had sought to accommodate and channel the automobile. In the 1930s and afterward, the city simply changed its approach to financing improvements by seeking outside funds. The most important change was an accentuation of the tendency to accommodate CBD traffic rather than to channel traffic citywide, for now the city worked for expressways aimed at the heart of the city.

Section III:
The Motorist Comes of Age:
Traffic Regulation, 1924–1930

> Rules which appear to be useless, arbitrary, or unnecessarily
> severe will inevitably be broken by many otherwise law-abiding
> citizens who feel that their individual freedom has been violated.
> . . . Traffic laws will be obeyed without compulsion in direct
> relation to their reasonableness.
>
> *Miller McClintock, 1926*[109]

> The police have our sympathy. God knows what they are up
> against.
>
> *Chicago Motor Club spokesman, 1925*[110]

A Business Problem:
Redefining Traffic Regulation

During the second half of the 1920s, Chicago's definition of reasonable
regulation came gradually to approximate that reflected in the behavior of
most motorists. At the same time, effective tools were fashioned for bringing
into line the driver whose habits did not conform to the norm. Several means
were found by which the motorist's behavior could be altered in ways which
would increase the usefulness of the automobile. Together, these seemingly
commonsense adjustments committed the city to what *Roads and Streets* in
1926 described as "the most far-reaching and inclusive program of traffic
control ever effected in the United States."[111]

Two major factors combined with the growing traffic crisis to bring about
a wave of innovation in traffic management after 1924. The first was the
interest on the part of the Chicago Association of Commerce and of Mayor
William Dever in the creation of a uniform traffic code and the rationalization
of downtown street use. The second, closely related to the first, was the
definition of CBD traffic as a business problem rather than one of law enforcement or overall transportation policy.

The CAC's standing Committee on Traffic and Public Safety carried on
the Association's tradition of involvement with the Loop traffic problem—a
tradition which dated from its sponsorship of the police mounted traffic squad
in 1910. When Mayor Dever, shortly after his election in 1923, proposed a
study of the entire city's traffic situation, the downtown business community
backed him.[112]

The CAC's Committee on Traffic and Safety thereafter blossomed into a
forum through which ideas from all concerned parties could be brought to

public attention and presented with authority to the city council. In 1925, the Committee took on the formal role of screening and formulating plans for traffic relief. In the same year the CAC, in cooperation with the city council, commissioned a study of the local traffic situation by a nationally known engineer, Miller McClintock.[113] The McClintock study, because of the scientific sanction it gave to the modernization of traffic law, constituted an outstanding achievement of business and political leaders working to in renovate Chicago's policy toward the automobile.

Superficially, McClintock's study occupies a position in the Chicago history of the automobile similar to that held by the first Arnold report in the history of local transit. As the Arnold report, taken together with the report of the Street Railway Commission of 1900, constituted a point of reference for the city's traction policy above the political battles of the day, so McClintock's study set forth a program which could be looked to with confidence by all concerned as a disinterested judgment of what was necessary to restore freedom of movement to the city's streets.

Here, however, the similarity ends. A number of the points in the Arnold report were revelations to the city's leaders. In particular his advocacy of through-routes, universal transfers, and CBD streetcar subways offered a precise vision of possibilities theretofore only hazily conceived. The McClintock report, though a very real engineering achievement, suggested little of importance which had not already been seriously considered by the CAC and other organizations. Indeed McClintock's 1925 book on street traffic control contains most of the recommendations which the engineer made one year later for Chicago. The CAC presumably knew in advance approximately what McClintock would say. Even more than in the case of transit, then, the real role of the engineer in the traffic problem *as a public figure* was the ratification of programs which already had the support of the ruling community.[114]

The traffic engineer took the role he did because Chicago in the 1920s defined the traffic problem as a business problem. This second major element in the 1920s traffic reform has two distinct aspects. First, during the 1920s, Chicago largely accepted the idea that the function of commercial streets was to facilitate movement to and from places of business. Parking became a privilege, not a right, and the motorist's behavior was regulated in order to increase facility of movement, even if this meant, for example, that a given driver had to circle a block instead of making a single left turn. Second, Chicago stopped viewing traffic as a moral or legal problem. Penalties for traffic offenses were lowered, enforcement procedures were simplified, and the motorist's experience of dealing with law enforcement agencies was made far less embarrassing.

Technological and Legal Innovations

The ideas behind the new traffic control and the difficulties involved in its implementation are best illustrated through three examples. These epitomize the changes of approach to street use taking place in the years after 1924. "Progressive" traffic signals mark the city's first effort to control the motorists' movements within a large area according to an overall plan. The prohibition of Loop parking during daylight hours made commercial streets pure access routes and the storage of automobiles more nearly a private responsibility. The modernization of enforcement procedure made it much easier to hold the motorist accountable for his behavior and at the same time increased the likelihood that he would respect the law.

Traffic Signals

Chicago experimented freely with traffic signals before 1924. Through 1925, however, engineers, police, and the Chicago Surface Lines opposed traffic signals on the ground that they slowed traffic, since they were unable to adjust to changing conditions and did not command the motorist's respect. Yet the Yellow Cab Company showed faith in the signal idea even before 1925: the company paid for the installation of synchronized traffic lights on Michigan Boulevard in 1924 and pronounced the system a great success.[115]

A movement began in 1924, under pressure from the CAC, Yellow Cab, and the local motor clubs, to extend such lights throughout the city. Study consumed most of 1925, with the CAC coordinating the effort. Nonetheless, the traffic signal, as a simple mechanical device, did not appear in 1925 to have much to offer the science of traffic control.[116]

Synchronized traffic signals—grouped in sets which "turned green" simultaneously along an entire section of a street—could do little to instill order in the traffic of a busy commercial street. A series of synchronized lights simply invited the motorist to race; the faster he went, the more lights he could clear before the entire group "turned red." Furthermore, such lights did not necessarily give the longer "go" time to the street carrying the larger amount of traffic. Worse, since motorists were "rewarded" for moving quickly down the street, fast drivers would catch up with slower drivers ahead of them and "bunch" at intersections. Because all lights stayed green for an equal amount of time, this small herd of vehicles created by the conditions set up by the lights might not clear the intersection on the next green light.[117]

What was needed was a means of making traffic movement orderly. The city needed a device which could be flexible—as was the traditional mounted policeman—yet could coordinate traffic movements from one intersection to another. Experimental installations of sequentially timed traffic lights, keyed to the average speed of traffic, had been made in New York and Detroit by 1925. McClintock thought them the most promising innovation of recent years when he wrote his classic work on traffic control in that year. But there were serious difficulties. Finding the proper timing required great skill, and McClintock believed such signals could not be used on streets with mass transit, since the stopping times of streetcars were too unpredictable.[118] Since most important Chicago streets carried streetcars, the new form of signal seemed to have little applicability for Chicago.

Transit engineers, not professional traffic engineers, resolved this problem. In 1923, the Chicago Surface Lines had undergone a major managerial reorganization. Hoping to avoid the political consequences of an increasingly bad public image, the interests which controlled the CSL brought in engineering and management personnel from outside the city to modernize their operations and win public favor.

The Surface Lines Engineering Department was reorganized by Evan J. McIlraith, a mechanical engineer with long mass transit experience in Seattle and Philadelphia. McIlraith and the engineers working with him were given a free hand to explore all matters relating to the company's operations. Since street traffic constituted a very major part of the difficulty confronting the CSL's operation, the engineers concentrated much attention on the chaos in the Loop and elsewhere. They lobbied for and provided statistical evidence in support of improved traffic regulation and approached the problem of signal light timing from their own perspective.[119]

McIlraith had been dealing with signal timing before it was applied to traffic control. In working with the Philadelphia Rapid Transit Company he designed a system of timing subway movements which, instead of simply preventing trains from running into one another, structured the subway traffic into an even flow. It was this idea, greatly modified, which he applied to Chicago street traffic.[120]

McIlraith was included, with the city electrician and others, on a Chicago Association of Commerce Committee on Traffic, which explored the problems of CBD congestion in 1925/26. By means of charts and theoretical demonstrations, the transit engineer managed to convince the sceptical businessmen and amateur traffic experts that a comprehensive system of traffic signals, timed to the movements of streetcars, would speed, not hinder, the movement of traffic. The task was not an easy one, for McIlraith had to argue against what appeared to be common sense: that if automobiles were limited to the speed of a transit vehicle, fewer private cars would be able to enter the Loop.[121]

An initial system of progressive signal lights, all linked to an adjustable timer in city hall, went into use on February 1, 1926. Studies of the densities of traffic at various intersections, done in connection with the McClintock investigation then in progress, enabled CSL, which set the timing of the lights, to interlace traffic on all the major streets entering the CBD into a pattern; thus at each intersection at each hour of the day, the "green" time given each street was approximately proportional to the amount of traffic expected from that street. At the same time the traffic was segmented into waves which, as the motorists began to understand the operation of the system, could flow with relative ease through the once-tangled Loop. Most left turns having already been outlawed and streetcars rerouted to eliminate unnecessary turning movements, the motor and transit traffic found a relatively free path.[122]

Shortly after the installation of the system, tests with automobiles and streetcars showed speeds had risen 25 to 50 percent on various Loop streets. Approval of the new system was not, however, universal. Hundreds of motorists had complained of having to make more frequent stops with the new system than with the old—apparently because they refused to adjust to an even speed. Pedestrians, who continued to defy the lights and the police, often found themselves trapped by oncoming waves of traffic.[123]

Within six months, however, the city had adapted to the system. The CAC proudly took credit for the invention, the city officials soon basked in the approval it brought them. There can be little doubt, however, where the impetus for this innovation originated. McClintock, who in his *Report and Recommendations* avoided all mention of the origins of the new signaling system, later described McIlraith as having "developed the principles and plans used in the progressive system installed in the Chicago Loop district. . . ." Throughout the 1920s, CSL engineers arranged the timing for the city's progressive traffic signals.[124]

The reason for the CSL's concern with traffic was clear: in addition to the public relations advantages to be gained, the company's interest in moving its cars down the street impelled it to find ways to move all traffic more expeditiously. Further, like transit executives elsewhere, CSL officials were infused with optimism about the possibilities of traffic control and the limitations of auto use.[125]

Surface Lines executives saw a positive correlation between modernized traffic regulation and the prosperity of their own business. Loop merchants, their dealings with the street being less direct, did not always see the problem so. Parking regulation therefore proved to be much more difficult to implement than did modern traffic signals.

Parking Control

By 1925, the idea that two lanes of every street could be reserved for automobile storage had come under heavy attack. The CAC had supported parking restriction from 1920, but many of the Loop's smaller merchants did not. The Association knew that the parking problem was, as R. F. Kelker had said, "a matter of salesmanship."[126] Hugh E. Young of the Chicago Plan Commission expressed the fears of many merchants and building managers when he predicted that a no-parking rule without the provision of free storage space would "disrupt business conditions." Young cited a survey showing that 67 percent of the autos in the Loop were used for business and that half the surveyed owners would not pay a 50¢ storage fee. Merchants who appeared before the CLT in hearings during 1925 and 1926 expressed similar fears.

"You are driving the trade out of the Loop to the outside," a spokesman for one relatively small store complained.[127]

Owner of physically small establishments catering to an affluent trade were particularly concerned about parking in Chicago, as in other cities. Affluent customers could park, go in to buy a book or an order of expensive pastry, and depart without violating the existing thirty-minute parking limit. One bookstore owner, whose street had been the subject of an experimental parking ban, alleged that his establishment had "contributed . . . $1500 a day to traffic control in lost business."[128] In general it was the Loop's larger business establishments which, together with the CSL, pressed for greater efficiency in street use.[129] Indeed, for practical purposes, the parking issue had become a dispute between different classes of business.

What was needed was evidence that a ban on parking in the Loop would not damage the trade of certain kinds of Loop businesses and the attractiveness of downtown office buildings. This the McClintock survey provided. In 1926 McClintock reported that, of 68,621 department store patrons surveyed, just 6.7 percent had arrived by automobile and fewer than 1.5 percent had parked at the curb. Nearly 13 percent of the visitors to office buildings, 16 percent of restaurant patrons, and 18 percent of book buyers had used automobiles to come to the Loop, but in each case less than 2 percent had parked at the curb. Motorists, it seemed, were more willing than anyone had thought to pay a small parking fee and walk from a garage to their destination.[130]

Other factors combined to favor a parking ban. During the third week of June 1926, the Eucharistic Congress of the Roman Catholic Church brought together prelates and faithful from all over the United States and many foreign countries. The city banned all automobiles from the CBD and restricted deliveries to off-peak hours. The result was a heightened interest in traffic control: the elimination of autos from the Loop did not necessarily mean the elimination of shoppers. At the same time, summons arrest for traffic and parking violations, which had gone into effect in March 1926, appeared to have little impact on the parking problem.[131] Clearly something more was needed.

Small Loop merchants continued to oppose a blanket parking ban through most of 1927. By the end of that year, however, the city council proved willing to pass the long-discussed legislation. The results of the ban were much as its advocates had predicted.[132]

At the end of 1928, auto registrations had risen 10.6 percent above the levels of 1926. The number of private vehicles entering the Loop, however, had jumped by over 14 percent. The number of police required for parking law enforcement dropped, and the speeds of streetcars through the Loop rose significantly.[133] Despite early fears, downtown business prospered. A major Loop property management company found in a poll of its tenants that 674 of 864 surveyed favored the new rule.[134]

The parking ban had other significant results. Chicago was a leader in providing both public and private parking facilities during the later 1920s. In 1935 Chicago was still the only major U.S. city with a parking prohibition covering a large part of its CBD. It also continued to lead in the provision of off-street parking and, McClintock believed, Chicago's concern to provide the motorist with parking space had brought more autos into the Loop than would otherwise have been possible.[135]

The CBD parking ban was an affair of business: ultimately the CAC and its allies had convinced the Loop's smaller businessmen that a parking ban was to their advantage. The progressive traffic signal was likewise a business affair: the CSL convinced other elements of the business community that lights timed to the movement of its cars would promote the movement of traffic in the interest of all. These two movements for efficient use of street space (with a successful drive on the part of the same interests to outlaw left turns in the Loop) did much to rationalize the behavior of motorists in the CBD. Businessmen also led in the drive to improve citywide traffic control through modernized traffic law enforcement.

Traffic Laws: From Enforcement to Administration

During the years which followed 1924, Chicago altered the administration of its traffic code by means of three simple steps. Together these changes amounted to a decriminalization of traffic offenses. While they did not solve all the problems of traffic law enforcement, the changes made during this period opened a new era in the relations between the city and its motorists.

The city's first steps toward a business approach to traffic control were the reduction in 1926 of the minimum fine for many traffic offenses and, in 1927, the elimination of minimum fines for most infractions.[136] More far-reaching was another major change which also began early in 1926.

At the urging of the CAC, municipal traffic court was reorganized into two branches: one a traditional court to hear contested cases; the other a "Traffic Bureau" with a clerk to take "guilty" pleas and accept standardized fines. Violators—given a printed slip by the arresting officer—came to court knowing exactly what would result if they admitted guilt. This system streamlined the process of traffic law enforcement, especially in the case of parking violations. The 38,280 parking misdemeanors of 1925 dropped to 307 in 1927, with summonses ("parking tickets") making up most of the difference. Speeding charges dropped from 36,020 in 1925 to 12,757 in 1928 and 8,165 in 1930; as speed limits were raised, and, after 1929, enforcement adjusted to the difficulty of obtaining convictions.[137]

Serious problems remained. Judges continued to be lenient toward motorists who did ask for trials, and low fines meant that some motorists came, as a 1927 city council resolution noted, to "[look] upon them as an incidental expense" in connection with the operation of their cars.[138] The solution to this problem lay in a revocable driver's license—an approach which had had the support of some local politicians since 1923. Only the state could alter the driver licensing law, however, and the city was unable obtain the necessary legislation during the 1920s.[139]

By 1930 the city had, however, committed itself to a new, coherent approach to traffic law enforcement. Motorists were no longer treated as conventional lawbreakers, summons arrest had helped to cut the number of parking violations, the city's accident rate was below that of most other cities, and its progressive traffic signals and downtown parking policy were models for traffic engineers and planners elsewhere.[140] Nearly one third of ticketed offenders failed to appear in court in 1930, and law-abiding motorists still had to contend with different sets of regulations on city streets and park district boulevards, but the fundamental shift in the bases of traffic regulation had taken place.[141]

The Automobile in 1930

During the 1920s, Chicago's street and traffic policies increased the usefulness of the automobile. They helped to make the private car less of an intrusion upon urban life, more nearly a workable element in the city's transportation system. Yet by the late 1920s, private autos had only begun to transform the city's way of life.

Street widening and boulevard improvements increased the speed of urban traffic. In 1932, autos on the major west side thoroughfare, which carried streetcars, average 15 mph between the CBD and a cross-street five miles west. On adjacent boulevards, equipped with stop signs and free from streetcars, auto traffic speeds averaged 20 to 25 mph to within two miles of the CBD. On the south Outer Drive, motorists averaged 25 mph between the Loop and Hyde Park, six miles away. Speeds on widened commercial streets were also greater than those on unwidened streets. Streetcars, on the other hand, averaged 9 mph citywide—an improvement of just 1 mph over figures from the early 1920s.[142]

Better streets, parking facilities, and downtown traffic control helped make the private car and the CBD more nearly compatible. By 1931 the CBD had storage spaces—mostly off-street—for 23,217 automobiles, and by 1926, 33.1 percent of the persons crossing the CBD's boundaries were riding in private cars, though it is not known how many of these stopped in the CBD.[143]

As always, however, such simple statistics are misleading. In 1928, fewer than 7 percent of the persons in the CBD at any given time had come by private car. The use of the auto for the journey to work outside the CBD is even harder to estimate, but it is probably fair to say that by the end of the 1920s the automobile was only beginning to become important in this role.[144]

One study, done in connection with the 1926 traffic survey, showed that only 2.6 percent of nearly 38,000 workers in a major plant at the city's western fringe used the automobile for their journey to work. The same number used steam commuter trains, and nearly 18 percent still walked to work. Residence data for the same group were not much different from those found for workers in this factory in 1916.[145]

Chicago undertook no major journey-to-work studies comparable to that of 1916 during the 1920s or early 1930s. Data from other cities, however, suggest that more and more workers citywide were using automobiles for commutation. Twelve percent of Milwaukee's industrial workers were using the automobile for the journey to work in 1928, as were 20.9 percent of those in Washington, D.C., and 8.2 percent of the dock and factory workers of Baltimore. Of Washington's workers, 23.2 percent continued to walk to work, as did 26.9% of those surveyed in Milwaukee.[146]

Such data, of course, can give only the broadest indication of what was probably happening in Chicago. We do know that auto ownership had become diffused throughout the city by 1926. Only the inner west side showed substantially fewer auto registrations than its population would seem to warrant. It is not possible to say, however, how these cars were used. Nevertheless, something can be inferred from transit ridership data.[147]

Chicago transit rides per capita rose through 1926 and declined thereafter. An economic downtown in Chicago during 1927 had something to do with this falloff, but when business revived temporarily in 1929, transit ridership did not rebound to the same per capita levels it had held in 1926. Thereafter, with the collapse of the national economy after 1929, transit ridership collapsed, never again to attain the levels of the 1920s.[148]

More revealing was the change in weekend ridership. Saturdays were the heaviest traffic days for the CSL throughout the 1920s, and Saturday ridership increased through 1929, as many Chicagoans rode the trolley to both work and play. But Sunday ridership experienced a steady decline after 1926.[149]

The automobile, then, may have begun to make serious inroads on mass transit's "journey-to-work" business by 1929, but its most serious effect was on recreational travel. The private car was ready by 1930 to assume the dominant role in urban transportation—a role which public policy had helped make possible.

By 1930 basic tools for the accommodation of the automobile had been discovered—the expressway and positive, facilitative traffic control. Yet when traffic revived in the mid-1930s, it once again became clear that improved roads drew more autos and created mounting traffic problems. After 1930, too, it became clear that innovation in traffic control was at an end.[150]

What Chicago had achieved for good or ill, then, was the depoliticization of traffic control. Miller McClintock declared in 1926 that traffic was ". . . similar in its general characteristics, to other physical problems of municipal administration, such as water supply, sewage disposal . . . and city planning. . . ."[151] This approach defined traffic control as relatively value-free and thus, ironically, limited in its potential.

In the short run, however, Chicago's auto policy achieved results partly because it defined traffic problems in terms of commercial access, freedom of movement, or, on occasion, something akin to hydraulics. "There is no republican or democratic method of handling traffic . . . ," McClintock wrote in 1926. R. R. Kelker agreed that solving the traffic problem was "merely a matter of engineering purpose."[152]

During the 1920s, Chicago's business community worked with others to depoliticize mass transit and remove it from the arena of controversy over social policy. In this the city was partially successful, but the success came too late. By 1930, large amounts of private capital were not to be had for mass transit, and the automobile had already become an accepted and increasingly important part of the urban transit system—the standard by which mass transit would soon be judged.

Chapter 6:

Making Do: Mass Transit in the Motor Age, 1914–1930

Section I: Transit Policy: 1914 versus 1930

> The people of Chicago have been trained to suspicion. They respond to demagogy where streetcars are the issue. The hand of Yerkes is on our traction lines yet.
>
> *Chicago Tribune, 1924*[1]

On July 1, 1930, Chicagoans were faced, for the third time in a dozen years, with an important "traction refendum." This time every active politician, every major newspaper, every downtown and neighborhood business organization, and most of the city's labor unions supported the proposition. "Really everything looks so hopeless," wrote Clarence Darrow, declining to join the tiny band opposing the ordinance. "Everything is so fully greased. Everyone is lined up."[2] This show of unity might have seemed remarkable to an observer transported from the Chicago of 1914; the substance of the ordinance being voted upon he would have found simply astounding.

The range of support for this 1930 compromise was truly remarkable. The Chicago Federation of Labor, steadfast for municipal ownership since the days of Yerkes and Dunne, supported this, the most generous settlement yet offered to private capital. The once-powerful obstacle of regionalism had been overcome: Tomaz Deuther gave full backing to the new ordinance, even though it guaranteed CBD subways more effectively than it did any other

164

improvement.[3] The city council itself had made equally profound adjustments.

The 1930 proposition did away with the short-term franchise entirely, replacing it with what was called a "terminable permit" of no fixed duration. Under this device, the newly created surface and elevated company was guaranteed the right to operate unless or until the city could buy the lines itself or find another purchaser.

In 1930 these provisions amounted to nothing other than a permanent grant. The combined company was assured a valuation of nearly $400 million, and the new franchise, which contained no effective amortization provision, made it certain that the valuation would rise. The city itself was effectively in receivership at this time, and there was no reason to expect that either it or some outside company would ever offer half a billion dollars for the transit properties.[4]

Despite this permanent franchise, the ordinance did not provide for effective regulation. It assumed a local regulatory commission, but the legality of this provision was untested, and the state had shown no willingness during the previous decade and a half to grant the city home rule. Further, the ordinance left the rulings of this commission open to court review, and the experience of the 1920s suggested that the courts would be sympathetic with the company.[5]

The old commitment to fixed fares and compensation was missing from the 1930 ordinance as well. The new company was to be guaranteed a "reasonable return"—the rate to be set by the local regulatory commission, subject to court review. This "Transit Commission" was defined in the ordinance only as "the body which by statute shall have jurisdiction and power to regulate the company. . . ." Whatever the "reasonable return" set by this Commission—if it came into existence—that rate would determine the level of fares. Compensation to the city, even in case the new company defaulted on its obligations under the agreement, was secondary not only to the return on the valuation but to specific securities of the new company.[6]

In return for all this, the city was promised $200 million in improvements and extensions, and the long-sought unification of surface and elevated service, though transfers from surface to elevated cars would require the payment of 3¢ above the streetcar fare. There was no effective provision to guarantee that any of the new company's obligations under the ordinance would be carried out.[7]

If all this did not add up to the realization of the Straphanger's Dream, it looked a bit like what might have been Yerkes' dream. Yet, given the fact that transit had long since been defined as a business, and that municipal ownership in any meaningful sense had been ruled out, the 1930 agreement was probably the best the city could have obtained. It at least freed transit to operate like other businesses—adjusting fares to costs and making, if it could, an attractive return. If mass transit could not become, like the street, a genuine public responsibility, it might at least attract needed investment as a genuine private enterprise.

It had taken policymakers fifteen years to reach this resolution, the same fifteen years in which the automobile had begun to transform the city's commuting habits. The process of adjustment by the city to the realities of local transit policy options had been slow, painful, and fraught with diversions. When the adjustment finally came, the fate of mass transit had to a large extent been sealed.

•   •   •

The distance between the transit policy of 1914 and that of 1930 was great. In 1914, the rationalization and modernization of mass transit had been impossible for four major reasons. First, there was the nexus among valuation, profits, and the hypothetical commitment of many politicians to municipal ownership. Elected officials dared not accept an agreement which embodied high valuations because the higher the valuation, the more difficult and expensive municipal ownership would be and the more the companies would be able to extract under the "5 percent" clause. The companies, on the other hand, could not accept lower valuations because they had real, if worthless, securities to "cover," and because the 1907 formula made overcapitalization the best assurance of higher short-term profits.

The second set of issues centered around the question of control. Local tradition required short-term franchises on the theory that the companies, if their right to operate could be revoked after twenty years, would be more likely to give good service. By 1914, however, as the fate of Mayor Harrison's subway project revealed, financiers would not undertake a comprehensive transportation plan with a short-term franchise. In the years after 1914 declines in the size and predictability of transit earnings made long-term franchises all the more vital for the marketing of transit securities. Yet, because the creation of a state Public Utilities Commission had removed the city's power of regulation, short-term franchises seemed the only means by which a local government might retain some influence over a utility's behavior.

A third and similar nexus was that among profits, fares, and unification of service. The city's policy through 1914 had been to press for lower "real" fares through universal transfers and the unification of the surface and elevated lines. But in 1913 the streetcar companies were already warning that they would accept no agreement requiring them to underwrite the prosperity of the elevated lines.[8] Unsubsidized unification of the surface and elevated systems certainly meant that surface riders and investors would in some way "bail out" the rapid transit. Yet neither politicians nor Surface Lines executives could accept such an arrangement.

Finally, no new arrangement could be reached in 1914 because neither political nor business power was sufficiently centralized. Competition between Democrats and Republicans, and within the parties themselves, kept the transit issue "hot"; effective organization by both CBD and neighborhood business kept business as a whole from bringing united pressure on politicians and companies for a quick resolution of the transportation dilemma.

All of this had been true in 1914 and had continued to be true for more than a decade thereafter. Static mass transit policy largely accounted for the fact that Chicago's transit system was substantially the same in 1930 as it had been in 1914. Between 1914 and 1930, however, changes in public perceptions and in power realities made possible the complete removal of transportation decisionmaking from the public domain. Traffic and street improvement policies were made outside the political arena too, but the motorist influenced their development through his behavior and voters could, and did, reject bond issues. The straphanger had no such options. The changes which led to the 1930 resolution may be considered in four broad categories.

First, the fifteen-year transit stalemate showed that the most important transit improvements could not be made unless a compromise was reached with the financial community on its own terms. Subway and rapid transit improvements—even the widespread rational use of buses—were impossible under the system of mixed control created by 1907.

Second, changes in the financial position of transportation companies took decision-making power entirely away from transit executives themselves and centralized it in the hands of bankers and financiers to a greater degree than before.

Further, events between 1914 and 1930 eliminated old solutions. They made it evident to any observer that regulation and municipal ownership were completely impossible in the prevailing financial and political climate. In addition, engineers lost or forfeited their image or neutrality, and their public function was slowly taken over by the CBD business community.

Additional changes were two remarkable examples of convergency. Regional business interests were coopted by the CBD business community and by utility interests; and political parties, downtown business, financiers, and labor unions came to share goals and interests to a greater and greater degree. This last change, part of the foundation for the creation during the 1930s of a long-lasting, citywide political machine, took place in stages after 1914. It received its final impetus from the coming of the Great Depression and the Century of Progress exhibition of 1933.

Section II:
Progress and Profit:
Transit Improvement,
1914–1930

[At] 5:30 P.M. yesterday thirty and more people were waiting for a [street]car at Madison and Paulina [two miles directly west of the center of the CBD]. Two cars went by, already overloaded. . . . A car finally arrived. All hands stormed the rear platform. Women took their chances with the men. They weren't savage about it; just fiercely persistent. . . . The car started off before all had climbed on. It was full anyhow. As it swung away, three women, one grey-haired and frail, were standing on the step, kept off the platform by the crowd inside, yet held to the step by the men clinging behind them. . . . [At the next two streets] each corner had another small crowd of which only the strongest found clinging room. . . . The weaker ones waited. . . . It was cold outside too, and damp.

Chicago News, 1923[9]

The effect of regulated private ownership on transit service in Chicago between 1914 and 1930 may best be understood if the period is divided at 1924. Prior to that year, tradition-bound management and engineering played an important role in limiting the quality of service on the Surface Lines, and technology set strict limits to the potential of the great alternative means of surface transport, the gasoline bus.

Beginning in 1924, however, technical and structural limitations became less important. The city's commitment to profitable, unsubsidized mass transit, together with the snare of regulation, made comprehensive rapid transit and unified service impossible throughout the period. After 1924, however, the skills and technology became available which could have made possible a better adaptation of mass transit to the needs of an increasingly dispersed and motorized city. Especially after 1924, therefore, it was the complexities of regulation and the needs of capital which set narrow limits to what technology and engineering could do for the surface riders who made up 80 percent of the city's straphangers during the 1920s.

Short of subways and protransit traffic control, two major options were available for the improvement of surface mass transit during the 1920s.

Transit could be adjusted to changing traffic and demographic conditions through new kinds of equipment and by means of routing designed to accommodate heavy auto traffic and population dispersion. In these areas— the adjustment of service and equipment to conditions—Chicago went as far as the combined technology and politics of the period would permit. These limited improvements did not begin, however, until 1924.

Between 1914 and 1924, transit service deteriorated badly. On the surface lines, average speeds dropped by 25 percent between 1915 and 1921, and in congested districts conditions were even worse. The number of revenue passengers rose by 10 percent between 1918 and 1923, but only 3 percent more cars were added. No significant extensions were made during this period by the surface or elevated companies, despite repeated pleas from aldermen and their constituents. In 1922 Insull himself declared that rapid transit service had reached its nadir.[10]

An engineer brought to Chicago by the CSL in 1924 recalled the lot of the straphanger in that year: ". . . Day after day, on many of the busy lines, all that could find room . . . [were] hanging on the window guard bars with their feet on the trucks under the car bodies . . . the platforms were jammed to refusal." This, and like circumstances on the elevated, could be traced in part to the especially difficult financial situation in which transit companies found themselves during and after World War I. But poor routing and scheduling practices, increased auto traffic, restrictive state regulatory practices, and the time lost in making change when fares rose to 7¢, all contributed to the deterioration of transit service at the time when the automobile was enjoying its postwar boom.[11] For economic, political, and public relations reasons, all the companies began after 1923 to look for profitable ways to improve their service and their public images.

Samuel Insull's elevated lines had a head start on the surface companies in the cultivation of their public image and in the art of building employee "esprit." Insull himself made frequent addresses on the role of utilities as public servants, and he gave the city a new opera house in the Edison Company's skyscraper officebuilding. After 1922 the elevated companies built several suburban extensions and added a small but significant number of new cars (complete with plush seats, clocks, and glass lightshades.)[12]

But the net effect of all this was chiefly cosmetic. Wooden cars of late-Victorian vintage remained the mainstay of Chicago's rapid transit system through the 1940s. Insull's suburban extensions, and the cars of two of his interurban lines, all fed into the overcrowded two-track Loop. The extensions, mostly on ground level, led to Niles, Maywood, and Bellwood. They were always lightly used: William Dever observed that they served only to "strike terror into the hearts of holstein cows," and Insull himself declared that they were designed primarily to promote the settlement of new territories and so the market for electricity.[13]

On the Surface Lines, a public relations and service improvement drive began in 1923, when Insull's confident, Henry Blair, replaced Chicago City Railways' Leonard Busby as head of the unified operating company, the CSL.

The Blair ascendancy typified Chicago mass transit events in the 1920s in that its significance was obscured from public view. In reality it represented the final transfer of control over mass transit from businessmen to financiers— a process which had begun about 1903.[14] To the casual contemporary observer, however, the events of 1923 must have seemed to have just the opposite meaning.

Before 1923, the visible officials of the CSL were all bankers, men easily identified in the public mind with the reputation for poor service and abuse of public confidence which had attached itself to the financiers' management of transit utilities. Blair, confronted with public dissatisfaction over the deterioration of service and with stubborn financial difficulties, promptly turned the day-to-day operation of the Surface Lines over to professional transit managers and engineers for the first time in the city's history.

This group, headed by Guy A. Richardson, who had helped see the Seattle and Philadelphia systems through the difficult years during and after the war, began a modernization and public relations program which simultaneously demonstrated the potential of a scientific approach and the limitations of a modernization program under private ownership.[15]

Richardson, a man of imposing physical stature, was presented to the press as the harbinger of a new era. He asked for public suggestions, held neighborhood meetings, and in general worked to create the impression of an open, progressive management. Propaganda played a major part in the improvement drive which began in 1923, and the public was saturated with declarations that service was improving and that management was doing everything in its power to show that mass transit could be fully reformed from within.[16]

The service improvements with which the CSL sought to back these claims were made entirely within the traditional context of private ownership: everything which was done either increased or maintained the companies' operating efficiency and so their potential profit.

The CSL's engineering staff, headed by Evan J. McIlraith, proposed a number of innovations. Those which promised to improve the companies' operating efficiency and which cost little or nothing were undertaken, sometimes with striking success. Those which threatened maximum profitability or which conflicted with the interests of some securities holders came to nothing.

Technology: Visible and Invisible

One scholar has observed that Chicago transit companies made no technical innovations in their rolling stock during the 1920s; he concluded that service was therefore stagnant. The contemporary public, like its progeny, was probably most impressed by new cars, of which the Surface Lines bought 546 and the elevated 400 between 1922 and 1929. The new surface cars made possible the addition of over 2,000 runs by March 1924.[17] But the really important changes in transit service, as in traffic control, resulted from new efforts to adapt and regulate the flow of vehicles rather than from the addition of more or better facilities.

All these changes were really "free" to the surface companies. New surface equipment was bought from funds already set aside by the Illinois Commerce Commission: in effect, it did not constitute a voluntary expenditure on the part of the companies. The more important changes in routing and scheduling likewise cost the companies nothing, and probably saved them money.[18]

Modern scheduling, in particular, emphasized the adjustment of car flow to demand. In conjunction with improved traffic control, for which the Surface Lines were in part responsible, good scheduling could limit the number of streetcars required to serve a given population and at the same time reduce crowding.[19] The key to this achievement was the Surface Lines' Traffic and Scheduling Department, reorganized from scratch by McIlraith, which drew together information about employment and living patterns, ridership, and street traffic in a way which had not previously been attempted.

Before the 1920s, transit scheduling was usually carrried out according to the principles developed for steam railroads, which had more or less complete control over their rights-of-way and which operated on relatively long headways and to some degree influenced the timing of the demand for their service. The new Chicago system, on the other hand, took fuller account of the fact that street and transit traffic were autonomous and determining factors in transit scheduling.

University-trained engineers who joined the Surface Lines after 1923 spent the first year of their service as motormen, conductors, and traffic checkers. On graduating into the Scheduling Department, they adjust the service to fit the conditions they had encountered in the street. Schedules were arranged to place a maximum number of cars on specific lines at the moment when heavy demand was anticipated. Streetcars were timed to meet at major transfer points—a feat requiring precise knowledge of traffic speeds on the streets involved.[20]

In short, Chicago's transit system, unable to get free of the street and its traffic because of the lack of comprehensive rapid transit, did its best to adapt the fixed-rail vehicle to automobile-clogged thoroughfares. Unable for the most part to convince employers to stagger their hours of business, the system tried to place cars where they were needed on the closest possible headways.

Adjustment

Careful scheduling, combined with a thorough rerouting of cars in the CBD, seems to have done much to improve the companies' image. Complaints continued to come to the companies themselves, but the press reported favorably on their efforts, and some of the most serious overcrowding appears to have been eliminated.[21]

The companies' improving image was as much a result of public relations as of engineering. CSL acknowledged this fact by funding a newspaper advertising compaign and by substituting a cheerful red for the old institutional green as the company's color for standard cars. The press, most of which backed compromise with the companies and favored the mayor who was trying to deal with them, had political reasons to boost the companies' image during the mid-1920s. It shared with the electric railway industry a tendency to idealize the achievements of the Chicago management. But there were real achievements nonetheless. CSL adjusted well to its situation—not just to street traffic but to the decentralized nature of the route system forced upon it by past regulation. The company actively fostered neighborhood travel through scheduling and advertising. With reason the economist of the American Transit Association observed in 1933: "There is probably no city in the United States in which the surface system is so well adapted to serving the community [as in Chicago]."[22]

But the phrase chosen to describe the surface system was unintentionally telling: the CSL, like the city's policymakers, adjusted to a situation it could not dominate. In an era when the standard for automobile facilities was excellence, the mark of a successful street railway was adjustment—the ability to make do with crowded streets and obsolete technology. During the 1920s Chicago had the engineering skills necessary to help mass transit keep pace with the changing city. But these skills were not fully exploited. Instead, despite increasing per capita ridership and progressive, efficient operating management, transit in Chicago did not keep up with the spread of population or with the changing technology which the companies' own engineers helped to create. The reason, in the last analysis, was that transit was a business.

•   •   •

Service did not improve as it might have in three important ways. Extensions were not made to keep pace with dispersing population; crowding, while reduced in the most extreme cases, continued to make the straphanger's ride uncomfortable, and the only significant transit innovation of the first quarter of the twentieth century, the gasoline-powered "motorbus," was not utilized to its full potential.

Public demand for extensions into developing territories continued throughout the 1920s. Property owners in "better" sections sometimes opposed streetcar extensions on the ground that their streets should be preserved for the automobile or that motorbuses were more compatible with residential districts, but business and real estate interests invariably lobbied for new lines.[23]

Citizens actively supported extralegal bus services run by neighborhood organizations or small private companies. They marched on the mayor's office to demand extensions and, in the summer of 1928, gave overwhelming support to a referendum proposition—not supported by any political machine—in favor of city-owned bus extensions where private companies had failed to give new service.[24] Many citizens in Chicago's newer, quasi-surburban areas clearly wanted mass transit service throughout the 1920s. They did not get it.

Extensions were not built because extensions were rarely profitable. Insull constructed long suburban extensions because he anticipated profits from the sale of electricity and, perhaps, of real estate. In the city, where such nontransit inducements were lacking, the elevated lines made no additions, though they had the franchise rights for over 100 miles of such extensions.[25]

The CSL added just thirteen miles of track between 1920 and 1929. Bus extensions, for political and financial reasons, were unattractive to the companies and difficult to achieve because of state regulation. Streetcar extensions were expensive. Further, by 1922, 30 percent of the companies' lines were not meeting expenses, and in something of a reversal of the 1914 situation, many of these were old, inner-city lines, not extensions. State regulation kept in service a number of lines which siphoned off money for useful extensions. This, to the companies, provided painful evidence that a line once placed in operation could not be discontinued, no matter how unprofitable it might become.[26]

Finally, the companies continued to use extensions as a bargaining tool. Once an agreement had been reached with the city on a new and favorable franchise, many long-sought bus and trolley bus extensions appeared, despite the effects of the Great Depression.[27]

Outlying riders, then, received little from the improvement campaign of the 1920s. But inner-city riders also had cause for complaint. Before the war, Arnold had estimated a comfortable rush hour load for a standard streetcar at 85 passengers. In 1923 CSL representatives contended that a car of the same dimensions was "not overcrowded" with 100 passengers. Through the 1920s, 100 passengers per car remained the CSL's accepted maximum load. CSL continued to add new cars including—in 1929—100 with leatherette "semibucket" seats; but for too many riders the hope of a seat of any kind remained an unrealized dream.[28]

In the 1920s, the causes of crowding were as complex as ever. Political regulation gave the surface companies a fare lower than that of most other cities—notably so if the effect of universal transfers is taken into account. And wages were higher than those in many other cities. Not surprisingly, the companies' operating ratio was slightly higher than the national average in 1927.[29] All these facts meant that the companies had fewer resources to allocate to service improvements than might otherwise have been the case.

The companies also suffered from inefficiencies which were not related to the wage–fare structure. Regulation retarded the introduction of "one-man" cars—trolleys on which the motorman collected the fares and the conductor was dispensed with. Lightly used lines were thus made less efficient than they might have been. But the most revealing inefficiency of Chicago mass transit in the late 1920s was the fashion in which the motorbus was introduced. Here regulation combined with the interests of those controlling the transit companies to determine how an important new technology was used.[30]

Politics and finance, as will be seen, continued to make it impossible to remove the streetcar from the surface of the street. In this situation, the motorbus provided one more means of adjusting mass transit to the automobile age. The bus, at least, could maneuver around some obstacles and could make a better showing in the competition for street space. Yet the most politically and financially viable use of this new vehicle was neither the most efficient nor the most rational from a transportation point of view.

The Bus: A Casualty of Regulation

T. E. Mitten, the outspoken president of the Philadelphia Rapid Transit, accurately described the role of the bus in the 1920s. "In democratic America we don't like to admit any class difference in anything," Mitten said in 1927, "but we have it never the less. . . . We have three classes of service at three different rates. . . . The bus is the middle class vehicle. It fits between the streetcar and the individual automobile or taxi. When combined these three agencies form a complete local transportation system. . . ."[31] The bus in Chicago was just such a "middle class" vehicle, though by 1927 it had developed the potential to become much more. This potential was not realized—and the bus was not fully integrated into the city's transit system—until after World War II. Full utilization of the bus's potential did not fit the needs of transit businessmen, and it was impossible as long as city and state regulation remained at odds.

In Chicago, the bus made its debut less as a mass transit vehicle than as an oversized automobile. In the years just prior to World War I, its role seemed self-defined. The city's business streets were amply filled with trolley tracks, but the miles of smooth-surfaced boulevards built for pleasure vehicles were accessible only to the owners of carriages and automobiles. Between 1914 and 1917, several companies made abortive attempts to open the boulevards to the general public by means of gasoline-powered buses.[32]

No doubt their primary motive was profit, but their most vocal public advocate was unpaid and unsolicited—a woman who saw the bus as a means of democratizing the parks. The introduction of the motorbus into Chicago would probably have proceeded at about the same pace had there been no Emily Larned, but this persistent north shore society woman, who filled the

city's newspaper columns with some of the most delightfully ridiculous dog-
gerel to survive from the dreary history of traction, pointed up the implications of the use of the bus in its early days.

> John has his car and Jim has his car
> A thousand kinds there are
> But the great big splendid motor bus
> Is everybody's car.

So Larned wrote to the *Chicago Tribune* some time in 1915. This was
her persistent cry throughout the pre-1917 years when park commissioners
were working to exclude the bus from parks. The parks, she argued, were not
alone for "Mr. Millionaire and Mrs. Goldmesh Bag." Without the bus, "the
masses of people can reach the parks . . . [only] provided they walk blocks and
blocks, after submitting to jolting, overcrowding, and hanging onto straps,"
wrote Emily Larned.[33]

Park commissioners proved wary of the innovation. One by one, however, the boards relented. By 1917, when bus service was established on
north and west side boulevards by the Chicago Motor Bus Company, the
Tribune had long since picked up Larned's rhetoric and was campaigning for
more buses on the ground that they were a social good. The bus company itself
promised to bring the parks into the daily lives of the people, and on the day
when the first motorbus rolled, the *Examiner* ran its picture with the caption
"The poor man's pleasure car." "Now that we have the motor bus, our family
car is su-per-flu-ous," wrote Emily Larned.[34] The story was really not so
simple.

The movement for buses in the parks was an affair of the progressive
upper classes, not the working poor. From the first petition for lakeshore
buses in 1913, the names of the city's most prominent society women
appeared at the forefront of those advocating change. The most prestigious
streets near the city's center did oppose the bus, but the north and south shore
boulevards appear to have been areas of strong support for its introduction.[35]

By the early 1920s, successful bus lines had also sprung up to connect
important outlying hotels with the Loop, and certain suburbs with the outer
city and some rapid transit stations. When a reorganized Chicago Motor Bus
Company went before the state Commerce Commission in 1922 to ask for new
routes on the city's south side, it faced no objection from the surface or
elevated companies, which apparently accepted the bus company's argument
that it would attract a class of riders—chiefly pleasure seekers—who would
not otherwise use traditional mass transit. Vigorous opposition came instead
from the Illinois automobile clubs, which objected to the obstruction of
automobile thoroughfares by the lumbering two-decked buses.[36]

By this time the Chicago Motor Bus Company, which had dominated
the major boulevard routes since 1917, had survived challenges to its right to

operate and had emerged from a receivership brought on by its rapid expansion in a period of economic recession. By 1923, the outlines of Chicago's independent boulevard bus system had taken shape.[37]

The most important fact about the bus in Chicago during the first half of the twentieth century was, not a technological, but a legal one. The companies which operated buses were offspring of the Illinois Commerce Commission. The city tried repeatedly to regulate the bus companies and resorted at one point to the arrest of drivers, but Chicago could not exert authority over these firms.[38] As a result, the bus developed a separate and rather narrow niche within the transit system—one defined as much by politics as by anything inherent in its own technology.

Bus lines constituted a completely separate transit system until 1935, when the Commerce Commission ordered limited transfers among the bus, elevated, and streetcar companies. Thus the bus operators escaped most of the social burdens placed by the city council on streetcar riders. While the companies did pay substantial compensation to the park boards, they were never asked to pave streets, provide night service, or undertake unprofitable extensions.[39]

Confined largely to the boulevards, buses served a limited public with well-scheduled rush hour and midday service which—especially after the development of the Outer Drive—provided an alternate form of rapid transit service with higher fares but less crowding. By 1925, the major bus company had captured just over 5 percent of the local transit market. Its share, apparently taken from the Rapid Transit rather than the Surface Lines, hovered between 5 and 6 percent throughout the 1930s.[40]

From early 1923, the major company was operated by a national corporation which included the Fifth Avenue Coach Company of New York and was controlled by a Chicagoan, John A. Hertz. Hertz had entered the transportation field through his Yellow Cab Company, which rose quickly to predominance in the Chicago jitney business through vertical integration and a system of direct employment of drivers under an open shop, as opposed to the leasing of cars or licenses to independent drivers. Hertz manufactured his own buses and developed a profitable sideline in the Yellow Truck and Coach Company, which was ultimately absorbed into General Motors.[41]

Chicago Motor Coach, as the operating company was called from 1923, was thus an independent and unpredictable factor in the local transit puzzle. Bus building was as much a part of its business as was transportation. The bus executives might therefore have had an interest in the expansion of bus service somewhat akin to Insull's concern with rapid transit extensions: the profits to be had from the provision of equipment—or, in Insull's case, electric power—could be a strong motivation for fostering the growth of transit service. On the other hand, in order to serve nonboulevard streets, the company would have to make some sort of arrangement with the city in which it was likely to lose the immunities it enjoyed while regulated solely by the Illinois Commerce Commission.

By 1927, technology had opened new possibilities for the bus. Earlier buses could not compete with the streetcar in any but the most specialized service. They were too light and too slow to load. The CMC admitted as much and, until 1926, advocated only one role for their vehicle. They saw it simply as a provider of fast service over boulevard routes free for the most part from parked cars and business traffic.[42] By 1927, however, the bus had begun to change, and policy rather than technology alone began to determine the place it would occupy.

A new sort of bus emerged in the later 1920s. This bus was developed by a variety of builders—that of the Fagol Coach Company was built to CSL specifications and with the advice of CSL engineers. The resulting bus was closer to a free-wheeling streetcar than to the oversized automobile which had been cruising the boulevards for over a decade. This blunt-nosed, rear-engined, single-decked bus was the direct ancestor of the vehicles which ultimately replaced the streetcar throughout most of North America. By 1928, the CMC had a plan before the city council to do away with trolleys altogether in favor of a single, city-wide bus system. With hindsight, CSL's McIlraith contended that such buses could have replaced streetcars on all but a dozen major lines by the mid 1930s.[43]

The "energy crisis" of the 1970s popularized the view that corporate conspiracy—largely on the part of General Motors and some oil companies—was responsible for the replacement of electric transit during the period after 1935. One recent analysis appears to show that the streetcar of the late 1930s was really more efficient from the operating company's point of view than the contemporary gasoline bus—assuming that both vehicles were filled to capacity.[44] Whatever the merits of the "conspiracy" view, it is interesting to note that in Chicago, in the later 1920s, political considerations and the interests of electric traction investors held back the rational use of the gasoline bus. Six factors helped to keep the streetcar running throughout the city, on light as well as heavy density lines, long after a viable alternative for at least some of its functions had developed in the form of the motor coach.

Sincere doubt about the viability of the bus for heavy mass transit was one factor. City council CLT's engineer R. F. Kelker long contended that buses could not do the work of streetcars. The city's commitment to underground transportation was a second factor. Buses, of course, were useless in subways, and Chicago's policy of facilitating ever more automobile traffic meant that streetcars, which *could* be put underground in the CBD, were a superior policy alternative for downtown transportation.[45] But factors which had nothing to do with traffic or transportation were also important.

Under the 1907 ordinances, streetcar riders paid for paving a substantial portion of the city's streets. Chicago Motor Coach showed no interest in assuming this burden, which would be greatly increased because streets with trolley tracks would have to be repaved to accommodate the company's solid-tired buses. As late as 1930, aldermen expressed reluctance to eliminate the paving requirements. The question was, as one councilman declared, "a hot one."[46] The bus also had another nontransportation drawback, one linked to its very efficiency.

From 1918 onward, the job-creating function of traction building programs had been among the arguments used for a succession of comprehensive settlement ordinances. In the later 1920s, when Chicago's unemployment situation became serious, job-creation grew more important as a reason for quick resolution of the traction problem. A bus system whose fixed plant consisted chiefly of garages and administrative facilities could not compete with an old-fashioned traction system as a job-producer.[47] The fact that bus transportation produced fewer nontransit jobs was never used against bus substitution, but the greater job-making potential of streetcar and rapid transit construction must have been apparent to policymakers.

Regulation and the city's perpetual war with the transit companies also held up bus extensions. The streetcar companies were reluctant to actually make bus extensions, but they and the elevated companies actively sought the *right* to do so. The state Commerce Commission, however, which had won in court the sole right to regulate bus transportation, wavered back and forth between the competing companies and helped create a jumble of permits and rights which held back bus extensions and prevented the creation of a logical bus system.[48]

Finally, the streetcar companies themselves seem to have opposed the new technology, not out of a misguided faith in the electric traction, but from a very rational calculation of their own economic interest.

Writing of the 1930s, Evan J. McIlraih recalled that "bankers of both companies' boards . . . insisted that the terms of the [1907] ordinance would remain valid until a new ordinance could be arranged. That ordinance allowed pyramiding of the purchase price, or so-called capital value, by the reconstruction costs. . . . The CSL board was [therefore] very reluctant to permit our [the company engineers'] wish to get bus operation."[49] The fact that the capital account could be maintained and even increased by the perpetuation of outmoded street railway service thus gave the companies' bankers a vested interest in obsolescence.

Backward-looking policies were not the result of indifference or ill-advised conservatism: they were profitable. They kept up the valuation on which the companies were guaranteed a 5 percent return and helped to assure that in the event of municipal purchase, the companies would receive the full face value of most of their securities. These were unforseen results of the city policy designed in 1907 to promote modernization.

Riders and property owners appear to have preferred buses to streetcars. While the rate of bus substitution in Chicago during the 1930's was about the same as that of other cities, Chicago's buses ran under three separate operating companies, and their role in the transit system was less important than the CSL's own engineers thought it should be.[50]

As in the cases of extensions and crowding, politics and private ownership had combined limit the impact of technological progress. Committed as the city and the companies were to electric traction, subways remained the most obvious means of transit improvement, at least for the downtown district. But neither subways nor extensions could be had without a new "settlement." This was one reason for the 1930 agreement: private ownership under the old compromise could not keep pace with its own technology or with the challenge of the changing city.

Section III:
Finance: The Strengths of Impotence

> You can write any kind of legislation you want, but you cannot get one damned bit of transportation . . . unless the people that own those properties are ready to accept what you hand them. They can last just as long as the city of Chicago can last.
>
> *Melvin Traylor (president, First National Bank), 1926*[51]

The financial weakness of the city's traction companies assured that any settlement would be on terms favorable to transit investors and the bankers who represented them. This was so because, well before the 1920s, mass transit had come to be more important to the city than it was to any major financial institution as a business proposition.

The convergence of traction investors which had taken place before World War I continued to be important throughout the 1920s. Insull maintained control of the elevated system, and after 1923 his friend Blair was the dominant figure in the CSL. Despite the role of Guy Richardson as CSL president, Blair controlled long-range planning, and major company policy statements were issued in his name.[52] To both Blair and Insull, however, transportation was a secondary interest; their major financial commitments lay elsewhere.

Thus despite those good "connections," surface and elevated companies had really become financial "orphans" by the 1920s. For practical purposes they had lost the ability to raise money for major improvements through traditional channels. Only major financial institutions could help the transit companies raise capital for a comprehensive modernization and unification of the system, and such institutions would lend a hand only if their terms were met. Banks with heavy investments in the CBD had much to lose if the system collapsed, but specific punitive measures taken against the companies by the city did them no harm. Unlike actual traction investors of former times, the banks in charge of transit bonds could simply withhold their aid and wait for a better political climate as long as the system continued to operate at a reasonable level of adequacy.

The traction companies had arrived at this position of financial impotence not through the failure of the transit market but through the effects of their own financial practices and the city's regulatory tradition.

In their period of rapid growth, transit companies had gathered capital mainly through the issuance of stock: stock which, because of the industry's record of rapid growth, appealed to the speculative investor. On the Yerkes lines in particular, speculative stock issues had led to serious overcapitalization. As the growth of the industry slowed, however, bonds became the accepted means of financing improvements. Chicago's street railways financed all their expenditures under the 1907 agreement in this manner, and by 1917 over half of the securities of U.S. transit companies were bonds.[53] Bonds appealed to an entirely different class of investor than did speculative stocks, and this difference ultimately changed power relationships between transit companies and their investors.

The chief appeal of bonds was their security. Traction bonds were attractive because public transit, like other utilities, appeared to have the advantage of an inexhaustible market and little rate or service competition. Thus traction bonds were undermined by any factor which cast doubt upon the long-term stability of the company which issued them.[54]

Overcapitalization undermined Chicago transit bonds well before World War I and the automobile had had their impact on the predictability of the traction industry: even before 1914 some bonds were bringing the companies only 75 percent of face value.[55]

Short-term franchises also contributed to the weakening of Chicago traction securities. Before the Great Depression, investors did not anticipate massive, long-lasting inflation. A forty-year bond thus sold at less of a discount and required a lower interest rate than did a twenty-year security. CSL companies had difficulty selling even twenty-year bonds because, after 1907, it was clear that their franchises would expire before the securities matured. When that happened, first mortgage bondholders would gain control of the properties and negotiate a new agreement—a procedure which could place all the junior securities of the companies in jeopardy.[56]

Wartime inflation, fixed fares, and automobile competition only exacerbated this preexisting situation. Thus the cost of money was always greater to the Chicago companies than to East Coast utilities with perpetual franchises.[57] Limited franchises also accentuated the need to market large blocks of securities at one time—a practice which forced the companies to deal through large brokerage firms which extracted a further toll on transit improvements through their commission.

World War I completed the destruction of the transit industry's credit. Between 1914 and 1918 the cost of new equipment more than doubled and the wages of CSL trainmen rose 150 percent. By 1922 federal courts and the state regulatory commission had forced fares up to a level which would provide a "fair return" on the valuation (roughly 6 percent), but without the possibility of long-term bonds this rate could not attract new capital.[58]

On the surface lines operating ratios remained at around 78 percent through the 1920s, fifteen points above the maximum level considered necessary to attract new investment. In 1921 the state commission, recognizing that the companies could no longer attract investment, set aside a special fund from their gross receipts for the purchase of new equipment, but it was clear that such a fund could contribute little to the creation of a comprehensive system.[59]

Substantial transit improvements would obviously require large amounts of new money. For private companies, this would mean complete reorganization and the issuance of new securities. Even with municipal ownership, the city would have to find a way to issue securities which used the transit properties or their fares as collateral, because it was unlikely that the city's debt limit could be raised for mass transit spending. All this meant that local banks would have tremendous power and responsibility.

It was local banks which would be called upon to endorse transit bonds, whether issued by the city or a reorganized company. The banks would have to vouch for the reliability of the securities, and assume a protective function should the company or city fail to meet its obligations. Bankers thus held veto power over any plan for a new transit system. It was they who had to judge whether any provisions of a new franchise would make bonds "unmarketable," and what new securities old investors should accept in return for their former holdings. In exercising these responsibilities, Chicago's bankers set new, strict limits to city transportation policy.

The city's bankers firmly rejected any possibility of municipal operation, against which one banker declared ". . . there is a natural prejudice among security holders. . . ."[60] Thus they determined that even should municipal ownership somehow come about, mass transit would for the foreseeable future be conducted as a private business. Public ownership, then, could mean little real change. It would simply be a device for eliminating the franchise question and for raising capital.

Bankers also demanded that the full BOSE valuation for the surface lines be accepted for any reorganized company and ruled out any effective provision for local regulation, which one important banker declared was "no better than municipal ownership." The terminable permit was another essential demand of the banking community, and one major banker even sought to end the tradition of referring all transit proposals to a public referendum.[61]

Thus the financial community, in passing judgment on the viability of traction securities, negated Chicago's tradition of local control of mass transit, helped change the meaning of public ownerhsip, assured that transit riders must pay the cost of overcapitalization, and demanded that transportation decisions be removed as far as possible from the democratic process.

•   •   •

While their power was apparent from at least 1922, the receivership of the Surface Lines, which began in 1927, confirmed the bankers' hegemony. Major banks became the "protectors" of existing traction securities, and, through the efforts of receivership court judge James Wilkerson (a solidly conservative jurist with a long pro-utility and anti-labor record), Chicago's bankers emerged as the negotiators for the company side in the creation of the 1930 ordinance.[62]

Another crucial figure in the creation of a financial compromise was Samuel Insull. Insull was much more than the head of the Chicago Elevated Railways; between 1914 and 1930 he helped determine the fate of three major traction plans. His influence was all the more ironic because, by all accounts including his own, Insull in the 1920s continued to care little for urban transit as a business—indeed he tried hard to avoid involvement in it.[63]

Insull helped kill two traction plans and made possible the passage of the third. A plan put before the public in 1918 proposed a valuation for the elevated companies some $20 million below the price of their securities. Fisher, who had engineered that compromise, observed that, in the 1911 unification of the rapid transit, "they . . . had this financial orgy . . . in which certain gentlemen were able to persuade very astute financiers that there was a lot of potential profit in the elevated properties. . . ." These financiers, who, as Fisher said, "voluntarily elected to be the goat. . . ," might have held Insull responsible if the new plan passed: it would have cost them about $15 million. Insull complained about the valuation and did nothing to prevent his political clients from engaging in a successful effort to defeat it.[64]

A staunch foe of public ownership, Insull worked actively during 1924 to undermine the credibility of a muncipalization plan brought forth by the city's Democratic administration. He gave speeches reminiscent of turn-of-the-century attacks on municipal ownership as a form of socialism and, when a greatly modified plan was submitted to the voters, it was again the power magnate's political allies who led the successful fight against the compromise.[65]

Finally, it was Insull who made the arrangement of 1930 possible. Persuaded by his friends of the necessity of finding some solution to the traction impasse, Insull organized a committee of bankers to reconcile the conflicting interests of Surface Lines securities holders, took responsibility for the refinancing of the new company, and even arranged $500,000 in bribes to help assure the passage of the ordinance at referendum. In return, Insull received effective control of the new company and a valuation which covered all the securities of the old elevated companies. The new ordinance called for downtown streetcar subways, and so for the perpetuation of the streetcar market for Insull's electric power.[66]

There was more still to Insull's power and influence, for it was the utility empire builder who helped keep William Hale Thompson in the mayor's office for twelve of the sixteen years between 1914 and 1930. Thompson, as will be seen, carried on a particularly destructive traction policy between 1915 and 1923. Yet, ironically, it was also Insull who helped raise the public image of utility owners during the 1920s.

For the moment, the point is that important financial interests set the terms for transit policy during the 1920s precisely because mass transportation had ceased to be attractive to investors. Insull sold stock to elevated riders—to their eventual grief—but large-scale improvements required the cooperation of capitalists whose real concern was the viability of the city and especially of the downtown district.[67]

Thus the very weakness of mass transit as a business meant that most of the terms of the nineteenth-century "Traction Barons" would have to be met if public transportation was to keep pace with the needs of the twentieth-century city. The prospects of direct profit from transit service could no longer attract capital for major transit improvements. Instead investment would have to be provided with multiple guarantees, and even then it would come only with the aid of financial institutions interested in the economic health of the city.

Section IV:
The Myth of Objectivity:
From Engineer to Businessman

We [engineers] can do anything you want, if you will tell us what you want.

Bion J. Arnold, 1922[68]

[The Chicago Association of Commerce] by virtue of its numbers and representative character, receiving its authority from the people, takes the lead in urging the acceptance of this [1918 Traction] ordinance.

Chicago Association of Commerce, 1918[69]

By 1914, engineers had been discredited as neutral arbiters of transit policy. They failed in this role because of the heated political context in which they worked, and because the city's commitment to profitable mass transit prevented them from providing the answers the public and politicians wanted to hear. In the decade and a half after 1914, the CBD business community assumed the role formerly assigned to engineers as the arbiters of fairness and advocates of progress.

Business's public role was by no means new. It had led the drive against Yerkes and had organized the "selling" of the 1907 settlement. In later years, however, its role became more open and more formalized. As it had with traffic control and the implementation of the Chicago Plan, CBD business after 1917 openly assumed responsibility for compromising divergent interests and validating the objectivity of public policy. At the same time, engineers became something they had not been before—straightforward representatives either of corporations or of governmental bodies.

One aspect of the transit engineer's role which had always undermined their claim to objectivity was their function as legitimizers of utility valuations which were, in fact, political compromises. This role they continued to perform throughout the 1914–1930 period.[70] As before, too, engineering judgments which failed to meet political or CBD business needs were simply ignored. The Western Society of Engineers protested to no avail when the city decided to permit high downtown buildings in its 1923 zoning ordinances. The argument of one WSE member, that the building of subways at the riders' expense for the relief of auto traffic was unfair, also fell on deaf ears. The council included nearly 300 miles of politically necessary extensions in the 1930 ordinance over the protests of its own engineering staff.[71]

Neutral engineers were impotent. Indeed the pretense that engineers could or should be neutral vanished for all practical purposes. Insull openly urged the engineering establishment to support utility rate increases, and at least one engineer from the staff of Arnold's firm declared that he was glad to oblige.[72] More damaging was the increasingly obvious political use made of engineers and "experts" in various fields.

Mayor William Hale Thompson (1915–1923; 1927–1931) actively sought to have his chief sanitary district engineer made president of the Western Society of Engineers. The mayor's record in the use and appointment of engineers—or pseudo-experts—was notorious. By 1922 it was a matter of public knowledge, and one alderman could joke that the superiority European transit was traceable to the fact that ". . . they do not make engineers in 24 hours like you do in city hall." Thompson used "experts" to pad expenditures so flagrantly that one newspaper initiated a nearly successful suit to force him to return to the city treasury over $1 million in "engineering" appropriations.[73]

After 1915, the CLT employed a real and competent engineer of its own, R. F. Kelker, Jr. Kelker, an electrical engineer by training, headed an engineering firm which did transportation studies of New York, Los Angeles, Baltimore, and Saint Louis, among other cities. He was a sincere and effective advocate of mass transit. Yet Kelker was also a past master at producing engineering judgments which reflected the prevailing aldermanic view. As will be seen, he lent his name to at least one transit plan whose primary purpose seems to have been the depoliticization of mass transit on terms favorable to CBD business.[74]

Engineers, then, lost whatever policymaking role they had had in mass transit matters. On the other hand, civil engineers played a major part in the work of the Chicago Plan Commission. Hugh E. Young, in particular, designed most CPC improvements, and Joshua D'Esposito, among others, contributed to local expressway design. Young made statements on behalf of plan projects, and on occasion he advocated an approach different from that taken by policymakers. Plan projects, however, were not presented in his name, and the only really controversial engineering issue connected with such improvements, the valuation of property condemned for street improvements and the like, was dealt with by city agencies.[75]

Whatever the real influence of men like Young, they never assumed the public role once occupied by transit engineers. Some of the reasons transit engineers lost the position of apparent influence they once held can be discerned from a consideration of the most ambitious study of mass transit made in our period—that conducted by the Traction and Subway Commission of 1916.

．　．　．

The 1916 Commission was appointed at the behest of Mayor Thompson, apparently with the hope that it would make the traction issue easier to handle by resolving contested points such as valuation, the nature and routing of improvements, and the sort of franchise to be granted a new, combined surface and elevated company. The Commission included Bion J. Arnold, Henry Brinkerhoff, and Robert Ridgeway (an expert on building foundations). It was dominated by its chairman, William Barclay Parsons. A longtime rapid transit consultant and major figure in the history of New York City transit, Parsons had done traffic studies on five major cities before 1916. Parsons approached Chicago with an awareness of the local traction tradition, but produced a report which opened new vistas in at least two ways.[76]

The Commission's traffic study—the most intensive made during the 1900–1930 period—led it to devise a physical plan different in one important respect from previous Chicago traction plans. Parsons and his fellows called for a comprehensive rapid transit system focused on the Loop, but advised the immediate construction of a crosstown line two miles west of State Street.[77]

New lines outside the downtown area were to be elevateds, not subways. This, the Commission believed, was the fastest, cheapest, and most effective way to open new territories and to link the sections of the city to one another. This general approach (elevated rapid transit outside the Loop and early building of crosstown lines) was followed in all subsequent plans through 1930.[78]

Parsons also broke new ground in his suggested franchise. Here his commission faced squarely the problem of finding massive amounts of new capital for the proposed improvements. Reviewing the histories of New York, Boston, and Philadelphia, they pointed out that long-term franchises, flexible fares, and guaranteed return were the only alternatives to a public subsidy for mass transit. Accordingly they reported, "Your Commissioners have endeavored to work out a financial plan that will keep Chicago from following in the footsteps of Boston, New York and Philadelphia, and avoid imposing upon the taxpayers of the city any portion of the transportation burden."[79]

The insistence that mass transit be self-sustaining was, of course, not new. What *was* new was the suggestion that both the physical and the financial plan must be flexible, able to adjust both to the changing city and the changing financial position of mass transit. The only alternative to subsidy, the Commission told Chicago, would be a new traction policy. This meant a "terminable permit" for the operating company—not a fixed-term franchise but one which would be valid until the city purchased the lines. It also meant fares keyed to company profits, not tradition or social policy. If profits fell below a fixed amount (6 percent in the original plan), fares would automatically rise. If they exceeded 8 percent, fares would fall accordingly.[80]

Finally, the Commission advised that state law be changed to permit day-to-day regulation of transit service and finance by a "Board of Regulation and Control" which should be "constituted along the lines of the Board of Supervising Engineers, but with added duties and powers."[81] In the Commission's report, this board was evidently expected to be a body of transit experts, some specializing in finance and some in engineering. Thus, through the agency of an expanded and more powerful version of the old BOSE, the Commission attempted to make an indefinite-term franchise and flexible fares palatable to the public and regulation acceptable to investors.

As it turned out, no proposition containing such a "Board of Control" was ever placed before the voters. For political reasons the city failed to obtain the necessary legislation empowering it to exercise regulatory power, through a board of engineers or otherwise. Instead Walter Fisher, the father of the 1907 compromise, developed a formula which he hoped would make a form of regulation compatible with heavy new investment by private financiers. It was Fisher's plan which placed businessmen in the seat formerly occupied by engineers.

The formula was a "trusteeship plan": an idea originated in Boston in 1917 and pressed upon a reluctant Chicago city council by Fisher in 1918. In essence, this plan called for a new, unified company to be managed by a board of trustees representing jointly the security holders from the old companies, the CBD business and financial community, and the City of Chicago. Such a plan would obviate the state's monopoly of regulatory power by providing for joint control of finance and operations by the city and the owners of the old companies.

The companies promptly announced that they would consider the plan only if the city's representatives were, not politicians *or* engineers, but prominent local businessmen. Since the board was essentially self-perpetuating, once the initial representatives were chosen (by agreement among the CLT, Fisher, and company representatives), the board could be expected to continue as a businessmen's body indefinitely. [82]

The initial board agreed upon by companies and council included four direct representatives of the old companies and five "city members." The city's members were two bankers, a steam railroad executive, the president of the Chicago Association of Commerce, and one lonely former alderman. After 1928 the city would be able to appoint some new members, but these too would have to be men of proven business (not transit) experience. [83]

Business domination of the board was a major arguing point against this resolution of the traction problem. Nonetheless, the next major traction plan, in 1925, provided for a similar board. Both plans failed at referenda, despite strong business support led by the Chicago Association of Commerce. [84]

These defeats in no way diminished the determination of the CBD business community to control transit policy. The CAC and its allies lobbied for the state legislation to remove the prohibition on long-term utility franchises and steadfastly opposed municipal ownership. There was, the CAC declared in 1925, "a feeling of dread on the part of the business community that the operation [of mass transit] by municipal employees had been considered." The view that city employees were by nature either dishonest or incompetent prevailed in the business community and citywide press during the 1920s, as the performance of utility-backed politicians like Thompson ironically added weight to the prejudice against "mixing utilities with politics."[85]

• • •

Downtown business, led by the CAC, sought a solution to the mass transit problem for the same reason it pressed for modernized traffic control: the Loop itself seemed threatened by the city's inability to provide access to the CBD. Nothing better illustrates the role and the motivation of major business than does the successful united effort of the CAC, Democratic politicians, and engineers to undermine neighborhood resistance to a downtown subway and privately owned mass transit.

During the early 1920s, major CBD business and real estate interests shared a strong commitment to ever denser Loop land use. It was they who gained for the CBD a last-minute exemption from the height limitations in the city's first zoning ordinance. As George C. Sikes, a veteran of the anti-Yerkes campaigns, observed in 1923, "The high building policy and the traction program of the Loop interests go together."[86] More density meant more street traffic and, in the context of Chicago's policy tradition, this meant getting the streetcars off the street to make room for more automobiles and to increase the system's capacity. Neighborhood interests were a major obstacle to this program.

In theory these interests should have been satisfied by the 1916 plan, since it called for a genuine comprehensive transit system, most of it above ground as the Greater Chicago Federation had demanded. The heart of the regionalist position, however, insofar as it was not simply a desire to force the dispersion of business, lay in mistrust of large corporations and the belief that spending was not the ultimate solution to transportation problems. Tomaz Deuther and his allies hammered away at the bad record of private companies under the 1907 agreement and argued that the best engineering studies showed more transportation only meant more congestion. Local business, led by Deuther, played a major part in the defeat of the 1918 ordinance and thereafter organized to work for municipal ownership.[87]

To neutralize Deuther and his kind, the Chicago Association of Commerce in 1921 formed an alliance with CLT chairmen Ulysses S. Schwartz, a longtime public ownership advocate, and, it would appear, with the CLT's engineer, R. F. Kelker, Jr. The CAC could cooperate with Schwartz because, in the climate of transit economics in 1921, some form of public ownership appeared likely, and Schwartz' version was more palatable than that proposed by the mayor, William Hale Thompson. In addition, Schwartz was an active advocate of CBD subways. In their efforts to eliminate neighborhood organizations—a nuisance to politicians and CBD business alike, from the policy-making process, Schwartz and the CAC cooperated openly.[88]

Late in 1921, Edward Gore, the president of the Association of Commerce, led the formation of an "umbrella" organization which called itself the "All Chicago Council." Gore visited neighborhood business organizations and enlisted them in the larger entity. He then approached the CLT as the representative of the neighborhoods and asked that his organization be consulted in any future traction planning.[89]

Schwartz fell in wholeheartedly with the ruse. As the Association of Commerce noted, the CLT's treatment of the All Chicago Council "suggests that [it] is accepted, as it should be, as expressing the desires of a great proportion of the people of Chicago."[90] Here in perfection was the technique the Association of Commerce had introduced with the Citizens Non-Partisan Traction Settlement Association in 1907. The direct representative of the city's largest business interests set himself up as spokesman for small businessmen throughout the city, and his role was ratified at once by the appropriate political figure. The CLT deputized the All Chicago Council to draft a legal and financial plan and then set it to work at "discovering" the transportation needs of each community and constructing an outline for a comprehensive transportation system.[91]

The results, when they came in, were not surprising. The ACC's legal and financial plan was thoroughly compatible with the Schwartz plan, and its physical outline bore a remarkable resemblance to the widely publicized proposals of the 1916 Traction and Subway Commission which the ACC absurdly claimed not to have seen.[92] The CLT's engineer, R. F. Kelker, was then ordered by the aldermen to produce an updated physical plan. When the engineer reported in May 1923, the Association of Commerce recorded the expected result: "Singularly enough," *Chicago Commerce* told the business community, "the proposed Kelker plan includes practically *in toto* the recommendations of the All Chicago Council."[93]

Concessions on the part of business and utility representatives themselves completed the neutralization of neighborhood organizations. Insull declared that he would build a system of crosstown elevateds independent of any general traction settlement.[94] Marshal Field and Company agreed to the idea of special assessments of Loop property to pay 55 percent of the cost of any Loop subway. Deuther, after a long visit with Insull, came away claiming to have been convinced of the good intentions of utility companies and proclaiming that Yerkes had not been such a bad fellow after all. By 1928, four years after Deuther publicly concluded that the automobile would negate the centralizing effect of downtown subways, all but the old inner-city wards had voted in an advisory referendum in favor of using the Traction Fund—the city's share of streetcar receipts—for the building of CBD subways. The 1930 ordinance called for Loop subways to be built from the Traction Fund and assessments of CBD property.[95]

Neither Insull's promised crosstown elevateds nor the assessment of Loop property for subways came about, but by the end of the 1920s CBD business had come to dominate traction policy almost completely. In major traction settlement proposals, it representatives assumed the regulatory role formerly held by engineers, and some of their planning functions as well. Engineers had proposed plans which provided something for everyone, but only an organization like the CAC, in cooperation with the press and the city council leadership, could undermine organized neighborhood opposition and draw concessions from Loop businessmen as well.

It is hardly surprising, therefore, that the Citizens' Traction Settlement Committee of 1929, which worked out much of the 1930 ordinance, was headed by James Simpson, who was simultaneously president of Marshall Field and Company and head of the Chicago Plan Commission.[96] Major business interests had by this time completely taken over the policy role once assigned to engineers.

Businessmen replaced engineers in this position because they could successfully set themselves up as "neutrals," while engineers could not. Factual engineering judgments invariably served some side of a political controversy, if they were not too complex for political use. "Neutral" businessmen could produce not facts but compromises. They could act as power brokers and propagandists, as engineers could not. They could work with politicians too, without seeming to fall under their control. Because they were neither neutral nor expert in mass transit matters, businessmen could treat the transportation problem as one of political economy and so deal with the real sources of the city's transit policy impasse. For engineers, whose credentials rested on technical expertise, the roots of the transportation problem were and always had been out of reach.

Section V:
Courts, Business Values, And Public Policy: Eliminating the Issues of Valuation and Control

> Alderman Nestor: Some of the old horses are still in [the valuation].
>
> Alderman J. Bowler: We are hopeful that *that* is out, Charlie.
>
> Alderman Arvey: The water is out of the horses, anyhow.[97]

By 1930, perfectly serious Chicago aldermen were able to treat the once vital issue of overcapitalization as a joke. And regulation itself, the keystone of the 1907 policy, while not yet a subject for semiprivate jests, was represented in the 1930 ordinance by provisions that were no more than tokens. The issues of valuation and control were eliminated not by direct assumption of power by the business community but through the adoption of "business values" by state and federal courts.

The role of courts in undoing Progressive Era legislation has long been apparent.[98] Indeed their decisions on transit regulation and valuation occupy a central place in the story of the abandonment of the era's transit policy. In attempting to rescue the transit company's credit by defining the industry as private utility rather than a public service, jurists set their seal of approval on the policy definition which had always limited the potential of American urban transit.

The story of the demise of local regulation is extremely simple and parallels developments in other cities nationwide. In 1918 federal courts ruled that the creation of the state Public Utilities Commission had destroyed all city regulatory power. Through the decisions of the PUC on fares and routing, state power was extended beyond the day-to-day regulation of service and negated specific provisions of the 1907 ordinances. This meant that agreements between city and companies had no binding effect if the Commission wished to overturn them.[99]

When the state Commission set routes and fares for motorbuses, the city lost even the rudimentary right, given it in the 1871 state constitution, to determine whether mass transit might operate in a particular street. A long, loud, and apparently hypocritical legal campaign led by Mayor Thompson proved in the end that the city could not evict the companies for nonperformance nor, as it later turned out, could the city displace them when their franchises expired. The more the city tried to regain some vestige of regulatory power, the cleared it became that regulation by any local body was absolutely out of the question.[100] When the streetcar companies went into voluntary receivership after the expiration of their franchises in 1927, their position became stronger still; even the PUC could not regulate them. As one alderman observed, "They are under the supervision of the Federal court, and they can tell us to go to Hades as long as they are under that protection."[101]

Just as important were court decisions on the subject of utility valuation. This seemingly obscure issue continued to be vital because a transit company's valuation determined not only the basis upon which the state Commission was required to set rates, but, to an extent, the way in which the utility regarded its own plant and equipment. It also helped determine a company's attitude toward public ownership, for while a court-determined valuation could not set the price for public purchase, it could, through its influence on fares and so on profits, make continued private ownership more or less attractive. In general, higher valuation resulted in higher fares, less attractive municipalizaton, and to a smaller degree, less motivation for modernization.[102]

The crux of the valuation issue continued to be the fact that the city insisted that rates should be set and a company's purchase price determined on the basis of the depreciated value of the physcial plant—what might be termed "resale value"—while the companies, the courts, and the state regulatory commission favored the cost of reproduction of the properties as the valuation for all purposes. In 1921 James Wilkerson, as chairman of the Illinois Public Utilities Commission, stated frankly that this approach to valuation was favored by the Commission because it "aimed to restore credit and stabilize investment."[103]

When a transit utility was valued at its hypothetical reproduction cost, less very narrowly defined depreciation, the effects of wartime inflation produced valuations which covered most or all of a company's securities, whether or not the securities represented existing property. The problem—or virtue—of this kind of valuation was that it took no account either of obsolescence or of the market value of traction securities. Federal court decisions in other states made it clear by 1928 that Chicago's transit companies could count on court-imposed returns of about 8 percent on a combined valuation of roughly $300 million, even though the market value of their securities was about half that amount.[104]

All of this meant that the courts had created another rationale for the use of obsolete equipment—in addition to that provided by the 1907 ordinances. Because courts valued a company's physical plant according to its ability to carry passengers, not attract them, they placed a thirty-year-old streetcar on the same "level" as a new motorbus, except insofar as time had diminished the antiquated vehicle's ability to carry passengers.

The policy assumed a captive market which was, in fact, ceasing to exist. Such an approach might have made some sense when applied to a gas or electric utility: power from a forty-year-old generating station was no less usable or attractive than power from a newly built plant. Applied to mass transit, however, this standard became increasingly inappropriate because it disregarded progress. A growling, swaying 1907 trolley with wicker seats was simply not the same thing as a new, quiet, soft-seated motorbus. Despite present-day nostalgia for the streetcar, it is clear that contemporary riders preferred the bus, which more nearly approximated the ride given by the automobile.[105] Most important, for the patrons of an electric utility there was generally no alternative service. For an ever-growing number of straphangers, there *was* an alternative to obsolete mass transit.

Regulation by courts and the state PUC, thus, effectively held back progress in three important ways. First, by setting limited returns regulators made it impossible for the companies to accumulate the capital for major improvements. At the same time, by assuring a 6 to 8 percent return on inflated valuations courts lessened the pressure for modernization. Finally, this same guaranteed return made it possible for investors to hold out for a high valuation or purchase price in any new plan, whether for public or private ownership. In all these ways, the companies' financial weakness led courts and state regulatory commissions to take actions which strengthened the companies' hands in their dealings with the city.

Section VI:
The Festival of Hokum:
Chicago Transit Politics,
1914–1930

It is a perfectly silly and an asinine thing to do, to refer [a transportation plan] to the people of Chicago on a referendum, because they can't apprehend it; they cannot understand it.

Owen Wetmore (Chicago banker), 1926[106]

They think that working people have neither memory or intelligence.

Henry Mazursky (Local 504, Carpenters' Union), 1918[107]

Nonsense, irrelevancy, and obfuscation had long been major factors in public discussion of Chicago's transportation policy. The crusade which blamed Yerkes for all the city's political and transportation woes was laden with hyperbole and symbolic gestures.[108] Dunne's 1905 promise of "municipal ownership before the snow flies," the Citizen's Non-Partisan Traction Settlement Association's pledge that the 1907 agreements meant seats for all and public ownership at the city's option—both were almost certainly calculated distortions. Carter Harrison II's "Comprehensive Subway" plan of 1913 was a blatant sham. Still, before 1914, it was usually possible for the careful observer to discern the identities of the characters who shaped the city's policies and at least the general aims pursued by that policy.

After 1914 the city's complete loss of control in transit matters, the consolidation and shifting alliances of the city's political machines, and the growing influence of the downtown business community—whose power had always been great—combined to make the city's transportation policy something of a masked ball. Interests once opposed converged, and the meaning of important phrases changed.

Transit policy was being transformed in a way which is apparent only through hindsight. Mass transit slowly moved from the status of a vital public issue, the subject of strong, conflicting views, to that of a practical problem which, like traffic control, was to be resolved with the minimum of controversy or public participation. The work of the Chicago Plan Commission only partly masked the nature of the city's policy toward the automobile; but fifteen years of political rhetoric and maneuvering veiled the real changes in mass transit policy which were taking place, while the outdated public attitudes which still govern the popular understanding of transportation policy remained largely untouched.

· · ·

Much of the deception engaged in by politicians after 1914 resulted from the fact that they could not publicly admit that the city had lost all hope of regaining the power of regulation or eventual public ownership. Equally important was the fact that, after 1914, utility interests regained the sort of power over local politics they had held in the day of Yerkes.

Many progressives hoped that regulation would remove the influence of utilities from politics. Instead, the regulatory power of the city, and then of the state, drew transit and other utility interests into local politics by giving them a vital interest in it. "From the high point reached by [Carter] Harrison [II] at the end of his second term [1915]," complained reformer Charles Merriam, "Chicago descended to the lowest depths in its history."[109] Those depths were occupied by not one but two pro-utility political machines.

One was the Democratic faction identified with Roger Sullivan, the other the Republican machine of William Hale Thompson. The Sullivan Democrats found special favor with the Peoples Gas Light and Coke Company, although their relations with Commonwealth Edison, the electric utility, were also cordial. Of Thompson, Professor Forrest McDonald observed that ". . . his principal financial backer during his twelve years as mayor was Samuel Insull and, to make affairs more binding, Thompson's corporation counsel and closest henchman was Samuel Ettelson, the law partner of Daniel Schuyler, Insull's close personal friend and his political lawyer." Insull could also count on cooperation from Frank O. Lowden, governor of Illinois from 1917, and from most of his successors.[110]

And, even more confusing, genuine reform politicians sometimes entered into alliances of one sort or another with utility magnates. Thus Dunne, as governor in 1914, signed Insull's bill for a public utility regulatory agency, albeit his motives were not Insull's. Harrison in 1915, upon being defeated for renomination by the Sullivan machine, threw his support to Insull's man, William Hale Thompson. Finally, the post-1914 era's only reform mayor, William E. Dever, came to power in 1923 through the aid of George Brennan, who had taken over Sullivan's machine upon the latter's death. Dever, the erstwhile archfoe of private ownership, was thus beholden to a man who was well connected with all the local utilities and was Insull's "intimate friend."[111]

Nor were political machines the only forces to find common interests with the transit utilities after 1914. The Amalgamated Transit Workers' Union, once a staunch bastion of municipalization settlement, made a comprehensive peace agreement with the CSL management after a particularly bitter strike in 1919. Union leaders helped beat down demands for high wages, supported company appeals for fare increases, and, on one occasion when fares were cut, negotiated a matching wage reduction.[112] Tomaz Deuther's conversion to Insull's point of view has already been discussed. Thus some of the most irreconcilable interests and perceptions of the Progressive Era dissipated in the conservative climate and superficial prosperity of the 1920s.

Given such a confusion of interests and loyalties—and given too the fact that any comprehensive transit plan would have to include provisions violating the popular conception of what constituted fair and safe dealing with a traction company—it is easy to see why the three major plans presented to the public between 1914 and 1930 were neither honestly described nor discussed on their merits. Disillusionment with traction reform must have set in, entirely aside from any loss of interest in mass transit occasioned by the proliferation of automobiles. For the referendum of 1930, the first major one scheduled at a special election (without aldermanic or other contests), attracted fewer voters than any traction proposal since the debacle of 1914.[113]

• • •

The campaign over the ordinance of 1918 illustrates the confusion of interests and ideas which was rampant in the era after 1914 and the degree to which transportation issues were distorted in the attempt to assure the passage or defeat of this particular compromise.

The 1918 ordinance, as will be remembered, drew its inspiration from the 1916 Commission. In it the companies agreed to merge and to engage in a comprehensive building program based on that outlined by the 1916 Commission. In return, the city agreed to a long-term, renewable franchise, guaranteed returns of between 6 and 8 percent, and fare increases whenever profits fell below a designated minimum. Amortization, and thus eventual public ownership, was deemphasized so that fare increases could be minimized. Regulation, for practical purposes, was abanoned in favor of Fisher's trusteeship plan.[114]

The plan had definite flaws. There was no provision for holding the trustees accountable, and the state legislature could invalidate the trusteeship arrangement altogether, leaving the city without even the pretense of a voice in transportation policy and planning. Further, the lack of effective amortization provisions might have made the ordinance's long-term franchise a perpetual grant. Finally, the plan really relied upon the financial community, whose interests were represented in the board, to find a way to raise $91 million for improvements within the first six years.

The 1918 arrangement effectively enacted into law the transit program of CBD business, with significant concessions to outlying interests. Loop streetcar subways were to be built from the Traction Fund and CBD rapid transit subways were among its first priorities. As is evident from Map 6-1, however, had the program been carried out Chicago would have had, by

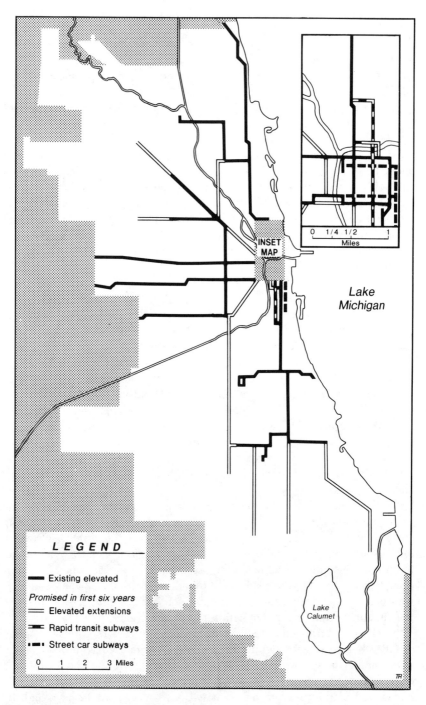

Map 6–1. 1918 transit plan.

1925, a fifteen-mile-long crosstown rapid transit line and substantial outlying rapid transit extensions. The plan's complicated financial arrangement also diminished the future effects of overcapitalization and, by allowing fares to rise and guaranteeing returns, might have helped restore mass transit's credit.[115]

The overcapitalization of the surface lines and the opposition of the business community had made real municipal ownership and control of transit impossible. State law had eliminated regulation which, in any case, could never have brought about major transit improvements requiring new capital. In the light of all this the 1918 plan, which effectively placed mass transit in the hands of the same interests which were to guide the city's traffic control and street improvement policies in the 1920s, had at least the virtue of making planning possible. Thus there were good arguments both for and against the ordinance. None of these real issues, however, featured prominently in the debate over the ordinance.

The lineup for and against the ordinance was itself confusing. Broadly speaking, the CBD business community and its press and political allies backed the agreement, while most other organized forces in the city opposed it. Thus Harrison and Dunne, longtime opponants of utility power, joined with Insull's ally, William Hale Thompson, who railed against the ordinance in part to take attention from his earlier ill-advised opposition to American participation in the war. With them were the Greater Chicago Federation, most of organized labor, and a number of figures from the turn-of-the-century battles. Hinky Dink and Bathhouse John also worked against the ordinance, but so did the council's only socialist, and a low tax reformer, Ulysses S. Schwartz.[116] All too often, arguments on both sides were nonsensical.

Mistrust of "traction barons" was one prime argument for defeat of the ordinance even though, in the absence of public ownership, the city had no alternative but to cooperate with utility owners. The state's attorney, a Thompson ally, charged massive bribery in the council passage of the ordinance and conducted a grand jury investigation which produced much publicity and no indictments.[117]

The Greater Chicago Federation made much of the companies' "betrayal" of the 1907 ordinances, which Carter Harrison II referred to as "Hun 'scraps of paper.'" John Fitzpatrick, head of the Chicago Federation of Labor, branded the agreement a plot by "wall street crooks." What was perhaps more culpable, opponents argued that *defeat* of the ordinance would preserve the 5¢ fare and that its *passage* would result in elevated lines being built down the middle of outlying commercial streets and streetcars being stopped only once every half-mile—all propositions completely without foundation.[118]

The claims of proponents were equally absurd. They promised that the ordinance would keep fares low, provide seats for all, and "enable working people to move to the outskirts of the city, where they can own their own homes."[119] Ordinance supporters engaged in a particularly revealing bit of opportunism when they attempted to exploit contemporary interest in the rights of women.

During the final week of the campaign, pro-settlement forces appealed to the city's female voters in a way that revealed not only the growing importance of women in politics but also the social class connotations of mass transit. Apparently inspired by the Association of Commerce's women's aux-iliary, groups of upper-class clubwomen and social workers rode the rush hour trolleys on the pretext of "discovering" whether the transit service was as bad as ordinance proponents claimed. Their reports, filled with complaints of the insults offered women's dignity and moral integrity by mobs aboard the cars, were prominently featured by the newspapers supporting the new agreement.

"It was awful," complained Mrs. Joseph Coleman, clubwoman, who was later reported confined to her home, "ill from the experience." "Such awful men on the back platform," she exclaimed to the *Daily News*. "Why are people allowed to sell such odoriferous whiskey, I wonder?" Other club-women and "suffragists" complained of being told to move forward in the car by conductors, suffering brushes with greasy strangers, and being unable to "free our persons from indecent contact with the struggling mob." "If Chicago does not pass the traction ordinance," the *Tribune* warned, "it will be commit-ting a crime against its womanhood."[120]

Mary Dobyne, secretary of the Chicago Political Equality League, observed on the day following the referendum vote that "business and social conservatives" had "particularly tried to delude women by arguments that did not hold water."[121] In fact, all the city's voters had been treated with some-thing less than respect by both sides in the traction debate.

"The truth is," the *Dziennik Zwiazkowy* lamented shortly after the ordinance was passed by the city council, "that only a few people know what the whole thing is about."[122] After two and a half months of campaigning, the *Dziennik's* assessment probably still held true. The results of the referendum, therefore, could show little about the population's opinions on the merits of public or private ownership, the importance of rapid transit, or even the old and now largely irrelevant issues of low fares and short-term franchises.

Chicagoans rejected the last real chance for a comprehensive trans-portation system by a vote of 243,351 to 210,611. Only the relatively affluent and independent Hyde Park and south shore areas, which included the University of Chicago, demonstrated great enthusiasm for the ordinance, although, as George Hooker had predicted, it received support all along the lakefront. The ordinance did worst in the inner northwest and southwest side districts, but the influence of political machines makes it impossible to inter-pret the vote in these most crowded sections of the city.[123]

The 1918 referendum was the only one for which the votes of men and women were tallied separately. Perhaps because of the clubwomen's efforts, more women than men seemed to favor the ordinance. With the exception of one ward, however, it was in outlying and gold coast districts that their votes differed most from those of their husbands, fathers, and brothers.[124]

• • •

Male and female alike, Chicago voters in the seven years after 1918 were exposed to more, not less, confusing rhetoric. During that period, much of the public attention to "traction" was taken up by two plans for municipal ownership. One, originated by Thompson and backed by most of the interests which had opposed the 1918 plan, was in all likelihood no more than a political stunt. The other, supported by most of the backers of the 1918 solution, was not municipal ownership at all in any meaningful sense. To make matters more confusing, some transit interests, beset by rising costs and tardy fare increases, began to look upon municipal ownership as a means of salvaging their investments.

Thompson announced his plan in 1919 with a typical rhetorical flourish. "We will fight the traction magnates now clutching at the public throat," proclaimed Samuel Insull's client, "until their greed is thwarted and their plans of extortion defeated."[125] Thereupon the mayor brought forward a plan which might be called extraordinarily farsighted were it not for the fact that even its creator must have known it had no real chance of being implemented.

Thompson's plan called for the creation of a new local government, a separate taxing body comparable to those that managed the city's parks. This "Transit District" would own surface transit in Chicago and some inner suburbs and was to have bonding and taxing power similar to that of other local governments. By negotiation or condemnation, it would be able to buy existing traction properties. The district was to be run by a board chosen at special, nonpartisan elections, and was to be required to charge no more than a 5¢ fare for transportation within Chicago. Any deficit was to made up from real estate taxes.[126]

The Thompson plan was courteously received by the city council perhaps because, as one alderman was reported to have said, "it's the mayor's funeral, not ours."[127] After an engineering commission duly confirmed the mayor's contention that the streetcars could be run profitably on a 5¢ fare, legislation based on the plan was submitted in Springfield on three separate occasions. Each time the legislation failed.[128] The Thompson plan had no tangible results, but the events surrounding its failure highlighted the changing attitudes of Chicagoans toward public ownership and helped to accentuate the change.

Thompson was the last Chicago mayor to speak of municipal ownership in the language of the turn-of-the-century radicals and the only mayor before the 1980s to suggest that transit fares be subsidized from general local taxes. Chester Cleveland, the attorney Thompson placed in charge of the plan, could sound remarkably like the urban populist Henry Demarest Lloyd. Cleveland spoke of the need to "accept as a vital fact in its full significance that the people are able through their general agents elected for the purpose to manage and operate the means of local transportation economically and efficiently."[129]

Thompson and his allies adopted the turn-of-the-century slogan "The Streets Belong to the People" and added to it a contemporary note: "The city has paid millions of dollars to provide boulevards and pavements over which vehicular traffic could operate . . . ," Cleveland told the CLT, ". . . and we do not complain because the people [with automobiles] go to and fro from work and we are taxed to maintain these boulevards." Why, then, Thompson and his spokesmen argued, should the city not own and operate mass transit with the support of general taxation? Was not the street railway as much a part of the street as was the pavement?[130]

These arguments would not be heard again in Chicago for over half a century. It was ironic—and damaging to their credibility—that they were aired at this time by the most ill-reputed politician to hold the mayoral office during the twentieth century.

The swift and sure reaction against Thompson's plan by the press and business community was based, not on the reputation of the mayor, however, but on the unpopularity of the ideas themselves among business-class reformers. To most of the press, the fact that the plan would raise taxes was sufficient evidence against its legitimacy. The CAC and Walter Fisher took much the same position.[131] The idea that good mass transit was profitable had, it appeared, risen from the level of an economic to that of a moral principle.

While many reacted aginst the superficial content of Thompson's plan, its real purpose was not hard to discern. As the *Chicago Journal* pointed out, the plan had the great advantage of being almost impossible to implement. It required state legislation and then, if the properties were to be purchased by condemnation, endless lawsuits. Furthermore, even if Thompson succeeded in forcing the sale of the surface lines, he could not purchase the elevated lines by condemnation, since they held state franchises which ran into the 1940s.[132] Thompson's "district plan" was probably little more than a talking point for his next mayoral campaign.

No doubt this was why the traction companies did nothing to oppose the plan, either privately or publicly. Yet behind the companies' seeming acquiescence may have been something more than mere complacency. Sources as different as Delos F. Wilcox and the Chicago Association of Commerce warned the public that municipal ownership could, in days of high legal valuations and low returns, be nothing more than a bailout for weary utility magnates. Something like this had occurred in Seattle in 1919. For a variety of reasons, some leaders of the auto industry supported municipalization of Detroit's street railways, which took place in 1922.[133] In Chicago, the potential of municipal ownership as a means of rescuing private investors and providing new capital for mass transit was one of the reasons why Ulysses S. Schwartz found a receptive audience for the plan he put foward to rival Thompson's.

Schwartz, who had a long record of interest in local transit reform and was a protege of Edward Dunne's political organization, had excellent reform credentials. When he became chairman of the Committee on Local Transportation in 1920, Schwartz set about trying to make municipal ownership a viable proposition. In doing so, however, he helped to change its fundamental meaning.[134]

Schwartz' plan went through several forms, but as it emerged toward the end of Thompson's second term, it was—as Schwartz liked to remind his audiences—a business proposition. The existing companies would be purchased by the city with certificates which would be a lien, not on the property itself, but on the fares collected from its operation. Title to the property would be vested in the city, and the property would be operated by a city-appointed board, thus restoring home rule.

The fact that the "Schwartz certificates" were a lien on the fares alone would obviate the city's debt limit problem, while provisions like those of the 1918 ordinance to guarantee fare increases when necessary to pay interest on the certificates would presumably restore to the "municipalized" system the credit private companies had lacked.[135] Transit, then, would be operated by the city under a lien agreement which guaranteed businesslike management. Composed though it was of old and borrowed elements, the plan was something new for Chicago.

Schwartz took care to assure all concerned that his idea did not depart from the old Chicago idea that transit must be profitable. "The traction system should pay for itself," he told the *Chicago Post*. "It is neither sound nor necessary to devote taxes to that purpose."[136] The heads of the existing companies received the new idea with apparent interest, and CSL President Busby went so far as to comment that continued operation under private ownerhsip appeared impossible. While the response to the plan was not universally friendly, the proposition *was* taken seriously, and bankers who had handled previous traction issues agreed to investigate its feasibility.[137]

As put forward by Schwartz, this plan seemed capable of satisfying the demands of a variety of interests in the historical traction community: it offered genuine public ownership while retaining the commitment to profitable operation; it restored mass transit's credit while proposing for traction investors a way out of their postwar dilemma. A plan, however, was not a working reality.

Before it could be used to resolve Chicago's transit impasse, Schwartz' idea had to find a political constituency and had to be transformed into a specific proposition which could win the assent of investors and financiers. As this process took place, two factors intervened which transformed Schwartz' plan from a practical tool for public ownership into the final proof that municipalization was meaningless and impossible. These factors were the suspicions and demands of the financial community on the one hand and the personality and political ties of Mayor William Dever on the other.

Dever came to power in 1923 as a result of a split within the Republican Party and a compromise between the Harrison (reform) and Brennan (utility) factions of the Democratic Party.[138] Dever was the choice of an incompatible coalition in which sooner or later one element or another was bound to dominate.

Dever was, in one sense, the representative of the more socially minded progressive elements of pre-1914 era. It was Dever—the aldermanic candidate of the northwest side settlement movement in 1903—who had championed public ownership as a means of improving the lot of the working poor. It was Dever who had turned Dunne against the 1907 ordinances and had led the futile fight against their adoption. When he returned to politics from the judiciary in 1923, he did so with the support of a wide range of old-line municipalizationalists such as Dunne, Sikes, Haley, and Merriam, and of Bullmoose Republicans like Harold Ickes.[139]

Yet the "respectable" press and business community also backed Dever as an alternative to "Thompsonism." Men and institutions who had always supported private ownership, the *Chicago American* complained in 1923, "are now talking glibly and garrulously about securing municipal ownership for the dear people."[140]

Dever's ties made the significance of his election uncertain. His own character in these later years made it certain that any plan he endorsed would meet opposition on nontransportation grounds. Once a champion of more positive local government, Dever as mayor lent credence to the cruel remark of one of his 1923 adversaries that "the Democrats are running an elderly gentleman long since dead."[141]

The mayor proved a stern enforcer of negative laws: home stills and speakeasies were shut down, boxing matches were banned, and "immoral" publications were confiscated from newsstands. Dever, who as alderman had represented a polyglot immigrant ward, showed little understanding of the life-styles of ethnics and was openly hostile to the aspirations of business and professional women (attacked for the absence of women on the proposed municipal traction board, he queried, "What do they think we ought to appoint, midwives?").[142]

One political old-timer summed up Dever's personal fate: "To give the impression that you've gone over to the silk hatters after you've been raised under a fedora is fatal in Chicago politics."[143]

But the same remark with which the political veteran dismissed Dever could also serve as the epitaph for public ownership in Chicago. Begun in part as a crusade against corporate domination of the city's life, public ownership in Dever's term completed its own transition from fedora to silk hat. Through no fault of the mayor's, municipalization by 1925 had become but another means of bailing out traction investors and ensuring control of mass transit policy by CBD business and the banking community.

Dever began his term with a commitment to the idea that city government could and must prove its own ability to manage transit as a business. He would not, he promised, ". . . agree to any settlement that at any time will obligate the general taxpayer to make up a deficit in streetcar earnings."[144] The mayor took special care in his approach to the downtown business community.

To the Association of Commerce, Dever explained that municipal ownership was pragmatically "the only way out." No one, he assured CAC members, was interested in confiscating the property of the thousands of small investors who had purchased traction securities when the great banks lost interest in them in 1916. He declared to all concerned that there would be no tax subsidy for transit under public ownership, and he opposed "improvidence in the building of additions and extensions . . . so that the transit system itself will be self-supporting." "The whole matter," he said, echoing Ulysses Schwartz, "is a straightforward business proposition."[145]

At the same time, Dever's commitment to real municipal ownership was evidently genuine. "No plan should be considered," the new mayor told the council in the summer of 1923, "which introduces a continuation of private ownership or private operation under the disguise of Municipal ownership and operation." Such a halfhearted arrangement, he told his audiences during the 1923 campaign, "is merely a camouflage for another extension ordinance, and that might become permanent." Private ownership was out of the question because "no franchise can be devised which will compel continuous good service and the building of adequate extensions."[146]

Valuation was the key to a successful municipal ownership program. A high purchase price had made something of a fiasco of municipalization in Seattle. Dever knew this and repeatedly attacked the valuation claims of the companies, which he described as "entirely too high."[147]

Here, however, Dever encountered a solid line of resistance. Neither the companies nor their bankers had any reason to accept lower valuations, and business and financial interests alike demanded *not* municipal operation but throughgoing business control of any city-owned transportation system. As a result, the ordinance which Dever endorsed after two years of struggle provided for public ownership in name only.[148]

The 1925 ordinance bound the city to pay the full BOSE value of over $162 million for the Surface Lines and $85 million—$5 million less than Insull's minimum price—for the elevated. The municipal company was to be operated by a Board of Control which would be dominated by appointees of the certificate holders until 51 percent of the securities had been amortized. Thereafter, the city could replace the company representatives with "person[s] of recognized character and standing who [have] had important and outstanding industrial, financial, legal, engineering or other business experience."[149]

The new "municipal" line was to operate much as the private corporation proposed in 1918 would have operated. A barometer fund was to be created to keep fares high enough to pay off the certificates, and an improvement program—substantially that of the 1916 plan—was agreed to. Altogether, the scheme ultimately pledged the car riders to pay off an investment of nearly $721 million, plus interest, which was to be expended in the twenty years following the passage of the ordinance.[150]

Mayor Dever backed this plan and helped to push it through the council. The ordinance for a "comprehensive municipal local transportation system" was submitted to a referendum on April 7, 1925, with the backing of the Chicago Association of Commerce, the city's two largest newspapers, and a somewhat reluctant Brennan machine. The surface companies gave the plan unenthusiastic approval and, late in the campaign, Insull made a radio address in its behalf—a move which, the *Tribune* suspected, might have been designed to turn voters against the plan while keeping Insull in the good graces of the CBD community.[151] Labor, regionalists, the Hearst press, most of the older municipal ownership advocates, and the followers of ex-Mayor Thompson worked against the ordinance. Their effort was successful. By a margin of over 20 percent, Chicago once again rejected compromise.[152]

On the surface, the 1925 vote was a straightforward rejection of the idea of municipal ownership by a solid majority of Chicago voters. A number of observers who opposed municipalization interpreted the vote this way.[153] In fact, what the voters rejected in 1925 was not a proposal for municipal ownership in any meaningful sense. For practical purposes, it was nothing more than a long-term private franchise with good amortization provisions.

The Municipal Traction Board provided for under the 1925 ordinance reflected the views of bankers, investors, and traction owners on the control of a municipally owned system. In essence, it required that investors maintain complete authority over the system until the bulk of their investment had been paid off; it was to be controlled by businessmen thereafter—this despite ironclad provisions in the ordinance requiring that fares be kept high enough to pay interest on and amortize the Schwartz certificates.[154]

In Seattle, a city with a much more violent history of anti-business agitation than Chicago, the corporation owning the street railways accepted certificates like those offered by Chicago without asking any share in control of the property. The 1925 Chicago plan, in contrast, not only kept the municipally owned lines safely out of the city's hands, but made the lines subject to no regulation whatever. As city property, the system would be free from state regulation, and the members of the Board of Control were removable only through court proceedings.[155] Clearly, in Chicago, it was the attitude of the financial community which made municipalization a totally unsatisfactory solution to the question of transit regulation and planning.

Public ownership as offered to the voters of 1925 would also have done nothing to lower the cost of local transportation. The city, under the plan, would have had to buy the entire load of "franchise values" and defunct cable cars represented in the companies' valuations. As it was, by 1925 the motorbus had begun to prove its worth, and one Chicago newspaper was crying for the replacement of all streetcars by the free-wheeling machines. The argument that municipal ownership meant buying "useless junk" was widely used by its opponents in the years before 1925.[156]

In short, the vote against the 1925 ordinance can hardly be taken as a reflection of public opinion on municipal ownership. All of the considerations cited above, together with Insull's offer to build comprehensive rapid transit at no apparent cost to the public, made the form of public ownership offered by the 1925 ordinance a poor bargain. To all this must be added the estimate of the *Tribune*'s veteran traction reporter that up to two thirds of the ballots against public ownership were in reality votes against Mayor Dever's policies toward "beer, booze, and gambling."[157]

· · ·

Thus the public ownership issue was virtually dead after the 1925 referendum, if only because of the attitudes taken by the city's banking community and by courts dealing with the question of valuation. The few remaining advocates of municipal ownership probably helped assure its speedy burial. And the scandals accompanying the city's roadbuilding program no doubt deepened old suspicions of local government.[158]

By the fall of 1925, Dever, Schwartz, and Jacob Arvey, one of the strongest Dever men in the council, all agreed that the only solution to the traction problem lay in a terminable permit. To them, the only remaining question was how to safeguard the interests of the city in the legislation which would empower it to give the necessary indefinite-term franchise. Among the city's major active political figures, only William Hale Thompson opposed the terminable-permit solution, and he did not attempt to make traction a major issue in his successful 1927 bid for reelection. Delos F. Wilcox and the Chicago Association of Commerce, who had not agreed on any traction measure since the 1907 ordinances, now concurred that a terminable permit with the proper safeguards might be the best solution.[159]

Most important, the Committee on Local Transportation appears to have followed Alderman Arvey's suggestion that all future traction measures be cooperatively worked out on a nonpartisan basis. The Committee heard many plans and propositions during the following four years, but none was the special province of a mayor or a political faction. Instead, the Committee developed its own statement of principles and sought a plan which would accommodate them. In the 1926 version of this statement, the CLT agreed, by an eleven to one margin which included Thompson Republicans, that a terminable permit with home rule and effective termination provisions was a satisfactory basis for a new comprehensive plan.[160]

By 1926, then, politicians had for all practical purposes agreed to forego the traction issue as a weapon in their battles with one another. What pressures were brought to bear upon them, it is not possible to determine. It *is* clear that, by allowing council unanimity on fundamentals, political leaders made it possible for themselves to accept a future solution without handing their rivals any credit for the achievement. In this one respect, transit improvement had finally reached the status long before attained by roadbuilding under the Chicago Plan: the general outlines of how it was to be carried out were the subject of a political consensus.

Convergence, then, together with the obvious absence of any alternative to dealing with the existing companies on their own terms, was a primary element in the creation of the 1930 transportation settlement. Active politicians agreed to depoliticize mass transit. Curiously, three of Dever's top aides were, by the mid-1930s, employees of the transit companies they had sought to curb. Dever himself ended his days as an investment banker.[161]

Much more significant, after 1929 Insull cooperated with the more staid elements of the business community for a traction settlement, and after 1924 the CSL worked hard through traffic control and public relations efforts to align itself with the dominant elements of the business community. In receivership, of course, the CSL was represented by major establishment banking figures.[162]

Businessmen and politicians, then, closed ranks among themselves and worked together after 1925 to see that the only possible traction settlement would be achieved. The 1928–1930 Citizens Traction Settlement Committee constituted the apogee of this cooperation. Appointed by the CSL's receivership judge and composed of leading members of the "civic-minded" business community, the CTSC relieved the city council of responsibility for unpopular decisions by setting its seal of approval on high valuations and flexible fares. The council cooperated by appointing its own valuation commission to certify the CTSC's recommendations, and Insull, as already noted, lent a vital hand by undertaking the refinancing of the proposed new company and by seeing to the provision of "grease" for the passage of necessary state legislation and of the 1930 referendum.[163]

Two more factors came together to help assure nearly total unanimity for the 1930 ordinance. One was the prospect of the upcoming 1933 World's Fair, the "Century of Progress." The city's pride was at stake as it had been in 1893. The desire that a modern transit system comparable with those of eastern cities would at least be under construction when the visitors arrived may have helped solidify support for this latest traction program.[164] But desperation as well as hope played a major part in bringing about almost universal support of the 1930 plan.

Labor, with the exception of Margaret Haley's Chicago Teachers' Federation, wholeheartedly supported the 1930 ordinance. To some extent such support was to have been expected: the CFL's vice-president and head of the powerful building trades union, Alderman Oscar Nelson, was a Thompson man with few qualms about how solidarity was to be achieved (at the labor meeting which approved the ordinance Nelson converted one objecting delegate by punching him in the jaw).[165] But John Fitzpatrick, the CFL's longtime president and backer of Dever in 1927, spoke out for the ordinance too. Almost sheepishly he explained to the CLT why he had abandoned his quarter-century devotion to publicly owned utilities: "We have had this situation. . . . I do not know that anybody is really able to define the causes or just how it happened. . . . All we know is we suddenly found ourselves in the midst of it, all at once, and now we are burdened with the responsibility of finding ways and means of getting out of it. So this is one of the ways."[166]

Fitzpatrick was referring to the Great Depression, which had begun to affect Chicago before of the rest of the nation. The housing industry had collapsed first in Chicago, and building trades unions, along with outlying real estate dealers who had opposed previous settlements, were among the most ardent advocates of quick action on a traction plan in 1930.[167]

The rationale for their support was expressed by Alderman Oscar Nelson. "People built homes 5 or 6 years ago on the outskirts of Chicago," the CFL vice-president said in 1930, "with the expectation that the traction question would be settled. The result [of the failure to settle the question and build extensions] is that there has been no real estate development in the last 12 months particularly. . . . I say the traction problem has got a stranglehold on Chicago."[168]

"Confidence in the financial centers that Chicago is progressing and that investments within this area are desirable" would be restored by a quick traction settlement, the Chicago Real Estate Board wrote to the CLT in 1930. "The expending of approximately $300 million . . . would not alone restore the prosperity of the city, but we are confident that it would be but a small part of the money released for wages and materials once this problem is settled. . . . The time has come when the city must take immediate and definite action in a manner that will restore prosperity."[169]

Transit building as an antidote for economic depression was not a new idea: advocates of the 1918 settlement had promised that it would assure jobs for veterans returning from World War I. But by 1930 the issue of jobs had taken on overwhelming importance. So the 1930 settlement was a part of Chicago's answer to the Great Depression—a gigiantic public works project paid for, not by the now reluctant owners of real estate, but by the city's transit-riding middle- and working-class residents.

A city which had exhausted the public's tolerance for government spending in necessary but often extravagantly financed roadbuilding projects, a city which would not have met its day-to-day expenses were it not for the charity of its wealthy citizens, could hardly be expected to reject a plan which promised over $300 million of construction. A city whose government and taxing system had become something of a national scandal was not likely to ask searching questions of its benefactors. Shortly before the referendum on the 1930 ordinance, the Chicago Real Estate Board sent 450 automobiles through the streets of Chicago bearing signs which read, "Vote for Prosperity."[170]

These signs summed up some of the nontransportation reasons for the success of the first traction ordinance to pass a referendum in twenty-three years. The depoliticization of mass transit, the degree of convergence on the council, and the fact that a "traction baron"—Insull—finally *had* taken a direct hand in seeing to the success of a specific plan—these factors largely completed the reasons for the apparent solution of the "traction problem."

Nontransportation issues had kept the city's transportation problem unsolved through the last period in American history when the large amounts of money needed to produce an efficient comprehensive system were available. Now, as that period passed away forever, nontransportation issues helped bring the city's politicians, businessmen, and traction owners together to eliminate the old set of obstacles. On July 1, 1930, the ordinance with which Chicago abandoned much of its reform tradition passed by an overwhelming vote of 325,837 to 56,690.[171]

By the summer of 1930, Chicago had come full circle. The reform movement which flowered in the wake of the 1893 depression had fixed on Charles Tyson Yerkes as the symbol of the city's ills and had sought to cure the metropolis by removing Chicago's first utility czar from his position of power. Thirty-seven years after the Colombian Exposition, Chicago looked to another "traction baron" and another Fair for salvation from another terrifying depression. The Fair did not save Chicago. Insull could not even save himself.[172] Half a century after the 1930 "solution" of Chicago's local transportation problem, the *Chicago Tribune*'s remark of 1924—"The hand of Yerkes is on our traction lines yet"—still went far toward explaining the persistence of the city's seemingly insoluble local transportation problem.

Epilogue

If the automobile industry has gotten into the local transportation industry, it has not been our fault.

[*C. F. Kettering (vice president, General Motors), 1928*][1]

. . . the effect of the automobile upon the public carrier has been largely overestimated.

[*E. J. McIlraith, 1931*][2]

C. F. Kettering, in the statement quoted above, was disclaiming responsibility, not for the use of automobiles for the journey to work, but for the rise of the gasoline bus. McIlraith, ever the optimist, underestimated the attractions of personal transportation. Yet their observations draw attention to an important fact: by the end of the 1920s the automobile had long been seen as a largely autonomous force, transforming the city for better or worse. The effect of the private car has indeed been overestimated, and the power of its makers to transform society, real though that power was, has obscured the central role of public policy, and the way that policy was made, in accommodating the transformations wrought by the private car.

Local transportation policies had significant impact on the lives of Chicagoans. That impact was evident in 1930, and continues to be felt today. The Chicago of 1930 reflected the street and transit policies of the previous quarter-century. As already noted, auto ownership by 1926 had spread far beyond the boulevard and lakeshore districts. Traffic studies of the 1930s show fairly heavy crosstown auto travel, though by far the largest number of cars on a single street continued to be found on the Lake Shore Drive. A large proportion of Chicago's rush hour auto traffic—38.8 percent on three major streets which ended in the Loop—was generated by the CBD. Policy had helped the Loop and the automobile adapt to one another.[3]

Surface mass transit continued to serve a dispersed population. Company transfer studies during the 1930s revealed that travel on major crosstown lines was scattered over a wide variety of routes: passengers did not simply ride to Loop-bound lines and transfer downtown. Several crosstown lines

209

carried two-car trains during the 1920s, and on one, six miles west of the CBD, the CSL found it impossible to place enough cars on the street to meet the demands of the rush hours. Neighborhood business centers continued to depend predominantly on mass transit for patron access.[4]

The thinning of inner-city population during the 1920s has already been discussed. The departure of 211,000 persons from a thirty-six-square-mile area of Chicago between 1920 and 1934 cannot be credited to any single factor. By 1930, most of the areas which received streetcar extensions in the prewar era had undergone residential development, and extensions were still in demand. On the other hand, while the city as a whole experienced a 25 percent population gain during the 1920s, its suburbs added 59 percent to their population. Here the automobile, which helped connect outlying areas with one another as well as with the city, may be assumed to have been—with the commuter railroad—the major transportation factor involved.[5]

Within the city too, the private car assumed an important function in the journey to work during the 1930s. Studies done by the CSL at the beginning of World War II revealed that both transportation and residence patterns had changed dramatically since 1916 and 1926. In 1916, 79.2 percent of workers in a southeast side steel plant had lived within walking distance of their jobs. By 1941, 27.4 percent walked to work, 34.9 percent used streetcars, and 35.2 percent rode in automobiles. At one major stockyards plant, a 1942 survey showed that only 7.8 pecent of the workers walked to work, while 28.8 percent drove and 62.5 percent used mass transit. In 1916 the proportion of presumed walkers in the same general area had been between 30 and 47 percent. Such figures suggest a dramatic change, not only from 1916, but from 1926 when, it will be remembered, Miller McClintock found less than 3 percent of the workers at a major west side plant commuting by private car.[6] The preference for the automobile was evident, rational and destined to grow. For this preference, however, neither the car nor its makers was solely responsible.

The automobile by itself did not determine the fate of urban transportation in Chicago. Long before the majority of Chicagoans could afford to own automobiles, misguided city policy had stifled the growth of public transit and made private transportation the rational, practical alternative for those who could afford it. The present-day visitor to Chicago can see all around him the monuments to Chicago's pre-Depression efforts on behalf of the automobile. These progressive and innovative street improvements—the products of collaboration among businessmen, reformers, engineers, and often-corrupt politicians—are still useful, though overloaded. One may search the city in vain for a single mass transit improvement constructed as a result of the city's policy of regulation. Chicago's transit system today depends upon structures which are either ancient relics of the days of unregulated private enterprise or more recent products of federal subsidies.

By inadvertence, city policy between 1900 and 1930 brought rapid transit development to a halt and directed the development of surface transit into a pattern which (in the absence of outside subsidies) seriously weakened the companies. They became, in the words of Alderman Ulysses Schwartz, "neither private nor public, but a curious hybrid. . . ."[7] By inadvertence too, Chicago's street and traffic policies fostered the use of the automobile.

The adversary relationship between Chicago and the organizations which provided its transit service—prevented the development of a modern, comprehensive system. Had Chicago either achieved municipal ownership or left its traction companies entirely free of oppressive regulation, a unified comprehensive transit network would probably have resulted. Throughout the period the companies showed interest in building such a system. The history of Chicago Plan improvements suggests that sectionalist politics would have demanded, if anything, overexpansion of a municipally owned system. Most Chicagoans of the 1900–1930 era would, of course, have responded that either unregulated private enterprise or municipal ownership meant corruption and waste. This fear, in the light of history, seems at once well grounded and of relatively little importance.

In fact, regulation gave Chicago most of the alleged disadvantages of municipal ownership with none of its attendant benefits. Chicago's street railways were the only companies to feel the full effects of the city's regulatory policies. By their own estimates, these companies contributed over $150 million in car riders' moneys to taxes and other public benefits between 1907 and 1935. If the $39,237,738 in profits extracted by the companies during the same years is added, the dollar-cost to the Chicago streetcar rider of regulated private ownership exceeds $188 million—an amount higher than the entire inflated valuation of the CSL in 1930. Even though most of the city's "traction fund" was eventually used in combination with federal funds for the building of a Loop rapid transit subway, the amount of transit riders' money diverted to nontransit purposes was in excess of $155 million.[8]

It is difficult to imagine that a municipally owned system could have stolen this much of the people's money without providing a more modern and comprehensive rapid transit system than the city received from regulated private enterprise. Graft there would no doubt have been, yet the private companies were engaged in their own form of cost-padding through overcapitalization. Wages under municipal ownership might have been high, but from the 1920s, the wages paid by the private companies were also high.

It has been pointed out by scholars that the municipal ownership and transit reform movements in other cities were closely tied to the drive by real estate interests to gain extensions which would raise the value of privately held land. This appears to have been the case with public ownership in Toronto.[9] But Chicago's regulatory policy produced an excess of extensions early in our period and failed to yield extensions in the 1920s, when they were again needed. Taxes did not pay for unnecessary extensions in Chicago: the fares of inner-city riders did.

Finally, by opting for regulated private ownership of transit utilities, Chicago cut mass transit off from city credit and left it outside the sphere of the city planning movement. That all these opportunities were relinquished in the name of separating "utilities" from "politics" was especially ironic, since regulatory power without responsibility led not to clean government and better public service but to ever-greater involvement of utility companies in politics. Impotence bred demagogy. Irresponsibility, not power, corrupted local government.

Nor did the effects of the adversary relationship between city and companies end with the compromise of 1930. That agreement collapsed after Insull's fall, and the city spent much of the following decade attempting to pressure the relatively solvent CSL into assuming the burdens of the bankrupt, CBD-oriented rapid transit company. The city continued to oppose fare increases through 1946, and helped keep Chicago's fares among the nation's lowest, thus making the financing of a consolidation or of new equipment all the more difficult.[10]

The same adversary relationship helped to doom surface mass transit, and particularly the streetcar. Harold Ickes, as head of the PWA, insisted that subways built by that agency be designed for rapid transit use only. His aim was to force the CSL to absorb the failing rapid transit company. In the absence of streetcar subways, the trolley lost whatever attraction it had continued to have for local planners. Traffic engineers had long looked upon the streetcar as a major obstacle to auto traffic, and had counted on its removal into subways to make possible the accommodation of Loop motor traffic.[11] Thus Ickes' decision to exclude streetcars from subways probably had more to do with the eventual elimination of light rail vehicles from Chicago's streets than did any efforts on the part of the makers of gasoline busses.

The legacy of 1907 continued to influence Chicago transportation policy in the 1940s. When municipalization was finally achieved in 1947, it was undertaken as a last resort. The Chicago Transit Authority, which thereafter operated the unified system, was governed by a board appointed by the mayor and the governor, composed for the most part of representatives of the business community and other organized interests, and effectively insulated from public pressure. The legislation creating the CTA required it to be self-supporting, and fares rose rapidly thereafter.[12]

This modified form of public ownership did restore transit's credit, and the CTA rapidly replaced late-Victorian wooden cars which had been the mainstay of elevated service for half a century. Streetcars too disappeared by the end of 1958 but, as had happened more than once before, increased street traffic neutralized the effects of the new equipment: the busses moved through the streets little faster than had the trolleys they replaced.[13]

There was real progress after 1947. By the 1970s three expressways had been built with rapid transit facilities in their median strips—a first, limited step toward unified transit and automobile planning. In 1974, despite strong resistance from most suburban districts, a Regional Transportation Authority was formed to coordinate and help finance mass transit in the six-county metropolitan area. Finally, in a step the significance of which was not widely appreciated, Chicago created a bus-only mall on State Street and reverse-direction bus lanes on several other CBD streets, and restored on-street parking in the CBD. These steps, undertaken in the late 1970s and early 1980s, removed much of the street space used for auto traffic in the Loop and seemed to reflect a traffic control policy shift in favor of mass transit in the CBD.[14]

Fundamental matters, however, had changed but little. Regional jealousies continued to be important. They were reflected in perennial complaints over the distribution of RTA funds, with some suburban representatives arguing, as had neighborhood organizations in earlier years, that transit funding and planning was disproportionately centered on the CBD. The allocation of service within the city was sometimes seen as being racially motivated, with black straphangers receiving the second-class service of which west side immigrants had formerly complained. Routing continued to reflect class feeling as well: when service cutbacks eliminated an inner-city bus line which ran one block from a parallel line, some riders complained loudly. The remaining route, on a business street, carried "roughnecks," one spokesman declared. "People, he observed, "have a way of sorting themselves out."[15]

The politics of transportation was familiar too. When Chicago's mayor, in 1981, proposed direct city ownership of the foundering CTA and a subsudy paid from a tax on professional services, vigorous opposition emerged from the business community. Suggestions were heard, too, for a return to private ownership. Most characteristically, the late twentieth-century debate over transit's future was structured largely in terms of political battles between the city's mayor and the governor of Illinois, with the press dwelling at length on alleged waste and corruption in the municipal system.[16]

One final continuity deserves attention: that in the pricing of transit service. At the end of 1982, the price of a transit ride (one dollar with transfer) bore roughly the same relationship to the price of a loaf of bread as it had in 1906. This fare, which many sought to raise, had not kept up with the cost of labor, and subsidies paid nearly half of the cost of the ride. Yet in 1982 the *Tribune* carried the story of a man who walked seven miles in search of work on one of the coldest days of a very cold winter. He could not afford the carfare.[17]

The parallels and differences between mass transit at the beginning and end of the twentieth century are far too complex to be explored in the space available here. It is best to end with this vignette. Heavily subsidized, transit fares still present a real hardship to more than a few. Transit cannot, and never could, erase the effects of the spatial segregation it helped to create or the economic system which gave it birth.

But the cost and quality of transit service are of major importance to urban workers as well as to the urban poor. "I moved closer to work . . . because I just got tired of watching six [full] busses go by me every morning," one south sider told the *Chicago Sun-Times* in 1982. Upper-income Chicagoans, living in dense lakeshore areas ill-suited to the automobile, have reason to complain as well: in the early 1980s, the busses they rode were among the most crowded in the city. And the author can testify that, on west Division Street, the straphangers still fight for standing room as they did in the days of Tomaz Deuther.[18]

Mass Transit *has* improved, and the automobile has provided escape— and new dilemmas—for most urbanites. But the fundamental failure to treat the urban and regional transportation problem as a whole and to define mass transit as the responsibility of all who use and profit from the city—not disproportionately the burden of those who must travel on city busses and subway cars—continues to make urban life less attractive and more difficult than it need to be.

But Chicago's local transportation history reveals more than the impact of policy definitions on the shaping of public facilities in a single city. One final look backward at the story we have been tracing suggests important connections between Chicago's transportation history and a number of issues which have occupied historians in recent years.

Historians have long debated the degree of power held by local big business in Progressive Era cities. Did organized business lead the movement for structural reform in American cities? Did this movement give it throughgoing control of local government? Did major business interests instead involve themselves only in specific issues which were of direct concern to them? Alternatively, did the real power go to middle-class lawyers and administrators? Further, what role did neighborhood organizations have in the distribution of power and the allocation of resources?[19]

In Chicago, major downtown business fought to gain control of transportation policy in a time when transportation technology had helped to make a degree of dispersion possible and to create alternative centers of power in neighborhood business communities. Early political fragmentation reflected the city's diversity and the existence of numerous local political power bases. Further, straphangers, responding to job and housing opportunities, helped shape the transit system, while the driving habits of motorists strongly influenced both traffic regulation and street improvement policies.

CBD business achieved control of local transportation only when it was able to present a united front and only after political consensus had been achieved. Even then it had to make concessions to neighborhood interests, political machines, and other entities which had goals and concerns of their own. The CPC's failure to achieve its ends in the Union Station controversy, the difficulties encountered in the imposition of a Loop parking ban, and the rejection by voters of two major transit plans which had CBD business backing illustrate this point. Yet downtown business, when it did achieve consensus, was the most powerful single force in transportation policymaking. Engineers, lawyers, and urban bureaucrats had their influence, but only CBD business was able to manage the creation of a consensus on street and traffic policy and, ultimately, on mass transit.

The role of planners in policymaking is likewise debated, with some scholars finding the key to their effectiveness in the goals of the planners themselves, others in the wide range of factors which affected the policymaking process.[20] In Chicago, Burnham's great plan provided an important justification for traffic-serving street improvements. At no time, however, were planners or the CPC able to override the wishes of major elements of the CBD business community or cut through the city's tradition of mass transit policy. Particularly after the city's commitment to the superhighway, city engineers and consultants, not the Chicago Plan Commission or its staff, appear to have controlled expressway planning. The CPC's engineer, Hugh E. Young, played an important part in the gathering of housing and urban renewal data and in the defining of "blighted" areas, but traffic and roadbuilding policies became even more clearly the provinces of the city itself.[21]

Transit engineers, planners in their own fashion, also played a limited role. They helped set the parameters of early transit policy, but the attempt to give them an independent planning and regulatory function proved an almost unqualified failure. Public opinion helped to limit the role of the engineer as planner. Looking back on transportation improvements he had considered suggesting to the city, Bion Arnold observed in 1917 that he had backed away from ideas which seemed likely to arouse public opposition. In at least one case, he told the CLT, "I did not advocate [a specific design] because I knew that the people would object. . . ."[22] Even the plans of the 1916 Traction and Subway Commission, which did help to change the thinking of policymakers on some issues, met a political fate.

Some scholars ask, with reason, whether policy and planning made a significant difference in the life of the American city in any case. Changing technology and the constraints of economics did much to determine policy and to condition its outcomes.[23] Certainly a different local transit policy in Chicago would not have prevented the rise of the automobile. It might, however, have provided alternatives for the urban commuter. Nor was it economics which prevented the city from dealing as constructively with mass transit as it did with the automobile. Rather tradition, and the structure of power within the city, determined how Chicago's limited resources would be allocated.[24]

Transportation policy, then, was the result of a bargaining process which included politicians, large and small business and neighborhood interests, planners, and engineers. If CBD business dominated the policymaking process, it never fully controlled it: motorists and neighborhood groups in particular won important concessions. Rather downtown business acted as a broker among interests, and even in this role could not prevent the development of a strifling deadlock on mass transit policy.[25]

Together with all these factors, Chicago's environment helped shape its transportation history. The river accentuated the city's class and ethnic divisions and—in the nineteenth century—helped to create the divided surface transit system which inspired Chicago's devotion to the single fare and "universal" transfer. The river too helped confine the CBD and accentuate transit's centralization, thus giving focus to the regionalists' complaints. Lake Michigan's shore attracted much of the city's wealth as well as the attention of its planners. Wealth, combined with lakeshore boulevards, brought the quasi-evolutionary development of an automobile speedway along the lakefront. To the west and south, great areas of open land made possible the creation, in the nineteenth century, of a belt of parks and parkways.[26] These attracted wealth—and automobiles. Mass transit too was affected by the lakefront and parkways, for they fostered a separate class of service which included, by the 1930s, "expressway busses" on the Lake Shore Drive.[27]

Chicago's experience adds something to the understanding of utility regulation as well. In one sense, the regulation of the surface system presents a picture comparable to that of "enterprise denied" found by Albro Martin in the case of inter-city stream railroads.[28] The business efficiency of the CSL was clearly impaired by regulation. Innovation was retarded, and low fares and high service requirements helped repel investment.[29] Certainly, too, regulation of local transit did not represent the triumph of "the public interest." Rather it constituted an effort to subjugate the interests of transit investors to those of a variety of business, political, and—more rarely—neighborhood organizations.

Yet the story of local transit is different from that of America's railroads. A real social need existed, and continues to exist, for high levels of rush hour service combined with an affordable and attractive fare structure. It is unlikely that private enterprise could have met this need under any circumstances. Neither regulation nor corporate greed explains Chicago's transit history. When Yerkes declared that the straphangers paid the dividends, he revealed a truth the importance of which has not diminished since his time.

In the absence of a redistribution of wealth among urban citizens and a general redesign of the American city, low fares and frequent (and therefore expensive) service are preconditions of an accessible city and requirements of social justice. Zone fares will not assure that those more able to pay pay more. They will simply encourage auto use and, in a city in which minorities and low-income citizens are still confined to strictly delimited areas of residence, will place an additional burden on those who must live in the inner city in an age when job opportunities are increasingly dispersed.

Arnold and his contemporaries were correct in insisting that Chicago was one city, whose citizens had the right to travel its length and breath at a single, affordable fare. They erred only in believing that this right of access could be secured through a system supported only by user fees. The metropolitan region is now one city, and its ruling community has the benefit of history. It is evident that good transit cannot transform a city. It is equally evident that good mass transit can open opportunities for many and make life more bearable for all. Genuine public responsibility is therefore essential. The fact that urban government is fragmented and continues to represent the most powerful interests best does not diminish this reality. It only shows how much work remains to be done.

Notes

Author's note: The following notes represent approximately one half of the original documentation for this work. Much of the book's epilogue has also been eliminated. Readers wishing further information on the subjects treated here should contact the author through Temple University Press.

Introduction

1. On the adaptive nature of highway policy see, for example, Mark H. Rose, *Interstate: Express Highway Politics, 1941–1956* (Lawrence, 1979); Alan Altschuler, with James P. Womak and John R. Pucher, *The Urban Transportation System: Politics and Policy Innovation*, (Cambridge, Mass., 1980). See also Catherine G. Burke, *Innovation and Public Policy* (Lexington, Mass., 1979), and George A. Avery, "Breaking the Cycle: Regulation and Transportation Policy," *Urban Affairs Quarterly*, 8, no. 4 (June 1973): 423–38.

2. Charles W. Cheape, *Moving the Masses: Urban Public Transit in New York, Boston, and Philadelphia, 1880–1912* (Cambridge, Mass., 1980). Other recent works on the eastern big three ask substantially different questions. See, for example, Sam Bass Warner, *Streetcar Suburbs: The Process of Growth in Boston, 1870–1900* (Cambridge, Mass., 1962). The best book-length work dealing exclusively with mass transit away from the East Coast is Clay McShane's *Technology and Reform: Street Railways and the Growth of Milwaukee, 1887–1900* (Madison, 1974).

On mass transit in San Francisco, see Bion J. Arnold, *Report on the Improvement and Development of the Transportation Facilities of San Francisco* (submitted to the Mayor and the Board of Supervisors, San Francisco, 1913), pp. 39–41, 68, 225 ff., 420.

Extremely useful for comparison is John P. McKay, *Tramways and Trolleys: The Rise of Urban Mass Transit in Europe* (Princeton, N.J., 1976). Some of the vast array of other works, published and unpublished, are cited in the text as they become relevant.

3. Mark S. Foster, *From Streetcar to Superhighway: American City Planners and Urban Transportation, 1900–1940* (Philadelphia, 1981). See also Anthony J. Catanese, *Planners and Local Politics: Impossible Dreams* (Beverly Hills, Calif., 1974).

Also necessary for an understanding of the relationship among businessmen, planners, and the private car are the works of Blaine Brownell, including "The Commercial and Civic Elite: Planning in Atlanta, Memphis and New Orleans in the 1920's," *Journal of Southern History* 41, no. 3 (1975): 339–68; "A Symbol of Moder-

219

nity: Attitudes Toward the Automobile in Southern Cities, 1920's," *American Quarterly* 24, no. 1 (1972): 20–44; and *The Urban Ethos in the South: 1920–1930* (Boston, 1975). Brownwell describes the limitations within which planners worked in the South of the 1920s. Chicago, however, was essentially a transit metropolis, built by the age of rail technology, and so responded somewhat differently than the southern cities to the coming of the automobile. See also Howard L. Preston, *Automobile Age in Atlanta: The Making of a Southern Metropolis, 1900–1935* (Athens, 1979); Mark Foster, "The Model T, The Hard Sell, and Los Angeles' Urban Growth: The Decentralization of Los Angeles During the 1920's," *Pacific Historical Review* 44, no. 4 (1975): 459–84; Clay McShane, "Transforming the Use of Street Space: A Look at the Revolution in Street Pavements," *Journal of Urban History* 5 (May 1979): 291–96.

The literature on the automobile itself is large. Standard works include John Bell Rae, *The Road and Car in American Life* (Cambridge, Mass., 1971) and James Flink, *The Car Culture* (Cambridge, Mass., 1975) and *American Adopts the Automobile* (Cambridge, Mass., 1970). Pro–mass transit views of the auto's rise include James V. Cornehls and Delbert A. Taebel, *The Political Economy of Urban Transportation* (Port Washington, 1977), and Robert B. Carson, *What Ever Happened to the Trolley?* (Washington, D.C., 1978).

4. On the social burdens imposed upon mass transit, see Cornehls and Taebel, *Political Economy of Urban Transportation*, pp. 42 ff.

5. Useful comparisons may be made with Albro Martin, *Enterprise Denied: The Origins of the Decline of American Railroads, 1897–1917* (Cambridge, Mass., 1971), and Stanley Mallach, "The Origins and Decline of Urban Mass Transportation in the United States, 1890–1930," *Urban Past and Present* 8 (1979): 1–17.

6. The preceding two paragraphs utilize concepts developed by Samuel P. Hays, "The Changing Political Structures of the City in Industrial America," *Journal of Urban History* 1, no. 4 (Nov. 1974): 6–36.

7. On the background of the planning movement, see Stanley K. Schultz and Clay McShane, "To Engineer the Metropolis: Sewers, Sanitation and City Planning in Late-Nineteenth-Century America," *Journal of American History* 65, no. 2 (Sept. 1978): 389–401; Mark S. Foster, "City Planners and Urban Transportation: The American Response, 1900–1940," *Journal of Urban History* 5, no. 3 (May 1979): 365–96; Jon Peterson, "The Impact of Sanitary Reform upon American Urban Planning, 1840–1890," *Journal of Social History* 13 (Fall 1979): 83–103. See also Peterson's "The City Beautiful Movement: Forgotten Origins and Lost Meanings," *Journal of Urban History* 2, no. 4 (Aug. 1976): 415–34.

8. The view that America's adoption of the automobile was somehow the result of trickery or coercion by planners and/or the motor industry is evident in Emma Rothschild, *Paradise Lost: The Decline of the Auto-Industrial Age* (New York, 1973); Robert Goodman, *After the Planners* (New York, 1971); Kenneth R. Schneider, *Autokind vs. Mankind* (New York, 1972); Helen Leavitt, *Superhighway–Superhoax* (Garden City, N.Y., 1970). These works are merely suggestive; the literature is immense. See also Bradford C. Snell, "American Ground Transport: A Proposal for Restructuring the Automobile, Bus, and Rail Industries, February, 1974," in U.S. Congress, Senate Subcommittee on Antitrust and Monopoly of the Committee of the Judiciary, *Hearing on Bill 1167*, 93rd Cong., 2d sess., 1974, pt. 4, pp. A21–27. One good antidote is Herman Mertins, *Urban Transportation Policy: Fact or Fiction?* (Syracuse, 1970). On East Coast mass transit see Cheape, *Moving the Masses*.

Chapter 1

1. Clarence Darrow, "The Chicago Traction Question," *International Quarterly* 7 (Aug. 1905): 22.

2. The most thorough review of the evolving relationships between cities and their transit companies is Delos F. Wilcox, *Municipal Franchises: A Description of the Terms Upon Which Private Corporations Enjoy Special Privileges in the Streets of American Cities* (Chicago, 1911). See also Delos F. Wilcox, "The New York Subway Contracts," *National Municipal Review* 2, no. 3 (July 1913): 375–91; C. Norman Fay, "The City Gets 55%," *Outlook* 92 (June 19, 1909): 407–11; John A. Fairlie, "The Street Railway Question in Chicago," *Quarterly Journal of Political Economy* 21, no. 2 (May 1907): 371–404.

Most cities sought regulation by contractual agreement with their transit companies (as opposed to regulation by commission). Regulatory commissions, while more flexible, tended to be guided more by the economics of the transit industry and by changing political conditions than by planning goals. Companies which favored regulation generally sought a commission form of regulation. See Wilcox, "The New York Subway Contracts"; Forrest McDonald. "Samuel Insull and the Movement for State Utility Regulatory Commissions," *Business History Review* 32 (Autumn 1958): 241–54; and Forrest McDonald, *Insull* (Chicago, 1962), pp. 113–23.

3. On the roots of popular conceptions about mass transit in Chicago, see R. David Weber, "Rationalization and Reformers: Chicago Local Transportation in the Nineteenth Century" (Ph.D. diss., University of Wisconsin, 1971), esp. chaps. 3 and 4; Sidney I. Roberts, "Portrait of a Robber Baron, Charles Tyson Yerkes," *Business History Review* 35, no. 3 (Fall 1961): 344–72. On turn-of-the-century conditions, see Bion Joseph Arnold, *Report on the Engineering and Operating Features of the Chicago Transportation Problem* (submitted to the Committee on Local Transportation of the Chicago City Council, Nov. 1902) (New York, 1905), pp. 26, 32 (hereafter cited as Arnold, *1902 Report*).

Evaluations of Chicago's traction system as the nation's worst include Harry Pratt Judson, "The Municipal Situation in Chicago," *American Monthly Review of Reviews* (hereafter *AMRR*) 27, no. 4 (April 1903): 434–35; Victor Yarros, "Chicago's Significant Election and Referendum," *AMRR* 29, no. 5 (May 1904): 586; Graham Taylor, "Popular Value of a Supreme Court Decision," *Charities* 16, no. 1 (April 7, 1906): 9–10.

4. *Chicago Daily News*, Nov. 26, 1906; *Chicago Tribune*, Nov. 27, 1906; *Chicago Record-Herald*, Nov. 27, 1906.

5. *Chicago Daily News*, Nov. 26, 1906.

6. *Chicago Record-Herald*, Nov. 27, 1906. See also *Engineering Magazine and Industrial Record* 30, no. 7 (July 1905): 882.

7. Arthur Robinson, "Proposed Inner Circle System, "*Journal of the Western Society of Engineers* (hereafter *JWSE*) 11 (Fall 1906): 592–604, esp. p. 600.

8. The literature on the early development of urban mass transit is immense, though most of that produced before 1960 focused on the work of urban reformers. A good introduction to the subject is Glen E. Holt, "The Changing Perceptions of Urban Pathology: An Essay on the Development of Mass Transit in the United States," in *Cities in American History*, ed. Kenneth T. Jackson and Stanley K. Schultz, (New York, 1972), pp. 324–43, esp. pp. 324–33.

9. *Chicago Economist* supplement, *History and Statistics of Chicago Street Railway Companies* (Chicago, 1896), pp. 7–8; [John Maynard Harlan, William S. Jackson, Adolpuhs W. Maltby, William T. Maypole, Carter H. Harrison (mayor and chairman *ex-officio*)], Street Railway Committee of the City Council of the City of Chicago, *Report of the Street Railway Committee of the City Council of the City of Chicago . . . , March 28, 1898* (Chicago, 1898), pp. 167–70, 207–9, 231–33 (hereafter cited as Harlan, *Report*).

10. The average Chicagoan rode mass transit 164 times per year in 1890 and 320 times in 1910 (Chicago Traction and Subway Commission, *Report to the Honorable the Mayor and the City Council of the City of Chicago on a Unified System of Surface, Elevated and Subway Lines* [Chicago, 1916], p. 80 [hereafter cited as *1916 Report*]). By the roughest estimate, this means that about half the population could have used the transit system on each work day in 1910, about one-fourth in 1890. On this subject in general, see Theodore Hershberg, Harold L. Cox, Dale Light, Jr., and Richard R. Greenfield, "The 'Journey to Work': An Empirical Investigation of Work, Residence and Transportation, Philadelphia, 1850 and 1880," in *Philadelphia: Work, Space, Family and Group Experience in the Nineteenth Century*, ed. Theodore Hershberg (Oxford, 1981), p. 143. On the situation in Milwaukee, see McShane, *Technology and Reform*, p. 59.

By 1906, fares were probably not the chief obstacle to transit use for most citizens. Industrial wages in 1906 were 24% above the 1890–1899 average, and food prices had risen 16%, while transit fares remained stable. (U.S. Department of Commerce and Labor, Bureau of Labor Statistics, *Bulletin XV, 1907* [Washington, D.C., 1907], p. 6).

On Polonia, see Edward Kantowicz, "Polish Chicago: Survival Through Solidarity," *The Ethnic Frontier: Essays in the History of Group Survival in Chicago and the Midwest*, ed. Melvin G. Holli and Peter d'A. Jones (Grand Rapids, 1977), p. 183; on Packingtown, see Glen E. Holt and Dominic A. Pacyga, *Chicago: A Historical Guide to the Neighborhoods: The Loop and the South Side* (Chicago, 1979), pp. 121–32. Ethnic neighborhoods were by no means homogeneous; see Thomas Lee Philpott, *The Slum and the Ghetto: Neighborhood Deterioration and Middle Class Reform in Chicago, 1880–1930* (New York, 1978), pp. 22–23, 130–45.

11. *Chicago Economist*, p. 8; Bion J. Arnold, *1902 Report*, pp. 24–26, 31, 43–44; *Chicago Post*, Nov. 3, 1893; *Chicago Times-Herald*, Jan. 14, 1896; *Chicago Herald*, Sept. 17, 1894.

12. City of Chicago, Department of Public Works, *32nd Annual Report . . . 1907* (Chicago, 1907), p. 403.

13. Arnold, *1902 Report*, p. 32.

14. For example, the remarks of street railway representative E. R. Bliss in Chicago City Council, Committee on Local Transportation, *Proceedings* V (July 8, 1904): 43–44. The proceedings of this committee are hereafter referred to as CLT, *Proceedings*. These records (in the Chicago Municipal Reference Library) cover nearly 50 years and lack uniform pagination, since they were kept by a variety of secretaries. The highest number on a given page is always used, unless for some reason this would be confusing.

15. *Chicago Record-Herald*, Nov. 27, 1906. See also *Electric Railway Journal* 33, no. 1 (Jan. 2, 1909): 12 (hereafter cited as *ERJ*).

16. The literature is vast. Again, Glenn Holt provides a good synopsis in "Changing Perceptions," pp. 328–35. Early views of the traction industry were

conditioned by reformist disapproval of the fact of municipally granted monopolies of public service, as well as by the actual inadequacies of transit service.

Good examples of late nineteenth- and early twentieth-century assessments of the rise of urban street railways include Edward W. Bemis, ed., *Municipal Monopolies* (New York, 1899); Frank Parsons, "Municipal Ownership and Operation of Street Railways," *Arena* 25, no. 2 (Feb. 1901): 198–200; Frederick C. Howe, *The City: Hope of Democracy* (New York, 1905); Ida Tarbell, "How Chicago Is Finding Herself," *American Magazine* 67, no. 1 (Nov. 1908): 124–38. Two classic contemporary government-sponsored studies of the problem outside of Chicago are State of Massachusetts, General Court, Special Commission of Street Railways, *Special Commission appointed to Investigate the Relationship Between City Transportation and Street Railway Companies*: Report, February, 1898 (Boston, 1898), and Milo R. Maltbie, *Report on Indeterminate Franchises for Public Utilities* (submitted to Public Service Commission for First District, Dec. 29, 1908) (New York, 1908). On the role of consumer concerns in the rise of Progressivism and the demand for regulation see David P. Thelen, *The New Citizenship: Origins of Progressivism in Wisconsin, 1885–1900* (Columbia, 1972). On street railways as such, see Stanley Mallach, "The Origins and Decline of Urban Mass Transportation in the United States, 1890–1930," *Urbanism Past and Present* 8 (Spring and Summer 1979): 1–17.

17. *1916 Report*, pp. 70–78. Chicago's transportation problems were complicated by the fact that its population density was less than that of any other city of over 500,000.

18. Ibid., p. 80.

19. Chicago City Railway, *The Humphrey Bills and Comparison of American Street Railways* (Chicago, 1897), p. 17; *1916 Report*, p. 89.

20. Some typical examples only: Citizens Association of Chicago, *Report of the Committee on Bridges and Street Railways*, (Chicago, 1884), pp. 2, 3, 5; *Chicago Tribune*, Jan. 4, 7, 10, 1886; *Chicago Evening Post*, April 19, 1893; *Chicago Daily News*, Feb. 1893; *Chicago Tribune*, Feb. 24, 1893; *Chicago Tribune*, Jan. 26, 1896, July 5, 1897.

21. Early street railway contract ordinances include: City Council of the City of Chicago, "Ordinance to the Chicago City Railway, August 1, 1858" (manuscript in the Archives of the Chicago Surface Lines, Chicago Historical Society, box 1, folder 1, doc. 12, sec. 4–10); North Chicago City Railway, "Corporate Minutes" (manuscript in Chicago Surface Lines archives), p. 72; see also Harry P. Weber, *Outline History of Chicago Traction* (Chicago, 1936), pp. 3–7.

22. *Chicago Economist*, p. 8; on bribery, see Roberts, "Portrait of a Robber Baron"; on the continuing awareness of nineteenth-century corruptionism, see *Chicago Tribune*, Dec. 30, 1905 (obituary of traction magnate Charles Tyson Yerkes).

23. *Chicago Economist*, p. 8; R. David Weber, "Rationalization and Reformers," pp. 5–7, 17, 119–20, 242–47; Mayor Carter Henry Harrison II in *Chicago Tribune*, June 29, 1897.

24. Yerkes' strongest modern defender is Weber, "Rationalization and Reformers," pp. 242–51, 275–308. Weber contends that Yerkes' methods of financing improvements were not properly understood by the "old-fashioned" reformers. Nonetheless, Chicago's street railways by 1907 *were* more heavily capitalized than those of most other cities. See also George A. Schilling, *The Street Railways of Chicago and Other Cities* (Chicago, c. 1897), p. 45.

25. On the political and reform consensus for regulation, see George C. Sikes,

George Hooker, and the Chicago Federation of Labor (all on the pro-municipal ownership side of the traction reform movement) in *Chicago Chronicle*, Dec. 28, 1895; *Chicago Journal*, May 4, 13, 1897; John H. Hamline, *A. Prophecy, and Its Fulfillment by the Supporters of the Mueller Bill* (Chicago, 1903), p. 1. Even anarchist Lucy Parsons concurred (*The Liberator*, March 18, 1906). The mainstream view (for private ownership) is reflected in Chicago City Council Street Railway Commission, *Report of the Street Railway Bill as Submitted to the City Council of the City of Chicago, December, 1900* (Chicago, 1901), pp. 14–15.

26. *Chicago Daily News*, May 18, 1906; *Chicago Tribune*, June 4, 1906; *Chicago Journal*, Sept. 8, 10, 1906, and *Chicago Tribune*, July 30, Sept. 17. 1906. See also *Chicago Journal*, July 13, Aug. 22, 1906; *Chicago News*, June 9, 18, 29, 1906; *Chicago Tribune*, May 19, June 9, 18, 24, 29, July 4, 7, 23, 29, 30, Aug. 8, 13, 21, Sept. 17, 21, Oct. 25, 1906; *Chicago InterOcean*, Aug. 20, 1906.

Examples of accidents clearly resulting from negligence or impatience on the part of streetcar passengers include *Chicago Tribune*, June 4, 18, 29, 1906, Sept. 18, 1906; *Street Railway Journal* 27, no. 16 (Oct. 20, 1906).

On the number of accidents in Chicago in 1906, see *Chicago Tribune*, June 24, Dec. 20, 1906. A summary is given in Maurice F. Doty, "Dangers of Chicago Transportation" (undated flier, probably c. 1907, in Hooker Collection, University of Chicago, Chicago, Ill.). A total of 1,314 Chicagoans were injured between Aug. 9 and Dec. 31, 1906. The situation was not unique to Chicago: in New York City during 1907, 42 people died in one 27-day period as a result of mass transit accidents (Charles N. Glaab and A. Theodore Brown, *A History of Urban America* [New York, 1976], p. 144).

27. *Chicago Tribune*, Jan. 11, 1893. The remark became so famous that it was ultimately considered doubtful that Yerkes had ever said it. See Holt, "Changing Perceptions," p. 335.

28. Henry Demarest Lloyd, *The Chicago Traction Question* (Chicago, 1903), p. 41.

29. U.S. Department of Commerce and Labor, Bureau of the Census, *Special Reports: Street and Electric Railways, 1907* (Washington, D.C., 1907), pp. 331–85, esp. p. 337.

30. The documents are: *Chicago Economist* supplement; Harlan, *Report*; George Schilling, "Street Railways of Chicago and Other Cities" (part of *The People Against the Humphrey Bills*, issued by the Campaign Committee of the Committee of 100 Against the Humphrey Bills) (Chicago, 1897); Milo Ray Maltbie, *The Street Railways of Chicago: Report of the Civic Federation* (Chicago, 1901).

31. Yerkes—as presented by the press—was, if anything, too blunt in his dealings with the organized public. See, for example, *Chicago Tribune*, Jan. 12, 1893; Jan. 21, Feb. 7, 1894; *Chicago Evening Post*, Dec. 11, 22, 1893; *Chicago Chronicle*, Oct. 24, 1895.

Yerkes' dealings with fellow members of the business community were little more amicable. See, for example, the record of a long altercation, conducted by mail, between Yerkes and the president of the independent Chicago City Railway Company (letters in Chicago Surface Lines Archives [hereafter, CSL Archives], box 8, folder 53 of the North Chicago Street Railway files, dated between Oct. 18 and Dec. 23, 1889). See also H. Wayne Andrews, *The Battle for Chicago* (New York, 1946), p. 196.

32. *Chicago Tribune*, Nov. 27, 1906; *Chicago Record-Herald*, Nov. 27, 1906; *Chicago Journal*, Nov. 27, 1906.

33. See Robert Woodbury, "William Kent, Progressive Gadfly" (Ph.D. diss., Yale University, 1967), p. 85. On the varieties of urban reformers, see Melvin G. Holli, *Reform in Detroit: Hazen S. Pingree and Urban Politics* (New York, 1969), pp. 157–83. See also Edwin Burrett Smith, "Council Reform in Chicago," *Municipal Affairs*, June 1900, pp. 347–62.

The Humphrey Bills were the 50-year franchise legislation sponsored by the traction companies. The best, easily available account of their content and the struggle against their passage is in Roberts, "Portrait of a Robber Baron." McDonald (*Insull*, pp. 86–87) describes the Humphrey Bills as the first attempt at "modern commission-style regulation."

The bills themselves are in *The Humphrey Bills as They Are and Not as They Are Falsely Said to Be by the Press* (anonymous pamphlet, copy in Chicago Historical Society) (Chicago, 1897), pp. 7–16. The press labeled these measures "Yerkes Bills," although all the street railway companies suported them. The major non-Yerkes company, Chicago City Railway, included many city luminaries (Field,, Armour, Leiter, Allerton) among its large stockholders, and simply did not make an effective target (Chicago City Railway, Stock Ledgers, June 15, 1899, in CSL Archives, box 3, folder 19; Carter Henry Harrison II, *Stormy Years* [New York, 1935], p. 138). See also *Chicago Tribune*, Feb. 19, March 24, May 7, 1897; Roberts, "Portrait of a Robber Baron," p. 356; Chicago Federation of Labor, *Organized Labor Against the Humphrey Bills* (leaflet, April 4, 1897); Municipal Voters' League, *The Campaign of 1897* (Chicago, 1897).

The legislature shelved the Humphrey bills but promptly produced and passed new legislation authorizing the city council to grant franchise of up to 50 years' length (*Chicago Tribune*, May 25, 30, 1897; June 2, 3, 5, 1897). This new legislation, called the "Allen Law," thus placed the political struggle back in the city council and precipitated the MVL's campaign of 1898 described in the text below.

34. Smith, "Council Reform," pp. 3, 9; Samuel Wilbur Norton, *Chicago Traction* (Chicago, 1907), pp. 74–76; *Chicago Tribune*, Dec. 6–21, 1898; Independent Anti-Boddle league, *Listen to the Voice of the People* (leaflet, 1898, Chicago Historical Society).

Weber ("Rational and Reformers," pp. 336–44) presents the evidence for the superficiality of Harrison's commitment to traction and council reform. Yerkes' reaction to his defeat is in *Chicago Tribune*, Dec. 20, 1898.

35. Tarbell, "How Chicago Is Finding Herself," p. 516.

36. William Bennett Munroe, *The Government of American Cities* (New York, 1913), p. 370; Lincoln Steffens, *Shame of the Cities*, (New York, 1957), p. 165.

37. Steffens, *Shame of the Cities*, p. 165–67, 189; editorial, *AMRR* 23 (May 1901): 515–16; Hoyt King, "The Reform Movement in Chicago," *Annals of the American Academy of Political and Social Science* (hereafter *AAAPSS*) 21 (March 1905): 235–49, esp. p. 242.

38. *Chicago Tribune*, April 15, 1896. see also Joel Tarr, "William Kent to Lincoln Steffens: The Origins of Progressive Reform in Chicago," *Mid-America* 47, no. 1 (Jan. 1965): 55. Municipal Voter's League, *The Municipal Campaign of 1897* (Chicago, 1897).

39. A variety of Harrison positions on the traction situation: see *Chicago Tribune*, April 1, 1899; Carter Henry Harrison II, *Message to the City Council of the City of Chicago* (Chicago, Jan. 6, 1902); (Chicago City Council, *Journal . . . 1903–1904*, Aug. 24, 1904, pp. 1145–49). On Harrison's early ignorance of the traction issue,

see Harrison, *Stormy Years*, p. 101. On the unread engineering report, see Harrison's statement in CLT, *Proceedings* IV (Feb. 5, 1903): 22 of second numbered set of this meeting.

40. On Dever's background, see *Commons* (Feb. 1902).
41. *Chicago Tribune*, Feb. 12, 1906. The power of mass transit to improve urban life was a frequent theme among municipalizationists as among city planners. See, for example, John Burns (British labor advocate), "Municipal Ownership a Blessing," *Independent* 60 (Feb. 22, 1906): 449: "Men like manure are no good in heaps; they are only good when scattered over fresh fields." Representatives of the Chicago Federation of Labor quoted or paraphrased in *Chicago Tribune*, April 4, 1894, May 11, 1897, April 12, 13, 1906, Feb. 18, 1907; *Chicago Journal*, Feb. 10, 1896; CLT, *Proceedings* 7 (Jan. 1, 1907): 1941–65.

This position should be distingusihed from that grounded in the belief that urban utilities as "natural monopolies" were thus proper subjects for government ownership because they were not effectively regulated by the market. See Richard T. Ely, "The Growth and Future of Corporations," *Harper's New Monthly Magazine* 71 (1887): 259–66; Edward W. Bemis, "Municipal Ownership of Gas in the United States," *Publications of the American Economic Association* 11, no. 4 (1891): 1–185.

42. Charles Tyson Yerkes, *Investments in Street Railways: How Can They Be Made Secure and Remunerative?* (reprinted from the *Report* of the 18th Annual Meeting of the American Street Railway Association, Chicago, Oct. 17–20, 1899), p. 6. See also Foster, *Streetcar to Superhighway*, pp. 37–38; Tarr, "City to Suburb."

43. Citizens' Association of Chicago, *Report of the Committee on Tenement Houses* (Chicago, 1884), p. 4.
44. Robert Hunter, *Poverty* (Chicago, 1965), p. 343.
45. Adna Weber, *The Growth of Cities in the Nineteenth Century* (Ithaca, 1899), pp. 470–71; Delos F. Wilcox, *Municipal Franchises* (New York, 1911), 2: 7, Maltbie, *Report on Indeterminate Franchises . . .* , p. 9.
46. Richard McCulloch, "Notes on European Tramways," *JWSE*, 8 (1903): 293.
47. Wilcox, *Municipal Franchises*, 2: 8; Citizens' Association, *Tenement Houses*, pp. 18–19.
48. *Chicago Daily News*, Feb. 10, 1893; Harlan, *Report*, p. 71–73.
49. Arnold, *1902 Report* (from appendices A and B); Bion J. Arnold, *Maps on the Chicago Transportation Problem* (New York, 1905), maps A through C.
50. *Chicago Tribune*, Jan. 5, 1902, May 11, 1897; H. D. Lloyd in *Chicago Tribune*, June 2, 1903. Holli, *Reform in Detroit*, pp. 101–24, 50–52, 109–10.
51. U.S. Department of Commerce and Labor, Bureau of Labor Statistics, *Bulletin XV, 1907* (Washington, D.C., 1907), pp. 6, 141–42, 232, 237, 271; J. C. Kennedy, *Wages and Family Budgets in the Stock Yards District* (Chicago, 1914), p. 7; William C. Hard, "The Stock Yards Strike," *Outlook* 77 (Aug. 13, 1904): 887. (The 1904 strike centered on demands for 20¢ per hour wage for unskilled workers.)

Contemporary studies suggest that fare cuts were the least promising transit reform proposal in terms of their potential to raise the incomes of the poor. Income rather than transit was the key to dispersion (Roger Eugene Hamlin, "Transportation and Income Distribution: A Comparison of Policies" [Ph.D. diss., Syracuse University, 1973], pp. 267–72; Altschuler, *The Urban Transportation System*, p. 309).

52. *Chicago Tribune*, Jan. 10, 1886; Bion J. Arnold, "Presidential Address,"

JWSE 11 (Winter 1906): 124. Arnold pionerred the multiple unit-single control electric rapid transit train (*Scientific American* 88 (Nov. 15, 1902): 322.

53. CLT, *Proceedings* 15 (Dec. 2, 1908): 4424.

54. CLT, *Proceedings* 12 (July 3, 1907): 3143.

55. *Chicago Tribune*, July 22, 1906.

56. CLT *Proceedings* IX (Nov. 17, 1905): 6th page from end of meeting.

57. James Leslie Davis, *The Elevated System and the Growth of North West Chicago* (Evanston, Ill., 1965).

58. CLT, *Proceedings* 43 (June 24, 1913): 656; *ERJ* 36, no. 11 (Nov. 1910): 1148; C. V. Weston, "Action Necessary to Assure a Reasonable Return on Investment," *ERJ* (Dec. 26, 1908, Jan. 2, 1909) (reprinted as a pamphlet, New York, 1909. See also Edward Higgins, "The Larger Transportation Problem in Cities," *Municipal Affairs* 3, no. 10 (June 10, 1898): 238.

59. CLT, *Proceedings* 29 (Sept. 29, 1911): 8479–84.

60. *Chicago Tribune*, Dec. 17, 1898.

61. CLT, *Proceedings* V (July 8, 1904): 43–44.

62. Among the sources illustrating the range of interest in municipal ownership are *Municipal Ownership Bulletin* 1, no. 7 (March 11, 1899): 4 (Altgeld's campaign organ); *Chicago Tribune*, Feb. 10, 1898, April 2, 10, 1899, Jan. 5, April 2, 1902, Jan. 18, 1906; Clarence S. Darrow, "The Chicago Traction Question," *International Quarterly* 12 (Winter 1905): 13–22; Federation of Improvement Clubs of Chicago, "Open Letter," March 31, 1906, Hooker Collection, vol. 3; Chicago Federation of Labor, "Letter to Members," March 18, 1907, Hooker Collection, vol. 3; George C. Sikes, *The Street Railway Situation in Chicago: A Paper Read Before the National Municipal League* (Boston, 1902), pp. 15–16; Ralph E. Heilman, "Chicago Traction: A Study of the Efforts of the Public to Secure Good Service," *American Economic Association Quarterly* 9, no. 2 (1906): 93; Lloyd, *Chicago Traction*, p. 35.

63. *Municipal Affairs* IV, no. 12 (Dec. 1900): 788–90; *AMRR* 23, no. 5 (May 1901): 515; *Municipal Journal and Engineer* 12, no. 1 (Jan. 1902): 13.

64. [Chicago City Council Street Railway Commission] *Report of the Street Railway Commission to the City Council of the City of Chicago*, (Chicago, 1900) (Hereafter Street Railway Commission, *Report . . . 1900*) p. 11.

65. Ibid., pp. 14–16, 17–18.

66. Ibid., pp. 29–32, esp. p. 31.

67. Ibid., pp. 14, 19–23, esp. p. 22.

68. Ibid., pp. 24–29, 61. On the segregation of debt incurred for mass transit improvements, see p. 34.

69. Ibid., p. 50.

70. On Arnold's background, see *Natural Cyclopedia of Biography* 14 (New York, 1910), pp. 62–64.

71. Arnold, *1902 Report*, p. 14.

72. Ibid., pp. 24–26, 30.

73. Ibid., pp. 21, 30, 107, 112.

74. Ibid., pp. 22, 49, 107–14.

75. Lloyd, *Chicago Traction*, p. 40; see also CLT, *Proceedings* IV (Feb. 5, 1903): 2d page 5 (there are two sequences in this meeting). On the report's reputation, see Tarbell, "How Chicago Is Finding Herself," pp. 36–37; George C. Sikes, "The

228 *Notes to Chapter 1*

New Chicago," *Outlook* 92 (Sep. 28, 1909): 100–101; H. A. Millis, "The Present Street Railway Situation in Chicago," *AAAPSS*, Sept. 1902, p. 566.

76. CLT, *Proceedings* VI (Nov. 9, 1905): 42–46 of meeting; X (Nov. 28, 1905): 16 of meeting; 10 (Jan. 14, 1907): 2618.

77. See especially pp. 26–27 of *1902 Report*, which reflects the author's optimism about the power of engineering to resolve the transportation dilemma. See also Arnold's statement in Paul R. Leach, *Chicago's Traction Problem* (Chicago, 1925), p. 4.

78. *Chicago Daily News*, Nov. 27, 1906; *Chicago Tribune*, Jan. 11, 1899. George Hooker, "Paper read at the Merchants' Club of Chicago," Dec. 8, 1900; CLT, *Proceedings* IV (Feb. 4, 1903): 42 (Feb. 5, 1903): 17, 15, (Feb. 16, 1903): 27; 10 (Jan. 14, 1907): 2659.

From the outset, engineers were expected to produce results in keeping with council policy. See Arnold's instructions in *1902 Report*, p. 10, which called for him to "show the necessity for and entire practicability of the abandonment of operating cars in trains . . ." Arnold complied (p. 38).

79. *Chicago Tribune*, Jan. 6, April 2, 1902. The vote was 142,000 to 28,000.

80. *Chicago Tribune*, May 21, 1903. See also John Maynard Harlan in *Tribune*, May 15, 1903.

81. *Chicago Daily News*, Jan. 25, 1903; *Chicago Tribune*, March 20, April 8, 24, May 1, 2, 3, 4, 7, 8, 19, June 15, 1903.

82. "An Act to Authorize Cities to Acquire, Construct, Own and Operate and Lease Street Railways, and to Provide the Means, Therefore," in *Acts of February 14, 1859, February 21, 1861, February 6, 1865, and Other Legislative Acts and Constitutional Provisions Involved in Chicago Street Railway Litigation* (Chicago, n.d.), pp. 36–44 (copy in Chicago Historical Society).

83. This is the author's opinion, based on subsequent events. See also John H. Hamline, *A Prophecy and Its Fulfillment by the Authors of the Mueller Bills* (Chicago, n.d. [c. 1905], esp. pp. 28, 31, 44.

84. On the receivership of the former Yerkes companies, see *Chicago Tribune*, Jan. 22, Feb. 12, June 19, 1903. J. P. Morgan gained control of the Chicago City Railway: *Chicago Record-Herald*, Jan. 12, 1905; *Chicago Daily News*, March 27, 1905; *Electric Railway Journal* 24, no. 25 (Dec. 25, 1909): 1280.

On the state of Chicago's traction service between 1902 and 1907, see H. A. Millis, "The Present Street Railway Situation in Chicago," *AAAPSS* 14 (Sept. 1902): 556; editorial, *Scientific American* 96 (May 25, 1907): 430; editorial, *Street Railway Journal* (hereafter, *SRJ*) 25, no. 1, (April 15, 1905): 693. On the other hand, see *AMRR* 31, no. 4 (April 1905): 387–90, for photos and descriptions of New York City streetcar conditions that rival anything reported in Chicago.

85. The CLT's first major published document set forth what was to be the council's fundamental negotiating position through 1915 (*Report of the Special Committee on Local Transportation of the City Council of the City of Chicago* [Chicago, 1901]). See also CLT, Subcommittee on Chicago City Railway Extension Ordinance, *Report of the Subcommittee to the Committee on Local Transportation . . . of a Tentative Ordinance for the Chicago City Railway Company, with Exhibits* (Chicago, Nov. 14, 1903), esp. p. 30; *Chicago Record-Herald*, Aug. 1, 1903: *Chicago Tribune*, April 22, 1904. See also Heilman, "Chicago Traction," p. 57.

86. Arthur F. Bentley, "Municipal Ownership Interest Groups in Chicago: A Study of Referendum Votes, 1902–1907" (typescript in Regenstein Library, University of Chicago, Chicago, Ill., probably c. 1910).

87. See, for example, Albert Shaw, "A Discussion on Municipal Ownership," *Independent* 49, no. 2527 (May 6, 1897): 567; W. B. Munroe, "The Civil Federation Report on Public Ownership," *Quarterly Journal of Economics* 23, no. 4 (Nov. 1908): 161.

88. *Chicago Tribune*, Nov. 9, 1905. See also, for example, Hayes Robbins, "Public Ownership vs. Public Control," *American Journal of Sociology* 10, no. 5 (May 1905): 801–12; Frederick C. Howe, review of Hugo Richard Meyer's *Municipal Ownership in Great Britain* in *American Political Science Review* 1, no. 3 (May 1907): 475–77; E. W. Bemis, *Municipal Ownership of Gas in Great Britain* (Baltimore, 1891), 76–77, 150.

On corruption as a bar to public ownership in Chicago, see, for example, W. K. Ackerman, *Reporting of Investigation Into the Street Railways of Chicago Made to the Civic Federation* (Chicago, 1898), p. 28; *Chicago Tribune*, March 12, 1903; *Chicago Daily News*, March 2, 1907; *City Club Bulletin* 6, no. 3 (March 6, 1907): pp. 19, 20, 27, 34.

89. John Maynard Harlan and Edward F. Dunne, "Chicago's Campaign for Municipal Ownership," *Commons* 16, no. 5 (May 1905): 137–44. On Dunne's early support of public ownership, see *Chicago Tribune*, Dec. 10, 1898; E. F. Dunne, *The Facts on Municipal Ownership* (Chicago, 1903); Judge E. F. Dunne, *Municipal Ownership: How the People May Get Back Their Own* (address to the Men's Club of the Stewart Avenue Universalist Church of Englewood, Jan. 12, 1904) (Chicago, 1904).

90. The best source on Dunne is Richard Edward Becker, "Edward Dunne: Reform Mayor of Chicago" (Ph.D. diss., University of Chicago, 1971). On Dunne's early background, see pp. 14–15; see also "Municipal Ownership in Chicago," *Harper's Weekly* 49, no. 2522, (April 22, 1905): 568, 583. On the parallels and connections between Dunne's drive for public ownership and the national movement, see Becker, "Edward Dunne," pp. 87–88, and the sources there cited.

91. Becker, "Edward Dunne," p. 50; H. Webster, "From Yerkes to Dunne," *American Illustrated Magazine* 61, no. 4 (April 1906): 693–94; Dominic Candeloro, "The Chicago School Board Crisis of 1907," *Journal of the Illinois Historical Society* 69, no. 5 (Nov. 1975): 396–406; *Chicago Tribune*, Feb. 10, 18, 1907; *Chicago Tribune*, April 3, 1907, for the *Tribune*'s "long-haired men . . ." remark.

92. Dunne, *The Facts on Municipal Ownership*; E. F. Dunne, *Summary of Correspondence Between Judge E. F. Dunne and Officials of English Cities—Actual Municipal Ownership of Street Railways*, (Chicago, 1905); E. F. Dunne, *Address to the Democratic City Convention, February 25, 1905* (Chicago, 1925); Democratic Party Convention, *Platform*, Feb. 25, 1905 (Chicago, 1905); E. F. Dunne, "Chicago's Onward March," *Municipal Journal and Engineer* 21, no. 10 (Sept. 5, 1906): 232.

93. Charles S. Dunbar, *Busses, Trolleys and Trams"* (London, 1967), p. 48. Dunne's position on the profit motive and his misinterpretation of the Liverpool experience are in *Address to the Democratic City Convention*. See also Bernard Aspinwall, "Glasgow Trams and American Politics, 1894–1914," *Scottish Historical Review* 66, no. 161 (1977): 64–84.

94. *Chicago Tribune*, April 6, 1905.
95. Ibid., June 30, 1905, Feb. 1, 2, 1906; Becker, "Edward Dunne, pp. 69–73.
96. *Chicago Tribune*, March 19, 1906.
97. On Dunne's school board, see Candeloro, "Chicago School Board," pp. 397–401. On municipal ownership, see Eugene Prussing, *Municipal Ownership and Operation of Street Railways: An Address at a Meeting of the City Club of Chicago* (Elmira, 1906), p. 11. Prussing's argument was not an unusual one. See, for example, Robert P. Porter (director of the 11th U.S. Census), *The Dangers of Municipal Ownership* (New York, 1907), esp. p. 7, and chap. 3 (which treats Csarist Russia as an example of "municipal ownership"). In a similar vein is James Mavor, "Municipal Ownership of Public Utilities," *Publications of the Michigan Political Science Association* 5, no. 4 (March 1904): 2–6, 10. These citations are merely indicative.

A second contribution to the identification of public ownership of utilities with socialism was the use of the municipalization issue by socialists. See, for example, Paul Douglas, "The Socialist Vote in the Municipal Elections of 1917," *National Municipal Review* 7 (March 1918): 131–39. See also Chicago Federation of Labor President John Fitzpatrick, a former blacksmith and founder of the short-lived Illinois Labor Party (*Chicago Journal*, Nov. 3, 6, 1918).

Another factor was the rise of what may loosely be termed "ideological" debates over public ownership in Great Britain *after* many urban services there had been municipalized. See, for example, Hugo Richard Meyer, *Municipal Ownership in Great Britain* (New York, 1906), pp. 5–6, 25, 38–44, 98–99, 106-9, 193, 247–53, 273–77, and 309–14.

One more factor was the simple identity taken on by the two concepts in the public mind. See the *Reader's Guide to Periodical Literature*, 2: 1534, for the heading: "Municipal Socialism," which refers the reader to "Municipal Ownership and Control."

98. Ray Ginger, *Altgeld's America* (New York, 1973), pp. 292–93; *Chicago Chronicle*, Feb. 1, 1906; *Chicago Tribune*, Nov. 17, 1905.
99. *Chicago Tribune*, Nov. 2, Dec. 5, 1905, Feb. 1, 9, 26, 1906; James Dalrymple, *The Street Railways of Chicago* (Chicago, 1907), p. 3. Dunne had asked the Glasgow Tramways to send its manager to inspect the Chicago street railways at the suggestion of Tom Johnson. As it turned out, Dalrymple was not the man Johnson had had in mind.
100. *Chicago Tribune*, Feb. 23, March 19, Aug. 8, 1906.
101. *Chicago Tribune*, Aug. 17, 21, 25, 1905; CLT, *Proceedings* X (Nov. 28, 1905): 38, 40, 41–43 of meeting, first transcript (a second, censored transcript follows). *Chicago Daily News*, Jan. 12, 1906; *Chicago Tribune*, Jan. 16, 1906.
102. Becker, "Edward Dunne," p. 110; Millis, "Street Railway Situation in Chicago," p. 63.

On the referendum "deal" and reformers' reaction, see *Chicago Tribune*, Jan. 18, 19, April 11, 1906; *Outlook* 82 (Feb. 3, 1906): pp. 241–42.
103. An Impartial Observer, "Immediate Municipal Ownership in Chicago, a Year After," *AMRR* 32 (May 1906): 552–53; *Chicago Tribune*, March 19, April 5, 1906; *Outlook* 82 (April 14, 1906): 818–19. On the background of the "other Democrats"— the Sullivan-Hopkins Machine, see Joel A. Tarr, "The Urban Politician and Entrepreneur," in *Mid-America* 49, no. 1, (Jan. 1967): 65.

104. *Chicago Daily News*, Nov. 27, 1906; *Chicago Tribune*, Nov. 27, 1906.

105. *Chicago Tribune*, March 3, 1906.

106. On Fisher's background and on his appointment, see *Chicago Daily News*, April 11, 1906; *Chicago Tribune*, April 11, 1906; Alan B. Gould, "Walter Fisher: Profile of an Urban Reformer," *Mid-America* 57, no. 3 (July 1975): 157–73, esp. pp. 167–68.

107. *Chicago Tribune*, Jan. 8, 15, 27, Feb. 5, 1907.

108. Citizens' Non-Partisan Traction Settlement Association (hereafter CNPTSA), *The Traction Ordinances in a Nutshell* (flier, 1907, in Hooker Collection).

109. On the organization and backing of the CNPTSA, see *Chicago Tribune* (a friendly source), Feb. 9, 10, 12, 14, 15, 20, 26, 27, 28, March 2, 1907; *Chicago Record-Herald*, March 4, 1907; *Chicago Chronicle*, March 7, 8, 1907.

For a list of organizations supporting or invited to support the movement, see Citizens Non-Partisan Traction Settlement Committee [sic], letter dated Feb. 20, 1907, in Hooker Collection. Tomaz Deuther, head of a citywide organization of the neighborhood businessmen active after 1910 claimed in 1918 that most of the 100 organizations the CNPTSA listed as supporters did not in fact exist (Deuther, *Looking Backwards* [Chicago, 1918], also in Hooker Collection).

110. Republican Party Central Committee, *Platform* (Chicago, 1907), p. 1, in Hooker Collection. *Chicago Tribune*, Feb. 17, March 16, 1907. Walter Fisher worked actively for passage (*Chicago Tribune*, Jan. 19, Feb. 7, 8, 1907). For the Hearst position, see *Chicago American*, March 7, 1907.

111. CNPTSA, *The Traction Ordinances in a Nutshell*. See also *Chicago Tribune*, March 18, 1907.

112. *Chicago Examiner*, March 31, 1907; *Chicago Tribune*, Jan. 15, Feb. 18, March 20, 1907; *Chicago American*, March 7, 1907; *Democratic Municipal Platform of 1907* (Feb. 23, 1907), bottom of sheet; Chicago Federation of Labor, *Letter to Members* (March 18, 1907); Municipal Ownership Central Committee, *Notice to Friendly Storekeepers* (March 20, 1907). *Chicago Daily News*, March 3, 1906 (letter from Mayor Brad Whitlock of Toledo).

113. *Chicago Tribune*, March 16, 17, 1907.

114. The "outside" origin of these ideas is evident in a discussion in CLT, *Proceedings* VIII (April 22, 1905): 4, and in *Chicago Tribune*, May 18, 1906. Delos F. Wilcox traces the links in *Municipal Franchises*, p. 49.

On the claim that the ordinances allowed the city to buy the companies at will, see *Chicago Tribune*, Jan. 13, 1907; *Chicago Daily News*, March 2, 1907.

The opposite position is evident in an advertisement of Harris Trust and Savings reported by Willard Hotchkiss, "Recent Phases of Chicago's Traction Problem," *AAAPSS* 31 (May 1908): 627. *Municipal Journal and Engineer* 22, no. 7 (Feb. 13, 1907): 105. On the New York lawyers' view, see CLT, *Proceedings* 41 (April 26, 1913): 244.

115. *Municipal Journal and Engineer* 21, no. 16 (Oct. 17, 1906): *Chicago Tribune*, Oct. 2, 1906. CLT, *Report of the Committee on Local Transportation to the City Council of the City of Chicago, January 15, 1907, and the Ordinances to the Chicago City Railway Company and Chicago Railways Company, Passed by the City Council. February 11, 1907* (Chicago, 1907) (hereafter, *1907 Ordinances*), see pp. 61–67.

116. *1907 Ordinances*, pp. 35–38, 65–68; CLT, *Proceedings* 6 (Oct. 29, 1906): 860–61, (Oct. 31, 1906): 896–902. See also Walter Fisher, "The Traction Question, 1907–1920," *Chicago City Club Bulletin* 13, no. 5 (Dec. 20, 1920): 145.

117. CNPTSA, *We Are a Proud City: Help Us Become Proud of Our Streetcars* (flier, Hooker Collection).

118. Harlan *Report*, p. 65. See also Schilling, *Street Railways of Chicago*, pp. 27, 34.

119. Arnold's 1902 valuation is in *1902 Report*, p. 204. On securities outstanding in 1906, see U.S. Department of Commerce and Labor, *Street and Electric Railways* [1907] (Washington, D.C., 1908), pp. 80, 337. The companies' negotiating position is in *Chicago Tribune*, Dec. 11, 1906.

The valuation commission's report is Bion J. Arnold, Mortimer Cooley, and A. B. DuPont, *Report on the Values, Tangible and Intangible, of the Chicago City Railway Company and the Chicago Union Traction Company* (submitted to the Committee on Local Transportation of the City Council of the City of Chicago, Oct. 10, 1906) (Chicago, 1906). See especially pp. 11, 14, 31–33, 37, 49.

The nature of the compromise is explained by Arnold in "Transportation Problems in Major American Cities," *California Progressive*, May 1, 1911, pp. 12–13. See also CLT, *Proceedings* 28 (June 27, 1911): 7612.

120. *1907 Ordinances*, pp. 87–88, 90.

121. *Chicago Tribune*, Dec. 16, 1906, Jan. 3, 1907; *Chicago Record-Herald*, March 10, 1907; Walter Fisher, "The Traction Ordinances," *City Club Bulletin* 1, no. 3 (March 3, 1907): 34. On the history of the idea of a profit limitation, see U.S. Department of Commerce and Labor, *Street and Electric Railways*, p. 134.

122. *Chicago Tribune*, Jan. 3, 11, 12, 1907; George Hooker, "The Traction Ordinances," *City Club Bulletin* 1, no. 4 (March 13, 1907): 9.

123. *1907 Ordinances*, pp. 52–55, 69–73; *Scientific American* 96 (May 25, 1907): 430; CLT, *Proceedings* VI (Nov. 9, 1905): 42.

124. *1907 Ordinances*, pp. 61 ff.

125. Ibid.

126. Ibid., p. 89. The companies' New York financial backers resisted this concession for two months (*Street Railway Journal* 27, no. 17 [Oct. 27, 1906]: 863; *Chicago Tribune*, Dec. 16, 1906). Their resistance was softened, reported the *Tribune*, when they heard of the condition of Chicago's street railway service and nearly "wept with compassion" (*Chicago Tribune*, Dec. 29, 1906).

127. Strong support came from CBD merchants (*Chicago Daily News*, Dec. 29, 1906; *Chicago Tribune*, Dec. 29, 1906), and the Chicago Real Estate Board (*Chicago Tribune*, Dec. 26, 1906, Jan. 3, 18, Feb. 10, 12, 1907).

128. *Chicago Tribune*, Jan. 9, 1907; *Chicago Tribune*, Dec. 20, 21, 22, 23, 31, 1906, Jan. 3, 4, 5, 8, 11, 12, 1907. Foreman's remark is in CLT, *Proceedings* X (Jan. 14, 1907): 2737–38; see also *Chicago Tribune*, Jan. 8, 13, 14, Feb. 5, 11, 1907.

129. Clipping in Hooker Collection, vol. 3 (undated).

130. Sample card is in Hooker Collection, vol. 3. Fisher's remark is in *Chicago Record-Herald*, March 8, 1907.

131. All Chicago papers, April 3, 1907. See also Bentley, "Municipal Ownership Interest Groups," chap. 3, pp. 2–3; chap. 6, pp. 22–23; chap. 7, pp. 1–6.

132. *Chicago American*, April 3, 1907; *Chicago Examiner*, April 3, 1907; *Chicago Tribune*, April 3, 1907; *Chicago InterOcean*, April 3, 1907; *Chicago Record-*

Herald, April 3, 1907. *Outlook* 85, no. 10 (March 9, 1907): 537–38; "An Impartial Observer" (probably Victor Yarros), "Chicago Election and the City's Traction Outlook, by an Impartial Observer," *AMRR* 35, no. 5 (May 1907): 583.

133. *Chicago Daily News,* March 4, 1909, May 5, June 16, 1913. Bentley, "Municipal Ownership Interest Groups," chap. 6, p. 22; chap. 7, pp. 1–6; *Chicago Examiner,* March 1, 1907, and "Impartial Observer," "Chicago's Election," p. 582.

Chapter 2

1. Cited in George Hooker, "Traffic and the City Plan," *Charities and the Commons* 19 (Feb. 19, 1908): 1491.

2. *Cassier's Magazine* 31 (Feb. 1907): 456.

3. *Chicago Tribune,* Oct. 22, 1906. See also (*Chicago Daily News,* Nov. 28, 1906), and Board of Supervising Engineers, Chicago Traction, *6th Annual Report of the Board of Supervising Engineers, Chicago Traction, for the Year Ended January 31, 1913* (Chicago, 1915), p. 51; *9th Annual Report . . . 1916* (Chicago, 1917), pp. 191–193 (hereafter, BOSE, *Annual Report*[s]).

4. Lyman Cooley, "The Ownership of Public Utilities," *JWSE* 11, no. 1 (Feb. 1906): 4. Cooley was a Chicago engineer specializing in canal and sanitation projects.

5. *Chicago Tribune,* Jan. 12, Feb. 16, 19, April 15, Sept. 17, 1856.

[City of Chicago], *Ordinances of the City of Chicago* (Chicago, 1858), p. 375 (ordinance of Feb. 15, 1851).

6. *Chicago Tribune,* June 27, 1906. On the history of the Board of Local Improvement (BLI), see *BLI Staff Report #27* (Chicago, Oct. 26, 1923), p. 1. On the problem of street paving in general, see City Club of Chicago, *Civic Committee Reports* (1906–1907), p. 56; Pratt Judson, "The Municipal Situation in Chicago," *AMRR* 27 (June 1903): 434; *Chicago Journal,* May 26, 1906.

7. City Council, *Journal . . . 1907–1908* (Chicago, 1908), p. 3829; DPW, *33rd Annual Report . . . 1908,* (Chicago, 1909), p. 314; *Chicago Tribune,* Oct. 25, 1907.

On collections and resistance, see City of Chicago, *Vehicle Tax Fund: Statement Showing the Amount, Cost and Location of Street Repairs and Improvements Paid Out of the Fund Provided by Vehicle Licensees* (Chicago, Dec. 31, 1908), p. 1.

8. Ibid, pp. 1, 3 ff.; DPW, Bureau of Streets, *Report . . . 1909* (Chicago, 1909), table on p. 236; DPW, Bureau of Streets, *Report . . . 1913* (Chicago, 1913), p. 367; City Council, *Journal . . . 1910–1911* (Chicago, 1911), June 27, 1910, p. 943; *Chicago Commerce* (June 3, 1910), p. 3. For opposition to this policy, see Chicago Commission on City Expenditures, *Preliminary Report on Street Paving* (Chicago, 1910), pp. 32–33.

9. City Council, ordinance (dated Aug. 16, 1868), in CSL Archives, box 1, folder 12, item 2, p. 5 (hand copy in Chicago Historical Society); State of Illinois, *An Act to Promote the Construction of Horse Railways in the City of Chicago, February 14, 1859,* sec. 6 (in [Anonymous] *Acts of February 14, 1859, February 21, 1861, and February 6, 1865, and Other Legislative Acts and Constitutional Provisions Involved in the Chicago Street Railway Situation* (Chicago, c. 1904), p. 5. CLT, *Proceedings* VI (o.s.) (Nov. 9, 1905): 21 (on the history of streetcar right-of-way in Chicago ordinances. Arnold's statement is in Arnold, *1902 Report,* p. 48.

10. (*Swanson* v. *Chicago City Railway,* summarized in Frank B. Gilbert, *Street Railway Reports (Annotated), Reporting Electric Railway and Street Railway Decisions of Federal and State Courts in the United States* (Albany, 1904–14), 6: 456. Other

cases in which the same principle was applied are cited from Milwaukee (p. 464), Boston (p. 754), Birmingham (p. 235), Salt Lake City (p. 749), St. Paul (p. 713), Tacoma (p. 514), and three smaller cities (pp. 357, 543, 830).

11. *North Chicago Street Railway* v. *Cossar* (Gilbert, *Street Railway Reports*, 1: 94); *North Chicago Street Railway* v. *Johnson* (in ibid., 2: 91); *Chicago Union Traction* v. *Browdy* (in ibid., 2: 138); *Chicago City Railway* v. *O'Donnell* (in ibid., 2: 170–81); *Chicago City Railway* v. *Barker* (in ibid., 2: 182–89); *Chicago City Railway* v. *Kinnare* (in ibid., 4: 189).

12. BOSE, *9th Annual Report*, pp. 190–91.

13. On the degree of congestion, see DPW, 32nd Annual Report, (Chicago, 1908), p. 403; *Chicago Commerce* (July 3, 1908).

14. *Chicago Tribune*, Nov. 27, 28, 1906; *Chicago Record-Herald*, Nov. 27, 1906. See also Arnold *1902 Report*, p. 49. Simple overloading was also common (CLT *Proceedings* 30 [Jan. 17, 1912]: 9654). Wagon and express drivers commonly fed their horses in the street (*Chicago Commerce*, July 3, 1908).

On the attitudes of courts toward traffic obstructions, see CLT, *Proceedings* 30 (Jan. 24, 1912): 9696; 66 (Aug. 12, 1916): 33 (testimony of Police Superintendent Healey); and *Eckels* v. *Muttschall* (82 NE 872); *North Chicago Street Railway Company* v. *Smadraff* (59 NE 527).

15. Bion J. Arnold, *City Transportation: Subways and Railroad Terminals*, reprinted from *JWSE* 19, no. 4 (April 1914): 44.

16. Clay McShane, "American Cities and the Coming of the Automobile, 1870–1910" (Ph.D. diss., University of Wisconsin, Madison, 1975), pp. 111–13. On the complex transformation of the street, see Clay McShane, "Transforming the Use of Urban Space: A Look at the Revolution in Street Pavements, 1880–1924," *Journal of Urban History* 5 (May 1979): 291–96.

17. John W. Alvord, *The Street Paving Problem in Chicago: A Report to the Street Paving Committee of the Commercial Club* (Chicago, 1904), pp. 26–27; John W. Alvord, "Street Pavements in Chicago, "*JWSE* 10, no. 6 (Dec. 1905): 681–82; Fred Rex, "Municipal Paving: A Symposium," *AAAPSS* 29 (May 1907): pp. 565–66; *Municipal Journal and Engineer* 8, no. 3 (Nov. 1902): 225.

18. "A Former Commissioner of Public Works" (probably John McGann), "The Paving Question in the Chicago Teaming District," *American Team Owner* 1, no. 1 (Aug. 15, 1906): 7; CLT, *Proceedings* 13 (April 30, 1908): 400.

19. Alvord. *The Street Paving Problem*, pp. 26–30 and exhibit 20; *Chicago Tribune*, Jan. 1, 1906.

20. Alvord, *The Street Paving Problem*, exhibit 20.

21. Ibid., pp. 6–7, 8; Rex, "Municipal Paving," pp. 564–68; Chicago Commission on City Expenditures, *Preliminary Report on Street Paving* (Chicago 1910), pp. 32–33.

22. CLT, *Proceedings* 64 (Jan. 17, 1916): p. 34 of meeting.

23. In 1904, 11 percent of Chicago's streets were paved with asphalt or macadam, as compared with 41 percent of the streets of New York City. By 1915, smooth-surfaced materials made up about 65.4 percent of all Chicago pavements (Alvord, *The Street Paving Problem*, exhibit 2; Frances Eastman, *Chicago City Manual* [Chicago, 1915], pp. 166–67). On motor interests and paving, see, for example, Flink, *America Adopts the Automobile*, pp. 202–13).

In 1913 the streetcar line was the only pavement on major sections of an important diagonal street, Elston Avenue (CLT, *Proceedings* 45 [July 23, 1913]: 235. See also CLT, *Proceedings* 13 [March 16, 1908]: 390; 25 [Aug. 12, 1910]: 7341 of series.

24. Calculated from DPW, *32nd Annual Report*, p. 403.

25. Alvord, *The Street Paving Problem*, p. 9; City Club of Chicago, *Civic Committee Reports* (Chicago, 1906–7), p. 54; Chicago Bureau of Public Efficiency (hereafter CBPE), *Street Paving in the City of Chicago* (Chicago, 1911), p. 10.

26. DPW, Bureau of Engineering, *Report on Transportation Subways* (Chicago, 1909), 1: charts dated April 16 and 19, 1909, before p. 8; *American Team Owner* 1, no. 1 (Aug. 1906): 5. There were nearly 100,000 teams in Chicago in 1906, 686 of which were owned by major teaming contractors operating in the CBD.

27. *Chicago Commerce*, July 3, 1908; George Probst, "The Most Congested Street in the World," *American Team Owner* 1, no. 3 (March 1907): 5–6.

28. DPW, *32nd Annual Report*, p. 403; DPW Bureau of Streets, *Traffic Census of the Central Business District* (Chicago, 1915).

29. Sanding and salting were adopted by 1915 (CLT, *Proceedings* 59 (Jan. 23, 1915): 19–20; On rubber horseshoes, see CLT, *Proceedings* 64 (Nov. 23, 1915): 19. On the early motor truck and its advantages, see J. A. Kingnor, "The Automobile in Business," *AMRR* 29, no. 6 (June, 1904): 709–15. The number of motor trucks in Chicago rose from 3 in 1909 to 900 in 1911 (DPW, Bureau of Streets, *Traffic Census . . .* recapitulation table at end).

30. CLT, *Proceedings* LXIV (Jan. 17, 1916): 29–33, 35, 41–42; BOSE, *6th Annual Report*, pp. 450–51; Chicago Plan Commission, *Reclaim South Water Street for All the People* (Chicago, 1917), pp. 6–7; BOSE, *9th Annual Report*, 210; Commercial Club of Chicago, *Report on the Plan of Chicago* (Chicago, 1908), p. 14.

31. On The Illinois Tunnel Co., see Architectural League of America, *Report of the Committee on Civic Improvements* (Chicago, 1906), p. 2. The tunnels could handle little of the sort of freight which required wagons, and no railroad transfer business (BOSE, *9th Annual Report*, p. 157).

32. CLT, *Proceedings* 66 (Dec. 20, 1916): 22; 64 (Jan. 17, 1916): 29–31.

33. *Chicago Commerce* (May 19, 1911): 23–26, (April 25, 1913): 29–30.

34. BOSE, *9th Annual Report*, p. 223; DPW, Bureau of Engineering, *Report on Transportation Subways* (Chicago, 1909), unpaged table headed "Police Regulation of Surface Traffic."

35. *Chicago Commerce*, March 24, 1911, p. 6; March 29, 1912, p. 7; BOSE, *9th Annual Report*, pp. 226–28. The photographs of "staged" traffic jams are in *Chicago Commerce*, May 19, 1911, pp. 23–26.

36. On early cooperation between the CAC and the traffic control squad, see General Superintendent of Police (hereafter GSOP), *Report . . . 1908* (Chicago, 1908), pp. 81–82; *Chicago Commerce*, March 24, 1911, p. 6.

37. GSOP, *Report . . . 1909* (Chicago, 1909), pp. 80–81; GSOP, *Report . . . 1910* [Chicago, 1910], p. 31; *Report . . . 1912* [Chicago, 1912], p. 97. See also *Chicago Commerce*, Nov. 4, 1910, p. 8. The greatest inroads were made by the conversion of alleys to one-way thoroughfares in 1915; this allowed 75% of goods to be delivered by alley (CLT, *Proceedings* 66 [Aug. 12, 1916]: 27, 28).

38. GSOP, *Report . . . 1909*, p. 82; *Report . . . 1911* (Chicago, 1911), pp. 95–96 (the quotation).

39. Ibid., p. 91; *Chicago Commerce*, Nov. 4, 1910, pp. 8–18; March 24, 1911, p. 6; Oct. 20, 1911, p. 21.

40. George Elsworth Hooker, *Through Routes for Chicago Steam Railroads* (Chicago, 1914), p. 16.

41. DPW, *32nd Annual Report*, p. 403; DPW, Bureau of Streets, *Traffic Census . . . 1915*, recapitulation table.

42. On the planning response, see Foster, *Streetcar to Superhighway*, pp. 42–44. On the general reaction, see p. 20–23. See also Flink, *America Adopts the Automobile*, pp. 31–55; Rae, *The Road and Car*. There is a large body of literature on the impact of and response to the automobile in specific local areas.

43. Frank Munsey, "Impressions by the Way," *Munsey's Magazine* 29, no. 2 (May 1903): 181–83.

44. Joseph Pennell, "Motors and Cycles: The Transition Stage," *Contemporary Review* 81, no. 2 (Feb. 1902): 188.

45. Henry P. Norman, "The Coming of the Automobile," *World's Work* 27, no. 4 (April 1903): 481.

46. Ray Stannard Baker, "The Automobile in Common Use," *McClure's* 12, no. 3, (July 1899): 200–202; Winthrop Scarritt, "The Low Priced Automobile," *Munsey's Magazine* 29, no. 2 (May 1903): 177.

47. Scarritt, "Low Priced Automobile," p. 178; John Brisbin Walker, "The New Transportation," *The Independent* 55, no. 6 (June 1903): 1334–35.

48. *Chicago Tribune*, Jan. 11, 1907.

49. Norman, "Coming of the Automobile," p. 479; U.S. Department of Commerce and Labor, Bureau of Labor Statistics, *Bulletin Vol. XV*, pp. 238, 157. Statistics are for the three years after Norman's article. On used car prices, see advertisements in *Chicago Tribune*, Aug. 5–11, 1906.

50. *Chicago Tribune*, Sept. 17, 1906 (incident at Superior and Wells Streets); Aug. 29, 1906 (shooting incident); Feb. 1, 1906 (commissioner's remark).

51. Sixty-nine automobile licenses issued by the Chicago Bureau of Automobile Licenses in 1906 have been preserved in a manuscript labeled "Automobile Licenses 1–1,000 (Chicago Municipal Reference Library). The location by ward of automobiles paying the wheel tax in 1907 is reported in DPW, *32nd Annual Report*, p. 40. Tables are available from author.

52. *Chicago Tribune*, Feb. 4, Aug. 5, 26, 1906 (advertisements).

53. On the speed limit, see *Chicago Tribune*, Aug. 10, 1906. On average streetcar speed, see BOSE, *39th Annual Report*, exhibit T (1908 speed averaged 8.016 mph). By 1913, police were enforcing a 12 mph limit in Chicago's business districts and a 20 mph limit on the boulevards, although the law called for limits 3 to 5 mph slower (*Chicago Commerce*, Sept. 19, 1913, p. 15; *Chicago Tribune*, Sept. 18, 1913).

54. Board of Examiners of Operators of Automobiles, *Annual Report, 1900* pp. 1–13 (summary of law, procedures) On the concealment of vehicle tax tags, see *Chicago Tribune*, Feb. 1, 7, April 12, June 10, 1906.

55. Protests against speed limits include *Chicago Tribune*, Nov. 29, 1905, July 9, Sept. 20, Nov. 9, 1906. Similar material from nonautomobile sources includes *Municipal Journal and Engineer* 12, no. 5 (May 1902): 202.

On the difficulty of apprehending offenders, see *Chicago Tribune*, Feb. 1, April 12, June 3, July 29, 30, Aug. 16, 23, 1906. On police shooting tires of speeders, see *Chicago Tribune*, July 30, Aug. 16, 23, 24, 1906: *Chicago News*, July 29, 1906.

56. On the indifference of motorists to police authority, see Police Department, City of Chicago, *Annual Report for the Year Ending December 31, 1913* (hereafter

cited as CPD, *Annual Report . . . 1913*) (Chicago, 1914), p. 16. On the Evanston speeder who refused to attend court, see *Chicago Tribune*, Aug. 25, 1906. See also *Chicago Tribune*, Sept. 19, 1909.

American-born white males were far more likely to be arrested for speeding than for any other misdemeanor (based on CPD, *Annual Report[s] . . . 1907–1918*).

57. On the decline in convictions for speeding, see Table 2-1 below, prepared from CPD, *Annual Reports . . . 1913–1917*.

Table 2–1. Court Action in Traffic Cases, 1913–1917

Year	Misdemeanor convictions as a % of misdemeanor arrests	Speeding convictions as a % of speeding arrests	Parking convictions as a % of parking citations	"Passing streetcar" convictions as a % of "passing streetcar" citations
1913	45.8	91.4	—	—
1914	44.5	80.3	—	—
1915	39.7	73.9	—	—
1916	33.1	55.6	—	47.0
1917	33.7	50.2	52.1	44.5

58. Commissioners of Lincoln Park, *Report . . . 1908* (Chicago, 1908), p. 37.

59. *Chicago Commerce*, Dec. 22, 1916, p. 24.

60. Ibid., p. 24; Joseph Sabbath, "What a Traffic Court Can Do," *American City* 17, no. 8 (Aug. 1917): 126–27.

61. CLT, *Proceedings* 63 (Nov. 11, 1915): 1–3 of meeting; 64 (Nov. 24, 1915): 24, 34; 66 (Nov. 23, 1916): 54.

62. The 1915 DPW *Traffic Census* counted 6,396 automobiles crossing *north* over the Rush Street Bridge on July 13, 1915, with more than 40% of these vehicles making their crossing between 2 P.M. and 6 P.M.

On the number of parked cars, see Capt. Healey, *Chicago Tribune*, June 18, 1914. The 1,029 figure is from BOSE, *9th Annual Report*, p. 205.

63. BOSE, *8th Annual Report*, pp. 205, 206–10; Capt. Charles Healey in *Chicago Tribune*, June 18, 1914.

64. On cab owners' purchase of the right to park, see CLT, *Proceedings* 59 (Dec. 30, 1914): 91 of meeting. There were 9,175 public vehicle licenses of all types in 1914 (CPD, *Annual Report . . . 1914*, p. 21).

On the role of chauffeur-driven cars, see CLT, *Proceedings* 63 (Nov. 2, 1915): 32 of meeting.

65. *Chicago Tribune*, Feb. 9, 1906. Service station location data from *Chicago Classified Telephone Directory* (1906); ibid (1913). See Map 2-1.

66. CLT, *Proceedings* 62 (June 30, 1915); 48–53; *Chicago Post*, Aug. 14, 1905.

67. Hooker, *Through Routes*, p. 16; Inter Urban Roadways Committee of the Commercial Club of Chicago, *Inter Urban Roadways About Chicago* (Chicago, 1908), pp. 20–21.

68. CLT, *Proceedings* 40 (Subcommittee on Complaints) (March 17, 1913): 162 of volume. Many of the motorists cited for parking violations simply did not answer the summons (GSOP, *Report . . . 1912*, p. 97).

69. City of Chicago, *Ordinances of the City of Chicago* (Chicago, 1858), pp. 336–67 (chap. L, art. I—Obstruction in Streets and Public Places).

70. BOSE, *9th Annual Report*, p. 237.

71. *Chicago Tribune*, May 28, 1914.

72. CLT, *Proceedings* 66 (Dec. 8, 1916): 39.

73. CLT, *Proceedings* 66 (Aug. 12, 1916): 20–21, 30.

74. CLT, *Proceedings* 66 (Dec. 8, 1916): 47–48; 66, (Aug. 12, 1916): 30, 45, 47–48; CLT, *Proceedings* 60 (Jan. 23, 1915): 49. The tax problem developed when the Cook County Board of Review cut the city's real estate tax valuations (Delos F. Wilcox, *Great Cities in America: Their Problems and Government* [New York, 1913], pp. 189–90).

75. CLT, *Proceedings* 66 (Dec. 20, 1916): 8, 11, 28.

76. CLT, *Proceedings* 66 (Aug. 12, 1916): 14, 25, 30–31.

77. CLT, *Proceedings* 63 (Nov. 10, 1915):15, 18 of meeting; also see Ald. Capitain in CLT, *Proceedings* 64 (Nov. 23, 1915): 4.

78. CLT, *Proceedings* 63 (Nov. 10, 1915): 12, 20 of meeting; Richard Hall (CAC Director), in Commercial Club, *Report on the Plan of Chicago*, p. 115.

79. CLT, *Proceedings* 63 (Nov. 1, 1915): 42 of meeting; Arnold, "City Transportation," p. 44.

80. CLT, *Proceedings* 63 (Nov. 10, 1915): 23 of meeting.

81. Ibid., p. 16 of meeting, and CLT, *Proceedings* 64 (Nov. 23, 1915): 4–6 of meeting.

82. CLT, *Proceedings* 63 (Nov. 21, 1915): 29–30 of meeting; 60 (Jan. 23, 1915): 41–45 of meeting; 59 (Dec. 30, 1914): 91; 66 (Jan. 9, 1917): 2–3 of meeting; *Chicago Daily News*, March 14, 1916.

On the role of small business in the South, see Blaine Brownell, *Urban Ethos in the South*, p. 120. In southern cities during the 1920s small CBD businessmen were most seriously affected by automobile congestion. Howard L. Preston ("A New Kind of Horizontal City: Automobility in Atlanta, 1900–1930" [Ph.D. diss., Emory University, 1974], pp. 178–82) traces a long parking controversy in Atlanta and finds that CBD merchants opposed to the parking ban were trying to maintain CBD hegemony. In Chicago, always more a transit city, controversy followed other lines.

83. Chicago City Council, *Journal . . . 1917–1918* (Chicago, 1918), p. 3754; *Chicago Tribune*, March 11, 1917. The ordinance passed by a vote of 44 to 43.

84. From *Chicago Classified Telephone Directory* (1917), listing under "Garage"; R. F. Kelker, in American Electric Railway Transportation and Traffic Association, *Proceedings, 1922* (New York, 1922), p. 150.

85. CLT, *Proceedings* 56 (June 10, 1914): 388; 63 (Subcommittee on Service Betterments) (Nov. 1, 1915): 43 of meeting; 66 (Dec. 20, 1916): 29; *ERJ* 35, no. 20 (May 14, 1910): 870; BOSE, *9th Annual Report*, pp. 226–28.

86. See below, Chapter 4, pages 105–7, and Chapter 5, pages 154–55.

87. Daniel H. Burnham and Edward H. Bennett, *Plan of Chicago, Prepared Under the Direction of the Commercial Club During the Years MCMVI, MCMVII, and MCMVIII* (Chicago, 1909), p. 39.

88. Commercial Club of Chicago, Inter-Urban Roadways Committee, *Inter-Urban Roadways About Chicago* (Chicago, 1908), p. 12.

89. George Hooker, in City Club of Chicago, *The Railway Terminal Problem in Chicago* (Chicago, 1913), p. 86.

90. Arnold, *1902 Report*, p. 116; West Side Rapid Transit Association, *Facts and Figures About Electric Railways*, (Chicago, 1887), p. 3; Hempsted Washburn and William Alexander, *Prospectus and Subway Ordinance (n.p., n.d. [probably Chicago; "1903" penciled in])*, pp. 1–3; *CLT, Proceedings* IV (Feb. 5, 1903): p. 15 of second typed set (statement of Mayor Carter Harrison II that street congestion is the chief justification for subways); Robinson, "Proposed Inner Circle System," p. 586; Chicago Subway Arcade and Traction Company, *Prospectus* (Chicago, 1908), p. 1–3; CLT, *Report on the Traction Subway Systems of Boston, New York, Philadelphia, Paris and London* (Chicago, 1909), pp. 10–12; DPW, Bureau of Engineering, *Report on Transportation Subways*, pp. 10–12; Bion J. Arnold, *Recommendations and General Plans for a Comprehensive Passenger Subway System for the City of Chicago* submitted to the Mayor and the Committee on Local Transportation) (Chicago, 1911), pp. 3, 8; Charles K. Mohler, *Report on Passenger Subways and Rapid Transit Development in Chicago* (for the Committee on Traffic and Transit of the City Club of the City of Chicago) (Chicago, 1912), pp. 8–9.

91. Subways themselves were not expected to "pay." See, for example, the remarks of Bion J. Arnold in CLT, *Proceedings* 60 (Subcommittee on Subways) (Jan. 28, 1915): 51–52, and Bion J. Arnold, *Recommendations and General Plans . . .* p. 3. (The system as a whole was expected to subsidize the subways.)

92. CLT, *Proceedings* 36 (Oct. 2, 1912): 87; Bion J. Arnold, *City Transportation: Subways and Railroad Terminals*, reprinted from *JWSE* 19;, no. 9 (April 1914): 20.

93. CLT, *Proceedings* 36 (Oct. 2, 1912): 84.

94. BOSE, *7th Annual Report*, p. 30.

95. Arnold, *Recommendations and General Plans*, p. 3.

96. BOSE, *7th Annual Report*, p. 31.

97. On this disillusionment nationwide, see Mark Foster, *From Streetcar to Superhighway*, pp. 73–90. Its cause was the realization that off-street rapid transit *promoted* congestion. See, for example, Milo Maltbie, "Transportation and City Planning," *Proceedings of the Fifth National Conference on City Planning* (Boston, 1913), p. 115.

98. See, for example, Ballard Campbell, "The Good Roads Movement in Wisconsin" *Wisconsin Magazine of History* 49, no. 4 (1966): 273–93.

99. See, for example, N. S. Shaler, *American Highways: A Popular Account of Their Conditions and of the Means by Which They May Be Bettered* (New York, 1896), pp. 88–110, esp. pp. 88–92; George Chatburn, *Highways and Highway Transportation* (New York, 1924), pp. 147–49.

100. *JWSE* 11 (1906): 12.

101. *Chicago Commerce*. Dec. 6, 1912, pp. 19–20, June 2, 1916, p. 38.

102. Speech by Edward F. Dunne, May 30, 1916, reprinted in *Chicago Commerce*, June 2, 1916, pp. 18–19, 38–40 (statement cited is on p. 19); Chatburn, *Highways and Highway Transportation*, pp. 153–54.

103. DPW, *32nd Annual Report*, p. 411. See also Illinois Highway Commission, *1st Annual Report . . . 1906* (Springfield, 1907), pp. 22–30; Illinois Highway Commission, *4th Annual Report . . . 1910–1912* (Springfield, 1912), p. 288.

104. On parks and planning see Mel Scott, *American City Planning since 1890* (Berkeley, 1969), pp. 11–23. Recent scholarship has emphasized other sources for the planning movement. See, for example, Jon A. Peterson, "The Impact of Sanitary

Reform Upon American Planning," *Journal of Social History* 13 (Fall 1979): 83–103; Stanley K. Schultz and Clay McShane, "To Engineer the Metropolis: Sewers, Sanitation and City Planning in Late Nineteenth Century America," *Journal of American History* 65, no. 2 (Sept. 1978): 389–401.

On parks and "social uplift," see, for example, Dwight Perkins, "The Proposed Metropolitan Park System of Chicago," *JWSE* 10, no. 5 (Oct. 1905): 586; Henry Foreman, "Chicago's New Park Service," *Century* 69, (Feb. 1905): 611; Thomas Hawkes, "The Parks of Chicago," *Municipal Journal and Engineer* 14 (April 6, 1903): 157; City Club of Chicago, *Civic Committee Reports, 1907–1908*, pp. 213–15.

On concern for the relative decline in park acreage, see Dwight Perkins, "A Metropolitan Park System for Chicago," *World Today* 7, no. 3 (March 1905): 268; Andrew Crawford, "The Development of Park Systems in the American Cities," *AAAPSS* 25 (March 1905): 227; [Anonymous], "Parks and Public Playgrounds—The Results of a Year's Advance," *AAAPSS* 26 (Nov. 1905): 764–67; A. W. O'Niel, "Chicago's Park and Playground Extension," *Charities* 13, no. 2 (Aug. 1904): 798–800.

109. George C. Sikes, in Chicago Plan Commission, Executive Committee, *Proceedings*, June 25, 1914, pp. 726–27. See also *Engineering and Contracting* 28 (Aug. 7, 1907): 94.

106. *Chicago Journal*, July 29, 1897 (The quotation); West Chicago Park Commissioners, *Proceedings, 1906–1907*, pp. 4066, 4138–39; *Chicago Tribune*, Nov. 25, 1905. On speed tracks, see Board of West Chicago Park Commissioners, *Municipal Code*, April 28, 1903, p. 17.

107. *Chicago Chronicle*, Aug. 2, 1897; B. A. Eckhart, "Progress in the West Parks," in "The Chicago Park System," *World Today* 13 (Sept. 1907): 907–8; *Scientific American* 83 (July 28, 1900): 51.

108. *Chicago Tribune*, June 18, 1899; J. S. Foster, "Park Roads," reported in *JWSE*, 5, no. 3 (June 1900): 243.

109. West Chicago Park Commissioners, *Proceedings, 1906–1907*, Sept. 25, 1906, p. 4138; West Chicago Park Commissioners, *Proceedings, 1908*, p. 5076; Commissioners of Lincoln Park, *Report . . . 1908*, p. 29; South Park Commissioners, *Report . . . for the Period . . . December 1, 1906 to February 29, 1908*, p. 16, and Report . . . for the Period, March 1, 1909 to February 28, 1910, p. 20.

On the effect of automobiles on paving in general, see *Engineering and Contracting* 27, no. 24 (June 19, 1907): 269. On park traffic signals, see City Club of Chicago, *Civic Committee Reports, 1907–1908* (Chicago, 1908), p. 217; West Chicago Park Commissioners, *Proceedings, 1908*, Nov. 10, 1907, p. 4950. On park speed limits, see West Chicago Park Commissioners, *Municipal Code, 1916* (with amendment to 1919), Oct. 28, 1919, p. 36; South Park Commissioners, *Municipal Code, 1009*, p. 30; South Park Commissioners, *Municipal Code, 1919*, p. 241; South Park Commissioners, *Municipal Code, 1911*, p. 27.

110. City Council, *Journal . . . 1916–1917*, Jan. 19, 1917, p. 2949.

111. *Chicago Times-Herald*, April 13, 1897.

112. Arnold, in *Chicago Tribune*, Nov. 2, 1905; Ald. Thomas Hunter in ibid., opposing the idea.

The suggestion for a State Street boulevard is in *Chicago Tribune*, Oct. 27, 1906. On the proposed boulevarding of Halsted, *Chicago Tribune*, Dec. 15, 1906; *Chicago News*, Nov. 2, 1906. See also *Chicago Post*, Dec. 2, 1897; *Chicago Chronicle*, (Aug. 16, 1897); *Chicago Tribune*, March 8, 1906).

113. O'Neil, "Chicago's Park and Playground Extension," pp. 798–99; Foreman, in "The Chicago Park System," p. 904; "Parks and Public Playgrounds," p. 764–67; Perkins, "Proposed Metropolitan Park System," p. 600.

114. West Chicago Park Commissioners, *Report . . . 1915*, p. 9242; McClintock, *Street Traffic Survey*, p. 24.

115. CBPE, *The Park Governments of Chicago* (Chicago, c. 1911), pp. 5, 7, 8, 13. On the history of Chicago's debt, see Charles H. Wacker, *Chicago's First Needs* (Chicago, 1919), pp. 3–5.

On the business and political backing for park expansion, see Crawford, "Development of Park Systems," pp. 227, 233; Michael McCarthy, "Politics and Parks: Chicago Businessmen and the Recreation Movement," *Journal of the Illinois State Historical Society* 60, no. 2 (Summer 1972): 158–72.

116. On the influence of Arnold's report, see John Rothwell Slater, "Making a City into a Metropolis," *World Today* 13 (Sept. 1907): 885. The aldermanic "Palm Garden" remark is in *Chicago Tribune*, Nov. 2, 1905. See also editorial, *Chicago InterOcean*, May 3, 1905; *Chicago Tribune*, May 23, 2897.

117. Charles H. Wacker, *Gaining Public Support for a City Planning Movement* (Address to the 5th National Conference on City Planning, Chicago, May 1913) n.p., n.d.), p. 13.

118. CPC, *Address of Charles H. Wacker at the First Meeting of the Commission*, April 11, 1909, p. 1.

On the local background of the plan, see Frederick C. Howe, "The Remaking of the American City," *Harper's Monthly* 127, no. 15 (July 1913): 188. The standard work on Burnham is Thomas S. Hines, *Burnham of Chicago: Architect and Planner* (New York, 1974), esp. pp. 312-13.

119. Burnham and Bennett, *Plan of Chicago*, pp. 5–6.

120. Ibid., pp. 108–11; also pp. 34ˉ35, 86, 95.

121. Ibid., pp. 108ˉ09, and p. 88 on the necessity of cutting "ruthlessly" through existing buildings.

122. Ibid., pp. 42, 75–78,

123. Ibid., chap. 3.

124. Walter Moody, "Address to the 22nd Meeting of the Executive Committee, CPC" (hereafter, Moody, "Address"), in CPC, Executive Committee. *Proceedings*, Jan. 24, 1913, p. 16 of meeting; also CPC, Executive Committee, *Proceedings*, Jan. 13, 1911, p. 7 of meeting.

On the defeat of the Halsted Street widening, see *Chicago Tribune*, Oct. 15, 1906, and CPC Executive Committee in *Proceedings*, June 9, 1911, pp. 242–43.

125. Charles H. Wacker, *Address of Chairman Charles H. Wacker of the CPC Before the Secretary of War* (Washington, D.C., Nov. 20, 1913), pp. 1–2.

126. Moody, "Address," pp. 10–11; List of Members, CPC, *Proceedings*, 2nd meeting, Jan. 19, 1910, pp. 21–22; CPC, Executive Committee, *Proceedings*, March 22, 1911, p. 1 of meeting.

127. CPC, *Proceedings*, 2nd meeting, Jan. 19, 1910, pp. 22–23. See also Walter Moody, *Teacher's Handbook to Wacker's Manual of the Plan of Chicago* (Chicago, 1912), p. 53.

128. CPC, Executive Committee, *Proceedings*, Oct. 18, 1910, p. 17 of meeting. See also CPC, Executive Committee, *Proceedings*, April 22, 1910, p. 57 (Wacker: "We will not undertake to govern the City of Chicago").

On housing and transit, see CPC, Executive Committee, *Proceedings*, Oct. 18, 1910, pp. 19–23; CPC, *Proceedings*, 10th meeting, Jan. 26, 1914, p. 571; CPC, Executive Committee, *Proceedings*, Sept. 28, 1914, p. 3 of meeting.

129. On the CPC's consciousness of the importance of publicity, see for example, CPC, Executive Committee, *Proceedings*, April 22, 1910, pp. 59–62; Charles H. Wacker in *Chicago Examiner*, April 19, 1913.

On interest in the CPC's work on the part of other cities, see CPC, *Proceedings*, 5th meeting, March 19, 1912, pp. 325–26; Moody, "Address," pp. 41–42; CPC, Executive Committee, *Proceedings*, June 19, 1911, pp. 236–37.

130. CPC, *Proceedings*, 5th meeting, March 19, 1912, p. 326. Moody, *Wacker's Manual of the Plan of Chicago* (Chicago, 1912), p. 53.

131. Moody, "Address," p. 9 of meeting. See also Wacker, *Address . . . Before the Secretary of War*, p. 1; Moody, *Wacker's Manual*, p. 95; Moody, *Teacher's Handbook*, pp. 63, 100.

132. Ibid., pp. 12, 17.

133. Wacker, *Address . . . Before the Secretary of War*, p. 8; see also Benjamin C. Marsh, "City Planning in Justice to the Working Population," *City Club Bulletin* 7, no. 8 (1909): 327–28.

134. Moody, *Wacker's Manual*, p. 63.

135. Ibid., pp. 102, 108; Moody, *Teacher's Handbook*, pp. 65, 69, 73.

136. Moody, "Address," p. 12. See also *Wacker's Manual*, pp. 104, 106–12.

137. Moody, "Address," p. 13.

138. Walter Moody, "Address of Walter Moody to the Illinois House and Senate," April 29, 1919, in CPC. *Pull Chicago Out of the Hole* (Chicago, 1919), p. 18; CPC, *Proceedings*, 5th meeting, March 19, 1912, pp. 327–28; CPC, *Proceedings*, 10th meeting, Jan. 26, 1915, p. 579; Moody, *Wacker's Manual*, p. 114; Moody, *Teacher's Handbook*, p. 114; Moody, "Address," pp. 12–13.

139. CPC, Executive Committee, *Proceedings*, Jan. 19, 1910, pp. 31–32; Moody, *Wacker's Manual*, pp. 106–8.

140. CPC, Executive Committee, *Proceedings*, Jan. 13, 1911, p. 7; CPC, Executive Committee, *Proceedings*, March 22, 1911, p. 1–3, 10; Moody, "Address," p. 32.

141. CPC, Executive Committee, *Proceedings*, April 22, 1911, pp. 1–3, 8–9, 11–14. Bion J. Arnold wanted a streetcar-only right-of-way in the center of 12th Street (letter to CLT, *Proceedings* 62, [June 19, 1915]: 9). All aldermen opposed this as against the principle of a heavy-traffic street (pp. 10–12).

CPC, *Proceedings*, 5th meeting, March 19, 1912, pp. 6–7. See also CPC, Executive Committee, *Proceedings*, April 22, 1910, p. 54; March 22, 1911; pp. 3–5, 11, 15, 16–18.

142. CPC, *Proceedings*, 10th meeting, Jan. 26, 1914, pp. 571–72; CPC, Executive Committee, *Proceedings*, July 6, 1911, pp. 6–7, 12–13 of meeting.

143. Moody, "Address," p. 31. The outline of the plan is in Moody, *Wacker's Manual*, pp. 99–100, 106–11.

144. CPC, Executive Committee, *Proceedings*, Oct. 24, 1912, pp. 3–4 of meeting; Carl Condit, *Chicago, 1910–1929: Building, Planning, and Urban Technology* (Chicago, 1973) pp. 6, 71, 83, 249.

145. CPC, Executive Committee, *Proceedings*, Sept. 14, 1910, p. 76; CPC, Executive Committee, *Proceedings*, Oct. 18;, 1910, p. 6. Moody ridiculed this view. See his "Address," p. 35.

On the spreading of assessments and condemnation proceedures, see CPC, Executive Committee, *Proceedings*, Sept. 28, 1914, p. 739; BLI, *Report . . . 1918–1919*, p. 3; Robert H. Moulton, "Chicago Improvement Plans," *Review of Reviews* 57 (March 1918): 288.

146. CPC, Executive Committee, *Proceedings*, May 5, 1916, pp. 864–65.

147. Railway Terminal Commission of the City of Chicago, *Proceedings* 1 (1913): 198; ibid., p. 356.

For the CPC's position, see Moody, in City Club, *The Railway Terminal Problem*, pp. 10–11; CPC, Executive Committee, *Proceedings*, Oct. 14, 1913, p. 551.

148. Bion J. Arnold, *Report on the Rearrangement and Development of the Steam Railroad Terminals of the City of Chicago* (submitted to the Citizens Terminal Plan Committee) (Chicago, 1913), pp. 44, 81, 84, 86–87 (Arnold supported the railroad's plan).

The city's engineer concurred in most of Arnold's recommendations; see John F. Wallace, *Special Report of the Committee on Railway Terminals of the City Council of the City of Chicago* (Chicago, Dec. 11, 1913), pp. 11–12, 14–16.

On business opposition to the CPC's objectives, see Railway Terminal Commission, *Proceedings* 1 (May 23, 1913): 404–10.

149. CPC, Executive Committee, *Proceedings*, Oct. 14, 1913, p. 9 of meeting. Opposition to the CPC's lakefront arrangement was lead by the Greater Chicago Federation and the Chicago Federation of Labor (Henry W. Lee, C.E., *The Lakefront Steal* (Chicago, Jan. 13, 1913), pp. 10–11, 13. See also George E. Newcomb, *Chicago Replanned* (Chicago, 1911) (leaflet in Hooker Collection), which uses arguments comparable to those used by anti-subway regionalists (see esp. p. 16).

150. BOSE, *Recommendations of the BOSE for Initial Routes for Surface Line Subways under the 1907 Ordinances* (submitted to the CLT of the City Council of the City of Chicago) (Chicago, Oct. 29, 1913), supplemental map IV; CPC, Executive Committee, *Proceedings*, Jan. 22, 1914, pp. 565–66.

151. CLT, *Proceedings* 61 (Feb. 10, 1915): 37 of meeting; Moody, *Teacher's Handbook*, p. 5.

152. CPC, Executive Committee, *Proceedings*, March 13, 1912, p. 15 of meeting (on Burnham's automobile commuting habits).

Chapter 3

1. Delos F. Wilcox, *Municipal Franchises: A Description of the Terms and Conditions Upon Which Private Corporations Enjoy Special Privileges in the Streets of American Cities* (Chicago, 1911) 2: 141–42.

2. See Chapter 5, page 133.

3. Delos F. Wilcox, "Street Railway Re-Settlements and Municipal Ownership," *National Municipal Review* 3 (Oct. 1914): 745–51; T. L. Sildo, "Cleveland Street Railway Settlement," *American Political Science Review* 4, no. 2 (May 1910): 279–86; Delos F. Wilcox, *Analysis of the Electric Railway Problem* (New York, 1921), pp. 426–27.

4. Wilcox, *Municipal Franchises*, 2: 18, 50, 94–95, 152; *Street Railway Journal* 28, no. 13 (March 30, 1907): 556–57; *Scientific American* 96 (May 25, 1907): 430; John Fairlie, "The Street Railway Question in Chicago," *Quarterly Journal of Economics* 21 (May 1907): editorial, *Municipal Journal and Engineer* 22, no. 15 (April 10, 1907): 367; Edwin O. Lewis, "Philadelphia's Relation to Rapid Transit Company" [sic], *AAAPSS* 31 (1908): 605, 611; National Municipal League, *Proceedings of the Providence Con-*

ference for Good City Government and the 13th Annual Meeting of the National Municipal League, Held at Providence, R.I., November 19–22, 1907 (n.p., 1907), pp. 129–30; Bion J. Arnold, "The Urban Transportation Problem: A General Discussion," *AAAPSS* 37 (Jan. 1911; reprinted: New York, 1968), pt. 1, pp. 3, 6, 8.

5. See, for example, Robert H. Wiebe, *Businessmen and Reform* (Cambridge, Mass., 1962), pp. 179–224; David Thelan, *The New Citizenship: Origins of Progressivism in Wisconsin, 1885–1900* (Columbia, 1972), p. 223–90; Frederick C. Howe, *The City: The Hope of Democracy* (New York, 1906), p. 115.

6. Cheape, *Moving the Masses*, pp. 82, 97–100, 132–33, 176–77, 182, 206–8; 208–19; Federal Electric Railway Commission, *Proceedings . . . Together with Final Report of the Commission to the President* (hereafter FERC, *Proceedings*) (Washington, D.C., 1920), pp. 2099, 822–824, 1453–1455; Wilcox, *Municipal Franchises*, 1: 202–13; 252–55; 2: 101-427 (provisions of franchises in major American cities).

7. James D. Johnson, *A Century of Chicago Streetcars, 1858–1958* (Wheaton, 1964), tables, pp. 32–35; Chicago Transit Authority, *Historical Information* (Chicago, 1965), p. vii. On money spent in the first three years, see *Financial and Statistical Exhibits and Schedules of the Streetrailways Under the Supervision of This Board, for the Financial Year Ended January 31, 1910* (Chicago, 1910), exhibit A, schedule 1 (p. 7 of report.) On average citywide speed, see BOSE, *4th Annual Report*, pp. 319–20, and BOSE, *8th Annual Report*, exhibit T. On passengers per seat, see BOSE, *3rd Annual Report*, p. 278.

8. *Electric Traction* 9, no. 8 (Aug. 1913): 443; "Service, Equipment, and Operation in Chicago" (abstract of an otherwise unpreserved BOSE report to the Chicago City Council), *ERJ* 42, no. 7 (Aug. 16, 1913): 265–66.

On the elevated, by 1912, 151 miles of track fed the two-track CBD Loop, see U.S. Department of Commerce and Labor, Bureau of the Census, *Census of Electrical Industries* (Washington, D.C., 1917), p. 46; *Electric Traction Weekly* 6, no. 5 (Jan. 29, 1910): 121–22.

9. "Service, Equipment, and Operation in Chicago," p. 265.

10. Loop elevated congestion see CLT, *Proceedings* 32 (March 16, 1912): 63-64. On excess streetcars, see "Service, Equipment, and Operation in Chicago," p. 267.

11. BOSE, *6th Annual Report*, pp. 432-34; CLT, *Proceedings*, 27; (May 17, 1911): 7003–7, 7016, 7020.

12. CLT, *Proceedings* 21 (Dec. 15, 1909): 6281–90; 14 (Oct. 21, 1980): 4270–72, (Nov. 11, 1908): 4335–43; 33 (May 24, 1912): 544.

On long headways on through-routes, and of long walks between the places where one line ended and another began see CLT, *Proceedings* 28 (July 5, 1911): 7737–39, 7741–43. This citation is merely indicative; the complaint was perennial. See also, City of Chicago, Department of Public Service (hereafter DPS), Transportation Bureau, *Report . . . 1914*, pp. 5–6.

On the resistance of elevated companies to through-routes, see, for example, CLT, *Proceedings* 17 (Feb. 17, 1909): 5317, 5328–32, 5336–39; *Chicago Daily News*, Feb. 7, 8, 1910.

13. CLT, *Proceedings* 29 (Oct. 11, 1911): 8599 (the quotation); 30 (Jan. 17, 1912): 9631; 34 (June 21, 1912): 585; 60 (Jan. 22, 1915): 84–85; 62 (May 19, 1915): 65; BOSE, *20th Annual Report*, pp. 310, 312.

On company refusal to build extensions see, for example, CLT, *Proceedings* 60 (Jan. 22, 1915): 84–85; Leonard Busby to Chicago Surface Lines Board, CSL Archives, Corporate Record, vol. 1, pp. 98–100.

14. See below, Chapter 4, pages 114–15.

15. *1907 Ordinances*, sect. 7 and exhibit B.

16. BOSE *26th Annual Report*, p. 86 (exhibit X); CLT, *Proceedings* 41 (April 25, 1913): 131.

17. The effects of overcapitalization are discussed below, pages 121–26.

18. CLT, *Proceedings* 15 (Nov. 25, 1908): 4387–88; 12 (March 6, 1908): 3677–80.

19. CLT, *Proceedings* 12 (March 6, 1908): 3658 (Fisher's statement). On the demand for a single fare south of the end of the city, see, for example, CLT, *Proceedings* 26 (Nov. 30, 1910): 7739; 28 (Aug. 15, 1911): 8269–74.

On the intangibles in the valuations of the various smaller companies, see CLT, *Proceedings* 12 (March 16, 1908): 3918; BOSE, *Advance Report*, Jan. 13, 1910, pp. 113, 131; Bion J. Arnold, George Weston, and Glenn Plumb, *Detailed Exhibits of the Physical Properties and Intangible Values of the Southern Street Railway Company as of August 1, 1908* (Chicago, 1908), p. 14.

20. *ERJ* 35, no. 4 (Jan. 22, 1910): 164; CLT, Proceedings 22 (Jan. 22, 1910): 6307–8, 6309–13, esp. pp. 6311–12. See also *ERJ* 42, no. 1 (June 5, 1913): 11, 44.

On the terms of the city's agreement with the CCT, see, for example, CLT, *Proceedings* 25 (March 8, 1910): 6957–58, 6959, 6965.

21. On Evanston service, see CLT, *Proceedings* 47 (Oct. 27, 1913): 475. On suburban complaints, see, for example, CLT, *Proceedings* 27 (April 26, 1911): 6916–48. On the "cutting of lines" at the city's northern boundaries, see *Chicago Tribune*, Dec. 27, Aug. 11, 1910. On the suburban litigation for transfer privileges, see *ERJ* 41, no. 9 (March 1, 1913): 398; 43, no. 17 (April 25, 1914): 942.

22. See, for example, Frederick Whitridge, "Public Morality and Street railways," *The Century* 87, no. 5 (March 1909): 787–88; *ERJ* 33, no. 1 (Jan. 2, 1909): 26; CLT, *Proceedings* 46 (Oct. 24, 1913): 796–97.

23. The companies insisted that transfers issued on diagonal streets or on "jogging" routes be honored in only one direction. See, for example, CLT, *Proceedings* 46 (Oct. 15, 1913): 80–90). The BOSE upheld the company position (CLT, *Proceedings* 53 [Mar. 18, 1914]: 445–47).

On the role of universal transfers in slowing the rate of increase of the companies' gross, see CLT, *Proceedings* 64 (June 10, 1914): 383. The companies' transfer rules are summarized in DPS, Transportation Bureau, *Report* (1914), p. 7. On the increase in the proportion of Chicago riders using transfers, see BOSE, *37th Annual Report*, exhibit T. On the New York City Railways, see Dr. Thomas Conway, "Decreasing Financial Return Upon Urban Street Railway Properites," *AAAPSS* 37, no. 1 (Jan. 1911): 20. The bargaining which led to truly universal transfer privileges is thoroughly explained in Chicago Railways Company, *History of the Ordinance* (a pamphlet in the Gottlieb-Schwartz Papers, Chicago Historical Society).

24. CLT, *Proceedings* 31 (Subcommittee on Crowding) (Feb. 12, 1912): 10057.

25. Albro Martin, *Enterprise Denied: Origins of the Decline of American Railroads, 1897–1917 (New York, 1971); Stanley Mallach, "The Origins of the Decline of Urban Mass Transportation in the United States, 1890–1930," Urban Past and Present* 8 (summer 1979): 1–17.

26. See, for example, *ERJ* 32, no. 5 (July 3, 1908): 248; 32, no. 6 (July 11, 1908): 292); 40, no. 1 (July 6, 1912): 39; 32, no. 7 (July 18, 1908): 324).

27. On open-air cars, see *New York Times*, Oct. 29, 30, Nov. 2, 1915. On lower steps, see *Electric Traction Weekly* 6, no. 3 (Jan. 15, 1910): 76.

28. CLT, *Proceedings* 39 (Feb. 28, 1913): 834.

29. Thirty-six traffic checks were recorded between 1907 and 1914: 22 by the BOSE, 5 by the companies, 7 by CLT experts and aldermen, 1 by the DPS, and 1 by the City Club on its own initiative.

30. BOSE decisions which reduced the effectiveness of extensions are reported in CLT, *Proceedings* 61 (Feb. 17, 1915): 5–15; 64 (Dec. 8, 1915): 4.

31. CLT, *Proceedings* 31 (Feb. 7, 1912): 9889; 63 (Nov. 1, 1915): 37-42; 59 (Dec. 16, 1914): 40–41.

32. The 1907 ordinances provided for sufficient cars to handle loads "comfortably" (sec. 22), but the courts ruled this language vague and, therefore, unenforceable (CLT, *Proceedings* 44 [July 9, 1913]: 330).

33. Forrest McDonald, "Samuel Insull and the Movement for State Utility Regulatory Commissions," *Business History Review* 32 (Autumn 1958): 251–54. But Chicago Surface Lines apparently preferred local regulation. See the company's advertisement in *Chicago Daily News*, Feb. 14, 1915.

34. McDonald, "Samuel Insull," pp. 251–52; William L. Sullivan, *Dunne: Judge, Mayor, Governor* (Chicago, 1916), pp. 397–99, 439–40.

35. The PUC's assertion of authority is in State of Illinois, Public Utilities Commission, *Reports*, 2 (1915): Cases 3281, 3283, 8384, 3292, 3301, 3302, 3330, 3331, 3332, Consolidated, Cook County Real Estate Board v. Chicago Surface Lines *et al.*, pp. 291–92; *ERJ* 45, no. 4 (Jan. 23, 1915): 205.

On the creation of the city department, see CLT, *Proceedings* 50 (Dec. 30, 1912): 761–75, 827.

36. On the difficulty of fixing a crowding standard, see CLT, *Proceedings* 59 (Dec. 30, 1914): 5–28 (all citations in this volume by page of meeting), (Jan. 13, 1915): 1–33; 60 (Feb. 6, 1915): 9–33, 50–53.

BOSE engineers on the inevitability of crowding: CLT, *Proceedings* 50 (Dec. 29, 1913): 710-13; 59 (Jan. 13, 1915): 133.

37. Aldermanic fear of "voting for straphanging" is most evident in CLT, *Proceedings* 60 (Feb. 6, 1915): 49–60, 64, 68. For the BOSE position, see CLT, *Proceedings* 55 (May 13, 1915): 2–48; 56 (June 5, 1914): 54.

38. Thomas Cummerford Martin, "Bion Joseph Arnold: An Engineer Who Is the Right Hand Man of Public Service Commissions," *Scientific American* 105, no. 1 (Sept. 9, 1911): 223.

39. CLT, *Proceedings* 49, (Feb. 21, 1913): 608.

40. Bruce A. Sinclair, *A Centennial History of the American Society of Mechanical Engineers, 1880–1980* (Toronto, 1980), pp. 87–105; Edwin T. Layton, Jr., *The revolt of the Engineers* (Cleveland, 1971), esp. pp. vii, viii, 5, 54.

41. See Arnold's advice to the *City of San Francisco (Report on the Improvement and Development of the Transportation Facilities of San Francisco* [San Francisco, 1913]: 390–91). See also Arnold's polite but decidedly negative judgment on Chicago politicians as potential transit managers in Paul R. Leach, comp., *Chicago's Traction Problem (Chicago Daily News* reprint, 1925), pp. 4-5.

42. On the Weston brothers, see BOSE, *1st Annual Report*, p. 55; *ERJ* 32, no. 1 (June 6, 1908): 58; 33, no. 1 (Jan. 2, 1909): 12–14; *Chicago Daily News*, Feb. 5, 1910; George Weston, "Elements of Utility Valuation," *ERJ* 47, no. 6 (Feb. 5, 1916): 266–67.

43. Bion J. Arnold, *Report on the Pittsburgh Transportation Problem* (submitted to the Hon. Mayor William A. Magee, Mayor of the City of Pittsburgh) (Pittsburgh, 1910), pp. 6–7; Bion J. Arnold, *Report on the Traction Improvement and Development*

Within the Providence District, to the Joint Committee on Railroad Franchises, Providence City Council (Providence, 1911), pp. 1, 3–4. For Arnold's biography, see Albert Nelson Marquis, dir. *Who's Who in Chicago and Vicinity* (Chicago, 1931), p. 42.

44. Arnold, *Report . . . Pittsburgh*, p. 10; Arnold, *Report . . . San Francisco*, 390–91.

45. Citizens' Non-Partisan Traction Association, *We Are Proud of Our Chicago*, side 2 (leaflet in Hooker Collection); CLT. *Proceedings* 12 (Oct. 21, 1907): 3314. Dissent came chiefly from the Chicago Federation of Labor, whose animosity toward Arnold dated from the opposition of his 1906 valuation commission to a minimum wage for transit workers (*Chicago Tribune*, April 3, 1907; CLT, *Proceedings* 12 [Feb. 6, 1908]: 3510).

46. For a defense of overcapitalization in general by Arnold, see Bion J. Arnold, "The Transportation Problem in Major American Cities," *California Progressive*, April 1, 1911), p. 12. For his defense of the 1907 settlement, see ibid., p. 13, and Arnold *et al.*, "Phases in the Development of the Street Railways of Chicago," p. 642.

47. *Chicago Daily News*, July 6, 1912.

48. City of Chicago, DPW, *Report on Transportation Subways* (Chicago, 1909), p. 9; Bion J. Arnold, *Recommendations and General Plans for a Comprehensive Subway Passenger System for the City of Chicago* (Chicago, 1911), p. 8.

Attacks on Arnold include *Who Is Bion Arnold?* (Chicago, c. 1918), a leaflet probably produced by the Northwest Commercial Association (in Hooker Collection); [Mayor] William Hale Thompson, "To the voters of Chicago" (letter mailed to Chicago voters, Oct. 31, 1918, copy in Hooker Collection). See also Arnold's request to be excused from a valuation commission because his participation would "prejudice the reception of the report" (CLT, *Proceedings* 94 [Dec. 11, 1924]: 244).

49. A summary (without the political context) is in BOSE, *7th Annual Report*, pp. 28–29.

50. BOSE, *Recommendations . . . on Routes for Streetcar Line Subways Provided in the 1907 Traction Ordinances* (Chicago, Oct. 29, 1914), esp. pp. 6–9; *Chicago Commerce*, March 13, 1914, p. 8, March 20, 1914, pp. 17–21.

51. Arnold's statement is in CLT, *Proceedings* 27 (May 22, 1911): 7081–82.

52. *Chicago Tribune*, Jan. 15, 1912; *ERJ* 23, no. 12 (March 20, 1909): 522; 23, no. 17 (April 24, 1909): 788; 24, no. 10 (Sept. 4, 1909); BOSE, *3rd Annual Report*, p. 320. The audit was never printed.

53. Three of the two dozen recorded complaints by aldermen of pressure for better service placed on them by their constituents: CLT, *Proceedings* 27 (May 17, 1911): 7018; 28 (July 5, 1911): 7757; 53 (March 25, 1915); 598. The exchange between Block and Arnold is in CLT, *Proceedings* 60 (Feb. 6, 1915): 59.

54. CLT, *Proceedings* 60 (Feb. 10, 1915): 49, 70, (Feb. 6, 1915): 88–89.

55. See, for example, CLT, *Proceedings* 60 (Feb. 6, 1915): 68; 39 (Feb. 20, 1913): 564 ff. On BOSE negation of fines, see (*ERJ* 43, no. 21 [May 23, 1914]: 1152).

56. CLT, *Proceedings* 60 (Jan. 22, 1915): 2; 61 (Feb. 10, 1915): second session, pp. 114–17; E. W. Bemis to William Hale Thompson, Jan. 24, 1916 (in William E. Dever, Papers, box 4, folder 35). On Kelker's work with the BOSE, see *Electric Traction Weekly* 4, no. 40 (Oct. 3, 1908): 1001.

57. Martin, "Bion Joseph Arnold," p. 223; CLT, *Proceedings* 42 (May 24, 1914): 768; 41 (April 26, 1913): 247.

58. On subways, see Bion J. Arnold, *City Transportation: Subways and Rail-*

road Terminals (reprinted from *JWSE* 19, no. 4 [April 1914]: p. 20). On wave patterns in traffic, see BOSE, *9th Annual Report*, pp. 226–28. Arnold's earlier advocacy of underground parking is reported in *Chicago Tribune*, July 26, 1920. On his suggestion for combined mass transit and highway facilities, see Foster, "Western Response," p. 37.

On tax assessments for transit, see Arnold's remarks in Milo Maltbie, "Transportation and City Planning," in *Proceedings of the Fifth National Conference on City Planning* (Boston, 1913), pp. 129–30, 132.

59. David F. Noble, *America by Design: Science, Technology, and the Rise of Corporate Capitalism* (New York, 1977), p. 63. On Arnold's belief in Engineering neutrality, see *JWSE* 16, no. 5 (Oct. 1909): 641–42; CLT, *Proceedings* 41 (April 26, 1913); 227.

For Arnold's identification of himself with the ordinances and defence of the 1906 valuation, see *Chicago Tribune*, April 3, 1907; *Chicago Record-Herald*, Jan. 1, 1912; *ERJ* 39, no. 6 (Feb. 10, 1912): 251.

60. *Chicago Record-Herald*, April 6, 1914.

61. *1916 Report*, pp. 8, 138.

62. Cheape, *Moving the Masses*, pp. 202–12.

63. *ERJ* 37, no. 22 (June 3, 1911): 992; 37, no. 23 (June 10, 1911): 1035; James N. Hatch, "The Development of the Electric Railway," *JWSE* 12 (Summer 1908): 504–5; BOSE, *6th Annual Report*, p. 406; Montague Ferry, *Preliminary Report of the Department of Public Service Upon Interlocking Control of Public Utilities in the City of Chicago* (Chicago, July 1, 1914).

64. Insull, *Memoirs* (in Cudahy Library, Loyola University, Chicago), pp. 174, 191 ff.; Federal Electric Railway Commission, *Proceedings* (Washington, D.C., 1920), p. 2024; McDonald, *Insull*, pp. 96–97, 103, 156, 252; *ERJ* 37, no. 20 (May 20, 1911): 893.

65. *ERJ* 38, no. 22 (Nov. 25, 1911): 1089. See also McDonald, *Insull*, pp. 96–97, 103, 122–24, 214.

66. *ERJ* 38, no. 20 (Nov. 11, 1911): 1036; CLT, *Proceedings* 29 (Nov. 2, 1911): 8850–51.

67. CLT, *Proceedings* 33 (Dec. 31, 1912): 550 ff.

68. After 1911, as the position of the transit companies deteriorated, they—and their creditors—attached more and more conditions to unification. See *Chicago Tribune*, Aug. 22, 1923.

69. CLT, *Proceedings* 46 (Oct. 24, 1913): 781; Chicago Railways Company, *History of the Ordinance*.

70. Altshuler, *Urban Transportation System*, pp. 33, 309–10, 395–404.

71. *1916 Report*, pp. 8, 10, 44, and Appendix I (pp. A-1–A-113).

72. Cheape, *Moving the Masses*, p. 127; FERC, *Proceedings* (1920), p. 2133; Sam Bass Warner, *The Private City: Philadelphia in Three Periods of Its Growth* (Philadelphia, 1968), p. 192.

73. *Chicago Commerce* (Dec. 4, 1914), pp. 40–42; *Electric Traction* 9 (Aug., 1913): 443. Arnold, *Report . . . San Francisco*, pp. 100, 113; *1916 Report*, pp. 8–9, 117, 23–41; *ERJ* 35, no. 20 (May 14, 1910): 870.

74. *Chicago Tribune*, July 1, 1913, April 6, 1914; *Chicago Record-Herald*, April 6, 1914.

75. Arnold, *1902 Report*, pp. 110–11. City of Chicago, DPW, *Report on Transportation Subways* (Chicago, 1909), p. 9; Arnold, *Recommendations and General Plans*, p. 8; BOSE, *Recommendations* (1914); Arnold, "City Transportation," pp. 21, 19–30; R. C. St. John, "Proposed Initial Passenger Subway System for the City of Chicago" (1911, mimeograph copy at Chicago Municipal Reference Library), pp. 5, 6.

76. Greater Chicago Federation, *Will This Condition Ever Change?* (flier, c. 1914, in Hooker Collection); Greater Chicago Federation to William Hale Thompson, Aug. 10, 1918 (on GCF letterhead in Hooker Collection); CLT, *Proceedings* 36 (Oct. 2, 1912): 116; 44 (July 9, 1913): 36; Benjamin Levering, *The Subway* (leaflet, 1911).

On the business character of the regionalists, see list of member organizations appended to Greater Chicago Federation to William Hale Thompson, Aug. 10, 1918 (Hooker Collection). See also CLT, *Proceedings* 44 (July 16, 1913): 596; Tomaz Deuther, *First Issue of Civic Questions Pertaining Entirely to Local Transportation in the City of Chicago* (Chicago, 1924), pp. 36–37, 58, 62; *15th Ward Optimist* 1, no. 1 (March 8, 1914) (socialist party ward organ).

77. Deuther's own story of the Association's history is woven through his *First Issue of Civic Questions*.

78. For example, see Northwest Side Commercial Association, *Manacled by the Loop* (flier dated March 12, 1910, in Hooker Collection).

79. CLT, *Proceedings* 61 (Feb. 10, 1915): 37–38; *Chicago Daily News*, April 6, 1914; *Chicago Record-Herald*, April 7, 1914; CLT *Proceedings* 81 (Jan. 15, 1921): 71–72. Col. (former alderman) Jacob Arvey (interviewed by the author, March 4, 1974).

80. CLT, *Proceedings* 16 (Jan. 6, 1909): 4778–79.

81. Deuther, *First Issue of Civic Questions*, p. 6; "Chicago Subways," *JWSE* 18, no. 9 (Nov., 1913): 920; CLT, *Proceedings* 92 (Jan. 25, 1924): 346–93.

82. Testimony before the CLT in *Proceedings* 36 (Oct. 2, 1912): 113; John Fox, "Relation Between Housing and Transit," *AAAPSS* 51 (Jan. 1914): 161.

83. "Chicago Subways," *JWSE*, pp. 921–22. See also CLT, *Proceedings* 36 (Oct. 23, 1912): 421.

84. "Chicago Subways," p. 922.

85. CLT, *Proceedings* 36 (Oct. 2, 1912): 97–99.

86. Tomaz Deuther, *Letter to the Honorable Mayor and the Members of the Chicago City Council*, Aug. 10, 1918 (open letter in Hooker Collection); Greater Chicago Federation, *Looking Backwards: Reminiscences on Traction* (flier, n.d. [1918]).

87. Ralph Heilman, "Chicago Subway Problem," *Journal of Political Economy* 22 (Dec., 1914): 1003. See also *Chicago Tribune*, April 8, 1914, April 8, 1925, and Cook County Board of Election Commissioners, "Vote Returns for October 6, 1918, Referendum," in Chicago Municipal Reference Library.

Chapter 4

1. CLT, *Proceedings* 44 (June 28, 1913): 59.
2. CLT, *Proceedings* 36 (Oct. 16, 1912): 270 of volume.
3. *1916 Report*, pp. 238–39, 240.

4. Ibid., pp. 238–39.

5. The Commission surveyed workers in establishments employing 100 persons or more. The data came from employers. Similar samples were used by Parsons's firm in studies of Cleveland and Detroit, though these studies surveyed only industrial workers. (Barclay, Parsons, and Klapp, *Report on a Rapid Transit System for the City of Cleveland* [made to Board of Rapid Transit Commissioners, City of Cleveland] [n.p., 1919], pp. 23–24; Barclay, Parsons, and Klapp, *Report on Detroit Street Railway Traffic and Proposed Subway* [made to Board of Street Railway Commissioners, City of Detroit] [n.p., 1915], pp. 45–46).

6. *1916 Report*, pp. 246–47.

7. Ibid., pp. 165–66, 208–9, Appendix I, pp. A-2–A-5, A-59–A-61, for example. Of workers surveyed, 52% did work in the 12.5-square-mile center city. While the CBD—broadly defined—did dominate the labor market, travel patterns varied considerably from line to line and from weekdays to weekends.

8. Barclay, Parsons, and Klapp, *Report . . . Cleveland*, pp. 23–24; Barclay, Parsons, and Klapp, *Report . . . Detroit*, p. 46; E. K. Morse, *Report of the Transit Commissioner to the Mayor [of Pittsburgh]* (Pittsburgh, 1917), pp. 67–69 (cited in Joel A. Tarr, *Transportation, Innovation, and Changing Spatial Patterns in Pittsburgh, 1850–1934* [Chicago, 1978], pp. 30–2). See also Edward Ewing Platt, *Industrial Causes of Population Congestion in New York City* (New York, 1911), pp. 116–88, on the correlation between long hours of work, low pay, and residence near work.

9. Malcolm J. Proudfoot, "The Outlying Business Centers of Chicago," *Journal of Land and Public Utility Economics* 13, no. 2 (Feb., 1937): 63–69; *1916 Report*, p. 14. Generalizations concerning weekend riding in the 1920s and after are based on the annual reports of the Traffic and Scheduling Department, Chicago Surface Lines, 1926–1935 (in the office of Chicago Transit Authority Managing Director Harold Hirsch).

10. The classic statement in Flink's *Car Culture*.

11. Wilcox, *Municipal Franchises*, p. 58.

12. CLT, *Proceedings* 69 (June 27, 1917): 31.

13. Deuther, *First Issue of Civic Questions*, p. 8; CTA, *Historical Information*, p. 10.

14. CLT, *Proceedings* 60 (Jan. 27, 1915): 2 P.M. session, p. 96. The factory line (Fulton Street) ran until 1947, although the City Club had recommended the abandonment of this and four other lines in 1910. See Alan Lind, *Chicago Surface Lines: An Illustrated History* (Park Forest, 1974), p. 258; CLT, *Proceedings* 26 (Oct. 3, 1910): 7566; 28 (July 26, 1911): 8163. On the "jogging" line (14th–16th), see James D. Johnson, *A Century of Chicago Streetcars 1858–1958* (Wheaton, 1964), pp. 29, 31. On the ethnic composition and varieties of land use in the areas in question, see George C. Olcott & Company, Inc., *Olcott's Land Values: Blue Book of Chicago* (Chicago, 1932).

On the "nicer class" of patrons, see CLT, *Proceedings* 45 (July 28, 1913): 360; 15 (Dec. 2, 1908): 4424.

15. The author has found three in the records of the CLT: CLT, *Proceedings* 41 (May 2, 1913): 456; 12 (Dec. 19, 1907): 3353; 37 (Dec. 18, 1912): 520.

16. See, for example, CLT, *Proceedings* 37 (Nov. 27, 1912): 150; 62 (July 14, 1915): 55; 44 (July 9, 1913): 312–16; 42 (May 21, 1913): 599 ff. On the decline in rapid transit business within the two-mile zone, see *1916 Report*, pp. 99–102.

17. CLT, *Proceedings* 36 (Oct. 7, 1912): 159; 20 (Dec. 18, 1909): 630–48.

18. *1916 Report*, p. 229.

19. CLT, *Proceedings* 13 (March 23, 1908): 3973; 25 (June 15, 1910):7249–52; 62 (June 30, 1915): 36; 27 (June 20, 1911): 367–89; 40 (March 12, 1913): 102–52. McChesney's complaint is in CLT, *Proceedings* 12 (March 3, 1908): 3564, 3582–93. See also Philpott, *Slum and Ghetto*, pp. 190–91.

20. CLT, *Proceedings* 18 (May 26, 1909): 5487–88; 18 (June 23, 1909): 5603–4; 44 (June 10, 1913): 407; 59 (Dec. 15, 1914): 8–13; 65 (March 22, 1916): 16.

21. See, for example, CLT, *Proceedings* 28 (Aug. 2, 1911): 8214; 31 (Feb. 12, 1912): 9964–65; 35 (July 2, 1912): 254 (the quotation in the text); 42 (May 20, 1913): 535. See also ibid. 21 (Dec. 18, 1909): 6323: "during the summer on Sunday nights . . . it is impossible to get on a car."

22. *1916 Report*, pp. 245–46, plates 9–20; CLT *Proceedings* 30 (Jan. 17, 1912): 9608.

23. *1916 Report*, pp. 132–33; 160–62, 195–99.

24. CLT, *Proceedings* 33 (May 29, 1912): 786; also George Weston of the BOSE: "Now . . . some run all over Robin Hood's farm in order to go by someone's house or factory" (i.e., are routed "politically") (32 [March 13, 1912]: 201).

25. CLT, *Proceedings* 29 (Oct. 11, 1911): 8600. See also CLT, *Proceedings* 30 (Jan. 17, 1912): 9664; 48 (Nov. 14, 1913): 77–80; 61 (March 31, 1915): 8.

26. CLT, *Proceedings* 66 (Dec. 2, 1916): 9; 45 (July 23, 1913): 264. Citations are merely indicative.

27. The account here is based largely on the transcripts of the CLT. The Committee in this period was a forum for interest groups as well as for its politician members. Nonetheless, to have the desired effect on those present, a complaint or description would have had to bear some resemblance to the realities evident in the street.

28. BOSE, *6th Annual Report*, pp. 436–37; CLT, *Proceedings* 31 (Feb. 17, 1912): 10410.

29. CLT, *Proceedings* 45 (July 23, 1913): 258. See also CLT, *Proceedings* 60 (Jan. 27, 1915): 66–68.

30. CLT, *Proceedings* 27 (May 17, 1911): 7008–9 (quotation in text). See also CLT, *Proceedings* 67 (Jan. 24, 1917): 165; 15 (Dec. 2, 1908): 4422; 25 (May 4, 1910): 7143, 7146, 7150; 30 (Dec. 13, 1911): 9113; 30 (Jan. 17, 1912): 9647–48.

31. On the public expectation of door-to-door service, see CLT, *Proceedings* 55 (May 13, 1914): 16; 31 (Feb. 13, 1912): 10150.

Requests which produced or would have produced service on streets less than a half-section apart include those in CLT, *Proceedings* 19 (Nov. 22, 1909): 6088; 34 (June 12, 1912): 300; 60 (Jan. 27, 1915): 94.

In newer areas, real estate developers' ideas of what walk prospective buyers might accept could be as important as the actual requirements of riders. The Polonia Line was on Noble Street. This line was clearly a community institution, according to one longtime community resident, literally picking up and letting off passengers in front of their homes (CLT, *Proceedings* 37 [Dec. 11, 1912]: 392; Johnson, *A Century of Chicago Streetcars*, p. 31; interview with Sister Lucille Marie, S.S.N.D., Principal, Saint Stanislaus Kostka High School, May 23, 1976).

32. CLT, *Proceedings* 34 (June 21, 1912): 611.

33. CLT, *Proceedings* 34 (July 8, 1912): 460; 33 (May 15, 1912): 59–63.

34. CLT, *Proceedings* 49 (Dec. 16, 1913): 594–95, 667–68; 50 (Jan. 7, 1914): 963–66. On the retention of lightly used stations, see *1916 Report*, exhibit A.

35. CLT, *Proceedings* 21 (Dec. 13, 1909) 6233; 21 (Dec. 15, 1909): 6235; 29 (Oct. 11, 1911): 8577.

36. CLT, *Proceedings* 66 (Dec. 2, 1916): 24; 31 (Feb. 12, 1912): 10065. See also CLT, *Proceedings* 42 (May 13, 1913): 27; 44 (June 28, 1913): 59; 64 (Dec. 8, 1915): 53.

37. CLT, *Proceedings* 65 (June 30, 1916): 14–15.

38. The proportion of transfers to fares paid was 61.8 in 1907 and 72.4 in 1917 (BOSE, *36th Annual Report*, exhibit T).

39. CLT, *Proceedings* 67 (Jan. 25, 1917): 38 (Kelker's statement). Long waits for cars were a favorite complaint. Extended discussions of the subject, initiated by aldermanic or citizen complaints, are found in a dozen places in the CLT proceedings between 1912 and 1917. Complaints usually referred to waits of 15 minutes or longer.

40. CLT, *Proceedings* 56 (June 9, 1914): 145; 67 (Jan. 25, 1917): 36.

41. CLT, *Proceedings* 52 (Feb. 25, 1914): 782; 33 (May 29, 1912): 789. Inadequate night service was taken up ten times between 1910 and 1914.

On the unavailability of schedules, see CLT, *Proceedings* 31 (Feb. 7, 1912): 9889; 53 (May 9, 1914): 852.

42. CLT, *Proceedings* 40 (March 19, 1913): 284–85; 59 (Dec. 16, 1914): 155; 60 (Jan. 22, 1915): 34–35.

43. CLT, *Proceedings* 12 (Dec. 19, 1907): 3368; 25 (May 18, 1910): 7176; 26 (Jan. 4, 1911): 7845; 37 (Jan. 8, 1913): 680.

44. CLT, *Proceedings* 37 (Dec. 18, 1912): 530; 42 (May 13, 1913): 28.

45. CLT, *Proceedings* 49 (Dec. 17, 1913): 780–81. The relationship of women to public places is a study in itself. Some further complaints of the effect of crowding on women, all voiced by men, are in CLT, *Proceedings* 30 (Jan. 7, 1912): 9641; 33 (May 22, 1912): 352; 37 (Dec. 18, 1912): 530; 41 (May 2, 1913): 456.

Other vicissitudes common to all transit riders were occasionally presented to the Committee in terms of their impact on women, perhaps simply for the sake of emphasis. Crowding and long waits were, on the other hand, certainly more difficult for women with children. One of nine such complaints is to be found in CLT, *Proceedings* 47 (Nov. 5, 1913): 624. None of these complaints was brought to the Committee by a woman. One complaint which does appear to have been the special concern of women was smoking. See, for example, CLT, *Proceedings* 37 (Jan. 8, 1913): 708.

One specific complaint suggests a class-linked approach to the relationship of women to mass transit: CLT, *Proceedings* 38 (Jan. 22, 1913): 436 (request for an elevated station one block from an existing station which was located in the "vice district" because "the stenographers and businesswomen who are obliged to go into that neighborhood do not like to get off at that station").

46. CLT, *Proceedings* 39 (Feb. 5, 1913): 59; 32 (March 13, 1912): 162. See also CLT, *Proceedings* 38 (Jan. 22, 1913): 453.

47. CLT, *Proceedings* 15 (Dec. 2, 1908): 4411; 37 (Jan. 8, 1913): 718.

48. CLT, *Proceedings* 18 (June 23, 1909): 5648; 19 (Nov. 10, 1909): 5974; 49 (Dec. 9, 1913): 324.

49. *Chicago Tribune*, March 22, 23, 24, 1907. Capitain's remark is in CLT, *Proceedings* 67 (Jan. 24, 1917): 162.

50. CLT, *Proceedings* 80 (July 9, 1919): 22; 22 (Feb. 9, 1910): 6728.

51. On the movement of the center of population and the character of population and industrial development of the far south side, see *1916 Report*. The judgment that transit access to the CBD influenced patterns of population growth between 1900 and 1920 is that of James Everett Clark ("Impact of Transportation Technology on Suburbanization in the Chicago Region, 1830–1920" [Ph.D. diss., Northwestern University, 1977], pp. 102–7). Map 4-1 shows the Surface transit system of 1926. Data are from BOSE, *Annual Reports*. See also Chicago Surface Lines, "Population for Census Years, 1910–1940," in the author's possession.

52. Between 1920 and 1934 a 30-square-mile area of the inner city experienced a population decline of 211,000. Most of this loss was from an oblong 5-square-mile slice of the inner west side (City of Chicago, Department of Superhighways, *A Comprehensive Superhighway Plan for the City of Chicago* [Chicago, 1939], p. 24). However, almost all of this inner area, and nearly half of the area in which population declined, was defined by the Chicago Plan Commission in 1943 as suffering from a "coincidence of factors indicative of blight" (50% or more of structures "substandard" and 20% or more in need of "major repair"). See CPC, *Master Plan of Residential Land Use of Chicago* (Chicago, 1943), p. 75. Whatever benefits followed from this thinning of population did not, predictably, accrue to the city's black population, which was largely confined to the near south side—by 1943 the worst of the "blighted" districts (p. 76).

53. Arnold, *City Transportation*, p. 32.

54. Based on 22 crowding complaint hearings in the CLT *Proceedings* between 1908 and 1916. Only complaints in which crowding was defined were considered.

55. CLT, *Proceedings* 49 (Dec. 9, 1913): 252; 41 (May 2, 1913): 443.

56. BOSE, "Report on Traffic and Service Investigations by the BOSE During January, 1915" (mimeographed, 1915), pp. 8–9 and exhibit B.

No crowding standard was ever established for the elevated lines. Calculating from a diagram of a fairly common type of "open platform" car, one finds 4 square feet of standing room per passenger with 95 persons aboard, and 3 square feet per passenger with 111 riders. Calculations are from a diagram in Central Electric Railfans, *Bulletin 113: Chicago's Rapid Transit* (Chicago, 1973), 1: 82.

57. BOSE, "Report on Traffic and Service Investigations" (1915), table 2a, table 2b, and p. 20 and exhibit B, table IV. Data from San Francisco and Cincinnati suggest that comparable crowding existed in those cities, but not over such a wide range of routes. (Arnold, *Report . . . San Francisco*, pp. 106–113; R. W. Harris, *Report on Cincinnati Traffic Conditions* [to the City Council of Cincinnati, Ohio] [Cincinnati, Sept. 1, 1912], p. 25, table IX).

58. By the time of the 1914 BOSE investigation, all of these measures had been undertaken to some degree.

59. See, for example, CLT, *Proceedings* 30 (Jan. 24, 1912): 9731; 31 (Feb. 13, 1912): 10156.

60. CLT, *Proceedings* 37 (Dec. 9, 1912): 326; 37 (Dec. 11, 1912): 388–91.

61. On elevated scheduling, see CLT, *Proceedings* 41 (May 2, 1913): 437-38. On Chicago City Railways, see CLT, *Proceedings* 49 (Dec. 3, 1913): 24; 63 (Oct. 20, 1915): 57; *ERJ* 43, no. 4 (Jan. 24, 1914): 210–11.

62. *ERJ* 43, no. 6 (Feb. 7, 1914): 336–37; CLT, *Proceedings* 31 (Feb. 13, 1912): 10148; 31 (Feb. 7, 1912): 9884.

63. CLT, *Proceedings* 31 (Feb. 7, 1912): 9876.

64. CLT, *Proceedings* 31 (Feb. 13, 1912): 10142; 44 (July 21, 1913): 731.

On the effect of the cuts discussed on lightly used lines, see CLT, *Proceedings* 31 (Feb. 13, 1912): 10142. Such a standard also made night service undesirable to the companies (CLT, *Proceedings* 37 [Dec. 9, 1912]: 332–35).

65. CLT, *Proceedings* 43 (May 24, 1913): 763–66 (Arnold's judgment); U.S. Department of Commerce, Bureau of the Census, *Electrical Industries* (Washington, D.C., 1917), p. 91; BOSE, *36th Annual Report,* exhibit T; *ERJ* 51, no. 7 (Feb. 16, 1918): 337.

66. BOSE, *36th Annual Report,* exhibit T. In 1910, Arnold found Chicago's system eighth among 16 major cities in gross receipts per mile of car operation and second highest in expenses, including taxes (Arnold, *Report . . . Pittsburgh,* pp. 119, 123, 136).

67. *1916 Report,* pp. 5, 163, Appendix I, pp. A-1–A-113.

68. *Dziennik Zwiakowy,* Aug. 20, 1918 (Works Progress Administration Foreign Language Press Translations, Chicago, 1942, typescripts in Chicago Public Library).

69. CLT, *Proceedings* 41 (April 26, 1913): 218–19.

70. Bureau of the Census, *Electrical Industries* (1917), p. 55; Chicago data calculated from BOSE, *36th Annual Report,* exhibit S. Arnold's statements are in CLT, *Proceedings* 41 (April 25, 1913): 131; 39 (Feb. 21, 1913): 627. On elevated securities, see *1916 Report,* p. 385.

71. Bureau of the Census, *Electrical Industries* (1917), pp. 58–59. Data are unreliable because, after 1907, the census reported surface and elevated companies together. Separate tables are kept for companies operating rapid transit, but here systems with and without subway mileage are mixed. The author has therefore compared combined surface and rapid transit figures with the return for all systems grossing over $1 million. *This produces a result which is too favorable to the Chicago companies generally.*

72. Arnold, *1902 Report,* pp. 83–98, plates 13, 14, 15; Phillip Harrington, R. F. Kelker, and Charles DeLeuw, *A Comprehensive Local Transportation Plan for the City of Chicago* [Chicago, 1937], fig. 59; *1916 Report,* pp. 3, 40–45.

73. FERC, *Proceedings* (1920), pp. 1268–69.

74. On the impossibility of amortizing a consolidated transit system in 20 years, see *1916 Report,* pp. 38–39; Arnold, describing the 1907 franchises to a non-Chicago readership, called them, "indeterminate . . . recoverable" (*Report . . . Pittsburgh,* Appendix B).

75. Arthur Stone Dewing, *The Financial Policy of Corporations* (New York, 1926): 592–93, 615–16. The proper conceptual framework for utility valuation was much debated. See Sinclair, *Centennial History,* p. 104. See also testimony of Mortimer Cooley and other engineers in FERC, *Proceedings* (1920), pp. 255–56, 261, 271.

76. See the observations of Walter Fisher, quoted in Edwin Lobdell, *The Chicago Transportation Problem in 1927* (Chicago, 1927), p. 17.

77. *The Fitch Bond Book: Describing All Important Corporation and Railroad Bond Issues of the United States and Canada* (New York, 1917), pp. 210–12. Lawrence Chamberlain (*The Principles of Bond Investment* [New York, 1911], pp. 329–37) reflects the suspicion of street railway securities due to overcapitalization.

78. CTA, *Historical Information,* p. vii; BOSE, *37th Annual Report,* exhibit T; Bureau of the Census, *Electrical Industries* (1917), pp. 24, 46.

79. FERC, *Proceedings* (1920), pp. 1205–6; Edward F. Dunne, *Mayor Dunne Tells the Truth About the Issues in the Municipal Campaign of 1907* (Chicago, 1907), pp. 11–13.

80. The valuation at the end of 1916 was $149,954,546. If $25 million is eliminated from the original 1907 valuation and 15% of all post-1907 additions is eliminated as the equivalent of the brokerage and construction allowances, a capital account of $109,961,364 remains. The lowest valuation located for this period is that made by Charles K. Mohler in 1918 ("Another Deal in Chicago Traction" [typescript paper dated Oct. 29, 1918, in Hooker Collection]). By accumulating 10% annual depreciation and discounting intangibles and scrapped equipment, Mohler arrived at a valuation of $64,165,500 for 1917. Figures in the (text) result from calculations based on these valuations.

81. BOSE, *37th Annual Report*, Exhibit T.

82. Kelker, Deleuw & Company, *Report and Recommendations on a Comprehensive Rapid Transit Plan for the City and County of Los Angeles* (to the City Council of Los Angeles and the Board of Supervisors of Los Angeles County) (Los Angeles, 1925), pp. 71–75.

83. Barclay, Parsons, and Klapp, *Report . . . Detroit*, pp. 7–8, 46; Barclay, Parsons, and Klapp, *Report . . . Cleveland*, pp. 23–24; Tarr, *Transportation, Innovation, and Changing Spatial Patterns*, pp. 30–32; *1916 Report*, pp. 235–236. The authors of the Pittsburgh study cited by Tarr considered "walking distance" to be up to 2 miles. The figure cited in the text is for those within one mile. The figure given for Chicago is, because of the way in which the Commission reported its data, highly speculative.

84. On the thinning of inner city population, see note 52, this chapter, and any edition of the City of Chicago's *Local Community Fact Book* (Chicago, 1940 ff.). On the role of belt railroad lines in industrial dispersion, see Howard Francis Newall, "Measurement of Urban Decentralization" (Ph.D. diss., Indiana University, 1974), pp. 130–153.

85. *1916 Report*, pp. 123, 138; U.S. Chamber of Commerce, Committee on Public Ultilities, *Conference on Street Railways*, 2d sess. (Washington, D.C., 1919), p. 49.

86. Bureau of the Census, *Electrical Industries* (1917), p. 46, tables 45 and 46; Arnold, *Report . . . Pittsburgh*, pp. 128–29 (in 1910 Chicago was second highest).

87. D. C. Jackson, *Street Railway Fares, Their Relation to Length of Haul and Cost of Service* (New York, 1917), pp. 50–51; *1916 Report*, p. 72.

88. *1916 Report*, p. 168; BOSE, *36th Annual Report*, exhibits S, T, X (author's calculation).

89. CLT, *Proceedings* 66 (Jan. 2, 1917): 204.

90. Ibid., pp. 9–27. On the fate of the 1918 ordinance, see Chapter 6.

91. Henry Brinkerhoff, who also conducted a work-residence study for Barclay, Parsons, and Klapp in Detroit, makes this observation in CLT, *Proceedings* 66 (Jan. 2, 1917): 16–17.

92. On the speed and flexibility of Chicago's system, see *1916 Report*, pp. 138–39. On the cost of equipment, see Central Electric Railfans Association, *Bulletin 113*: Chicago's Rapid Transit, p. 203. See also FERC, *Proceedings*, p. 400.

93. In Pittsburgh, public outcry over the apparent unfairness of a zone system

led to its abandonment (FERC, *Proceedings* [1920], pp. 1908–9). In New Jersey, zone fares met with a response which included public violence (Wilcox, *Analysis of the Electric Railway Problem*, pp. 220, 229–43, 668).

94. The companies paid an average of 7% per year over the period 1907–1917. This total includes only the traction fund; it excludes city and, after 1913, federal taxes.

95. Calculated from BOSE, *25th Annual Report*, exhibit W.

96. On Chicago wages, see *ERJ* 62, no. 1 (July 7, 1923): 31.

97. *Chicago Commercial and Financial Chronicle* 84, no. 2183 (April 27, 1907).

Chapter 5

1. Cited in John Bell Rae, *The Road and Car*, p. 276.

2. Evan J. McIlraith, "Keeping Traffic Arteries Open," American Transit Association *Proceedings* (1946): 149.

3. *Motor Age* 44, no. 18 (Nov. 1, 1923): 58.

4. Miller McClintock, *Report and Recommendations of the Metropolitan Street Traffic Survey Commission of the Chicago Association of Commerce* (Chicago, 1926), pp. 2, 73–74. On the united demand of Loop business for traffic relief in 1922, see *Chicago Commerce* (Dec. 13, 1922), p. 17.

5. CLT, *Proceedings* 80 (Dec. 16, 1920): 72; *Chicago Journal*, Jan. 31, 1924 (editorial).

6. George C. Sykes, *The Eternal Traction Question* (in Dever, *Papers*, box 5, folder 36), p. 3. The ordinance before and after alterations is in Chicago Zoning Commission, *Tentative Report and Proposed Zoning Ordinance* (Chicago, 1923), p. 6; Hon. William E. Dever, Administrative Office of Zoning, *Chicago Zoning Ordinance* (Chicago, 1924), p. 12. See also Andrew J. King, "Law and Land Use in Chicago: A Pre-history of Modern Zoning" (Ph.D. diss., University of Wisconsin, Madison, 1976), pp. 398–405.

7. *Engineering World* 17, no. 3 (Sept. 1920): 149–53; 21, no. 4 (Oct. 1922): 211–12. By 1924 the average Loop building had ten usable stories. The average height of the 104 tallest Loop buildings in 1926 was 14.6 stories (Hugh E. Young, "Pedestrian Traffic," *JWSE* 29, no. 3 [March 1924]: 71; McClintock, *Report and Recommendations*, p. 32).

8. McClintock, *Report and Recommendations*, p. 57 (table); R. F. Kelker, Jr., in *Chicago Commerce* (Dec. 28, 1922), p. 17; U. S. Schwartz, *The Anti-Parking Ordinance* (Chicago, 1920), p. 7; CPD, *Annual Report . . . 1919* (Chicago, 1920), p. 32; CPD, *Annual Report . . . 1923*, p. 32; CLT, *Proceedings* 92 (Nov. 23, 1923): 112, 114; 95 (Jan. 13, 1925): 284.

9. *Chicago Daily News*, Nov. 7, 1922.

10. *Chicago Commerce*, May 1, 1920, p. 45, Dec. 4, 1920, pp. 11, 43, 64, 71; CLT, *Proceedings* 91 (May 4, 1923): 127; Charles H. Wacker, *An S-O-S to the Public Spirited Citizens of Chicago* (Chicago, 1924), p. 18.

11. *Chicago Daily News*, Oct. 23, 1923; C. E. DeLeuw, "The Basic Elements of Traffic Control," *Roads and Streets* 69, no. 10 (Oct. 1929): 359.

12. C. Z. Elkins, "Private Auto Traffic," *JWSE* 29, no. 3 (March 1924): 104; *Chicago Commerce*, April 16, 1921, p. 13; *Chicago Herald and Examiner*, March 13, 1920. John R. Guilliams, "The Control of Street Use," American Electric Railway Association, *Proceedings, 1928* [New York, 1928], pp. 446, 476.

13. Among the half-dozen serious proposals for two-level streets in Chicago are J. H. Sawyer, "Second Story Sidewalks," *JWSE* 29, no. 3 (March 1924): 87–101; City Council, *Journal . . . 1920–1921*, pp. 1661, 1742, 1840–41; *Chicago Tribune*, March 29, 1924. U. R. Pfuhl's letter is in *Chicago American*, Dec. 12, 1922. Deuther's claim of authorship is in *First Issue of Civic Questions*, p. 73.

14. Young, "Pedestrian Traffic," p. 74; *Chicago American*, May 28, 1923.

15. Among the measures opposed by the Illinois auto trade was the renewable driver's license (*Motor Age*, 47, no. 13 [March 26, 1925]: 21). See also, for example, *Automobile Trade Journal* 28, no. 1 (July 1, 1923): 55; *Motor Age* 42, no. 15 (Oct. 12, 1922): 4; 47, no. 9 (Feb. 26, 1925): 36.

16. *Motor Age* 44, no. 20 (Nov. 15, 1923): 31. See also *Motor Age* 44, no. 25 (Dec. 20, 1923): 28.

Business and auto industry interests frequently pointed out during the 1920s that traffic was a business and engineering problem, not one of legal policy. See, for example, John Ihlder, "The Auto and the Community," *AAAPSS* 116 (Philadelphia, 1924): pp. 199–200; *Engineering News-Record* 90, no. 1 (Jan. 4, 1923): 30–31; *Public Works* 55, no. 9 (Sept. 1924): 285–86; *Scientific American* 130, no. 1 (Jan. 1924): 18–19, 65–66; Morris Knowles, "City Planning as a Permanent Solution to the Traffic Problem," *American City* 32, no. 4 (April 1925): 333. Citations are merely indicitive.

17. Ald. Francis Tomczak to Mayor William Dever, Nov. 20, 1923 (in Dever, *Papers*, box 5, folder 22); CLT, *Proceedings* 91 (Nov. 2, 1923): 643; 91 (Nov. 13, 1923): 751–55; 91 (Nov. 2, 1923): 651. The quotation in the text is in CLT, *Proceedings* 92 (Dec. 7, 1923): 181.

18. On the support for a parking ban, see *Chicago Daily News*, Sept. 20, Nov. 18, 1920; Ald. U. S. Schwartz, *Anti-Parking Ordinance*, pp. 1, 4, 6; CLT, *Proceedings* 80 (Nov. 18, 1920): 64–65; 92 (Dec. 7, 1923): 123–24, 179; Police Chief Morgan Collins, in *Chicago Daily News*, Dec. 28, 1925; Lieutenant John Martin (head of the police Loop traffic detail), "Obstruction of Street Traffic," in American Electric Railway Transportation and Traffic Association, *Proceedings, 1922*, (New York, 1922), pp. 148–49; *City Club Bulletin* 12, no. 27 [July 4, 1920]: 146.

19. Charles Hayes of the Chicago Motor Club, in CLT, *Proceedings* 80 (Nov. 19, 1920): 49.

20. CLT, *Proceedings* 80 (Nov. 18, 1920): 104; 80 (Nov. 16, 1920): 26, 29; 80 (Dec. 20, 1920), 14, 37–38; DPS, *Annual Reports . . . 1916–1920*, p. 14. Quotation is from CLT, *Proceedings* 80 (Dec. 16, 1920): 30.

Business opposition to parking restriction and its rationale are described by CPC engineer Hugh E. Young in "Day and Night Storage and Parking of Motor Vehicles," *American City* 29, no. 7 (July 1923): 44–45. See also CPC, *How the LaSalle Street Improvement Affects You* (Chicago, 1924), p. 16.

21. *Chicago Commerce*, Nov. 26, 1920, p. 12; Dec. 11, 1920, p. 12.

22. Elkins, "Private Auto Traffic," pp. 106–7. See also *Chicago Daily News*, April 18, 1924; CLT, *Proceedings* 81 (March 31, 1921): 17–19; *Chicago Commerce* (Jan. 14, 1922): 29; City Council, *Journal . . . 1920–1921*, p. 2212.

23. *Chicago Commerce*, Jan. 7, 1922, p. 16; June 28, 1924, p. 7; July 4, 1925, p. 14; South Park Commissioners, Edward J. Kelly, President, *Annual Report to the Judges of the Circuit Court of Cook County, December 31, 1924*, p. 19.

24. *City Club Bulletin* 14, no. 29 (July 18, 1921): 120; *Chicago Commerce*, June 12, 1926, p. 19.

25. *Motor Age* 42, no. 26 (Dec. 28, 1922): 38.

26. City Council, *Journal . . . 1920–1921*, pp. 1269–70.

27. CLT, *Proceedings* 92 (Jan. 25, 1924): 321. See also CLT, *Proceedings* 80 (Nov. 18, 1920): 106.

28. Citizens' Police Committee, *Chicago Police Problems* (Chicago, 1931), pp. 160–62, 165; McClintock, *Report and Recommendations*, pp. 101–103.

29. CLT, *Proceedings* 99 (Dec. 1, 1925): 335; CPD, *Annual Report . . . 1920*, p. 9; *Annual Report . . . 1921–1922*, pp. 9, 10; *Annual Report . . . 1923*, p. 7a; *Annual Report . . . 1924*, p. 9; *Annual Report . . . 1927*, p. 8a; *Annual Report . . . 1929*, p. 8; *Annual Report . . . 1930*, p. 12; Citizens' Police Committee, *Chicago Police Problems*, pp. 165–66; *Chicago Journal*, June 16, 1923, July 22, 1925.

30. Elie v. Adams Express Company, 133 NE 243. See also A. H. L. Street in "Limitations of Municipal Power to Control Street Traffic," *American City* 27, no. 10 (Oct. 1922): 343; Chicago Motor Coach v. City of Chicago, 169 NE 22; Haggenjos v. City of Chicago, 168 NE 661.

31. John Bell Rae *American Automobile*, p. 10; *Chicago Commerce*, Dec. 27, 1924, p. 8; McClintock, *Report and Recommendations*, p. 176.

32. *Chicago Daily News*, Nov. 1, 1924.

33. *Chicago Tribune*, May 22, 23, 1924; Elkins, "Private Auto Traffic," p. 107; *Chicago Herald and Examiner*, June 20, 1923. On the reaction to police inability to enforce parking restrictions, see *Chicago American*, Aug. 9, 1924.

On the "unreasonableness" of the 20 mph limit, see *Chicago Tribune*, May 22, 1924; *Chicago Daily News*, Oct. 31, 1924; McClintock, *Report and Recommendations*, p. 108.

34. West Chicago Park Commissioners, *52nd Annual Report . . . 1922*, p. 13; McClintock, *Report and Recommendations*, pp. 194, 203; CPD, *Annual Reports . . . 1920–1925*.

35. McClintock, *Report and Recommendations*, pp. 194–95; *Journal of Commerce*, Jan. 18, 1923; Ihdler, "The Auto and the Community," p. 199.

36. Dixon Merrit, "The Deadly Driver," *Outlook* 142, no. 14 (March 1926): 521–22; Francis Warfield, "America on Wheels," *Outlook and the Independant* 152, no. 15 (Aug. 14, 1929): 631.

37. *Chicago Daily News*, Oct. 23, 1923; Citizens' Police Commission, *Chicago Police Problems*, p. 158.

38. On painted safety islands, see *Chicago Commerce*, Nov. 10, 1929, p. 11; Victor F. Noonan (of Chicago Surface Lines) to Ald. Thomas Bowler, Oct. 14, 1926 (in Dever, *Papers*, box 5, folder 22). The quotation in the text is from CLT, *Proceedings* 91 (Nov. 13, 1923): 751–52. See also, for example, City Council, *Journal . . . 1925–1926*, p. 3938.

39. On poor law enforcement, see, for example, *Chicago Herald and Examiner*, Feb. 2, 1926; *Chicago Evening Post*, May 22, 1926. On variations in signal design, see CLT, *Proceedings* 90 (Nov. 23, 1922): 17, 27–28; City Council, *Journal . . . 1922–1923*, Nov. 29, 1922. On drivers unable to understand traffic lights, see *Chicago Commerce* (Oct. 12, 1929), p. 11; *Chicago Daily News*, Nov. 8, 1927.

40. Comments on John Martin, "Obstructions of Street Traffic," p. 151; McClintock, *Report and Recommendations*, p. 90.

41. CPD, *Annual Reports, 1920–1931* (tables available from the author). On

the changing role of police, see James Q. Wilson, "The Dilemma of the Urban Police," *The Atlantic* 223, no. 3 (March 1969): 129–35. See also Darwin L. Tielhart, "Is Motoring a Crime?" *Outlook and the Independant* 154, no. 7 (Feb. 12, 1930): 248–50, 275.

42. City Council, *Journal . . . 1922–1923*, p. 1766; *Journal . . . 1923–1924*, p. 377; *Chicago Tribune*, Sept. 27, Dec. 10, 1923; *Chicago Herald and Examiner*, June 2, 1923; William E. Dever to Police Chief Morgan Collins, Oct. 30, 1924 (in Dever, *Papers*, box 4, folder 27); *Chicago Daily News*, Nov. 1, 1924.

43. On the use of street space by contractors, see *Chicago Commerce*, Aug. 9, 1924, p. 58.

On the failure of the movement for night delivery, see CLT, *Proceedings* 99 (Dec. 1, 1925): 339, 451; *Chicago Commerce*, May 24, 1924, p. 60. On the underutilization of the Illinois tunnel system, see James Rowland Bibbins, "City Building and Transportation," *JWSE* 25, no. 12 (Aug. 20, 1920): 456; *Chicago Commerce*, April 3, 1926, p. 13–14.

On peddlers, see Ald. Guy Gurnsey to Mayor William E. Dever, June 7, July 24, 26, 1923; Dever to Gurnsey, July 30, 1923 (all in Dever, Papers, box 2, folder 18); *Chicago Daily News*, Dec. 19, 1924; *Chicago Commerce*, Jan. 25, 1930, p. 30.

44. C. Z. Elkins, "Private Auto Traffic," p. 105.

45. A. E. Burnett, *A Sixteen Year Record of Achievement, 1915–1931: The Board of Local Improvements* (Chicago, 1931), p. vi.

46. Auto registration in Chicago rose as follows: 1918, 59,965; 1921, 137,750; 1924, 260,887; 1927, 335,263; 1930, 406,916. See Chicago Department of Superhighways, *A Comprehensive Superhighway Plan for the City of Chicago* (submitted to the Mayor and the City Council of the City of Chicago, Oct. 30, 1939), table 1, p. 22.

On the hope that the auto would bring the benefits of rural life to city dwellers, see, for example, "The Auto Industry," *Outlook* 111 (Nov. 17, 1915): 757–62; Alonzo Barton Hepburn and Francis H. Sisson, quoted in National Automobile Chamber of Commerce, *Facts and Figures of the Automobile Industry: Statistics of Production, Export and Use of Motor Cars and Motor Trucks* (hereafter cited as NACC, *Facts and Figures*) 1 (1921): 11; Drew Pearson, "The City Passeth—An Exclusive Interview With Henry Ford," *Automobile Trade Journal* 29, no. 3 (Sept. 1, 1924): 27–29.

47. *Automobile Trade Journal* 25, no. 3 (Sept. 1923): 19–20; NACC, *Facts and Figures . . . 1921*, p. 16; *Facts and Figures . . . 1924*, p. 92; *Literary Digest* 75, no. 6 (Nov. 11, 1922): 78. On the correlation of auto ownership with other marks of status, see NACC, *Facts and Figures . . . 1924*, pp. 10, 12–13; *Facts and Figures . . . 1925*, pp. 42–43; *Facts and Figures . . . 1927*, p. 38.

48. Paul Nystrom, *Automobile Selling: A Manual for Dealers* (New York, 1919), p. 4 (on limits of use). On the cost of auto ownership versus other costs, see NACC, *Facts and Figures . . . 1925*, p. 73; *Engineering and Contracting* 68, no. 12 (Dec. 1929): 535; T. R. Agg and H. S. Carter, "Automobile Operating Costs," *Highway Engineer and Contractor* 35, no. 2 (Aug. 1929): 44–46. Agg and Carter found that the cost of operation on intermediate quality roads ranged between 6.02¢ and 9.45¢, depending on the size of the automobile. The NACC had claimed a 10¢ per mile figure earlier in the 1920s. Significantly, owners themselves put the cost of operation lower than any published estimate ("What Do You Pay to Ride in Your Car?" *AMRR* 75, no. 1 (Jan. 1927): 19.

On the role of the automobile industry in the development of time-payment plans, see *Motor Age* 42, no. 17 (Oct. 26, 1922): 30; Fred E. Clark, *Reading in Marketing* (New York, 1931), p. 292; Flink, *The Car Culture*, p. 149.

49. Nystrom, *Automobile Selling*, p. 64; advertisements in *Chicago Daily News*, May 20, 1921; *Chicago Evening Post*, March 27, 1919; *Chicago Tribune*, May 5, 1929.

50. Nystrom, *Automobile Selling*, p. 64; Lee J. Eastman, "The Closed Car: Its Development and Its Great Future," *Automobile Trade Journal* 28, no. 6 (Dec. 1923): 22–23; *Chicago Commerce*, Oct. 20, 1923, p. 16.

51. Walter Jackson, "The Sale of the Ride," in Illinois Electric Railway Association, *Papers, March 14, 1923*, p. 3; John A. Miller, "Increasing the Efficiency of Passenger Transportation in City Streets," in American Society of Civil Engineers, *Transactions* 89 (1926): 927.

52. American Electric Railway Association, Committee on Rapid Transit, *The Economics of Rapid Transit* (New York, n.d. [c. 1927]), p. 28; Subcommittee on Two Level Streets and Separated Grades of the Committee on Traffic and Public Safety of the City Council, City of Chicago, *Further Preliminary Report with Reference to Elevated Highways for the Chicago Metropolitan Area* (Chicago, May 1929), p. 5; John Burmeister & Company, *Burmeister's Chicago Transportation Directory* (Chicago, 1915), p. 54. Burmeister's time was presumably a minimum.

53. Eugene Taylor, "Progress on the Chicago Plan," *Engineering News-Record* 90, no. 12 (March 22, 1923): 528; Samuel Shelton, "The Greatness of Automotive Transportation," *Motor Age* 50, no. 5 (Aug. 5, 1926): 12; McClintock, *Report and Recommendations*, pp. 2, 70–71. Citations are merely indicative.

54. On planners and the automobile, see Foster, *Streetcar to Superhighway*, pp. 61–62. On southern cities, see, for example, Blaine Brownell, "A Symbol of Modernity: Attitudes Toward the Automobile in Southern Cities [during the] 1920s," *American Quarterly* 24, no. 1 (1972): 20–24; Brownell, "The Commercial and Civic Elite: Planning in Atlanta, Memphis and New Orleans in the 1920s," *Journal of Southern History* 41, no. 3 (Autumn 1975): 339–68. The phrase "Commercial and civic elite" is Brownell's. See also Preston, "A New Kind of Horizontal City," pp. 101–5, 152–233. On Detroit, see Donald F. Davis, "The City Remodelled: The Limits of Automotive Industry Leadership in Detroit, 1910–1929," *Histoire Sociale* [Canada] 13, no. 26 (Nov. 1980): 465–86.

55. Wacker, *An S-O-S*, p. 14. See also, for example, Eugene S. Taylor, "The Relation of Transportation to the Chicago Plan," *JWSE* 29, no. 3 (March 1924): 118; BLI, *The Chicago Plan by the Board of Local Improvements* (Chicago, 1922), p. 7; *City Club Bulletin* 12, no. 24 (Oct. 25, 1919): 214; South Park Commissioners, *Annual Report . . . 1924*, p. 18; Eugene S. Taylor, "The Plan of Chicago in 1924, with Special Reference to Traffic Problems and How They Are Being Met," *AAAPSS* 116 (Philadelphia, 1924): 224–31. These citations are merely indicative.

After the demands of auto traffic and the value of the plan for the city's economy as a whole, the promise that street improvements would create jobs was perhaps the third most important argument made in their favor during the 1920s. See, for example, Charles H. Wacker, *An Appeal to Businessmen: Provide Work Now for the Unemployed—The Relation of National Prosperity to City Planning, Business, and the Chicago Plan* (Chicago, 1921), pp. 8–9; CPC, *South Water Street Facts* (Chicago, 1922), p. 10; "A Historian," *Big Bill the Builder: A Chicago Epoch* (Chicago, 1927) (a campaign pamphlet for William Hale Thompson), p. 6; BLI advertisement in *Chicago Defender* (a newspaper with a largely black readership) April 7, 1928.

56. Compare CPC, *Ten Years' Work of the Chicago Plan Commission, 1909–1919: A Resume of the Work Done on the Plan of Chicago* (in CPC, *Proceedings* 5 [1919–1921]: 1034), with CPC, *West Side Superhighway* (Chicago, 1929), p. 21, and "A Chicago Traffic Problem Solved by a Weekend Job," *American City* 45, no. 12 (Dec. 31, 1931): 756.

57. *Chicago Commerce*, Oct. 4, 1924, p. 23; South Park Commissioners, *Annual Report . . . 1924*, pp. 4–5; *Chicago Herald and Examiner*, May 2, 1924, July 16, 1925; Eugene Taylor, *The Outer Drive Along the Lake Front* (Chicago, 1929), p. 27.

On streetcar-free streets for automobiles, see *Chicago Commerce*, Dec. 27, 1924, p. 9; similarly Linn White, "Parks and Parkways," in American Society of Civil Engineers, *Transactions* 86 (1923): 1347.

58. Taylor, "Relation of Transportation to the Chicago Plan," pp. 118, 120.

59. On planners and the automobile nationwide, see Foster, *Streetcar to Superhighway*, pp. 91–115. On the parameters within which southern planners had to work, see Brownell, *The Urban Ethos in the South*, pp. 171–80. Brownell has generously shared his insights on this subject with the author.

60. BLI, *A Sixteen Year Record of Achievement*, p. 80; CPC, *The Plan of Chicago in 1925* (Chicago, 1925), pp. 5, 12, 16.

61. BLI, *A Sixteen Year Record of Achievement*, pp. 80, 101; Harrington, Kelker, and De Leuw, *Comprehensive Local Transportation Plan*, p. 68.

62. Bruce Allen Hardy, "American Privatism and the Urban Fiscal Crisis in the Interwar Years: A Study of the Cities of New York, Chicago, Philadelphia, Detroit and Boston, 1915–1945" (Ph.D. diss., Wayne State University, 1977), pp. 98, 202, 215.

63. McClintock, *Report and Recommendations*, pp. 58–59; R. F. Kelker, Jr., in CLT, *Proceedings* 99 (Dec. 4, 1925): 506; Thomas MacDonald and H. B. Fairbank, "The Development of Improved Highways," *JWSE* 31, no. 4 (April 1926): 115.

64. H. J. Fixmer, (engineer, BLI), *Data on Overloading of Trucks* (report to the subcommittee of the City Council Committee on Efficiency, Economy, and Rehabilitation) (Chicago, 1928), p. 16; McClintock, *Report and Recommendations*, p. 58.

65. BLI, *A Sixteen Year Record of Achievement*, p. 80.

66. On the cost of South Water Street, see Hugh E. Young, "The South Water Street Improvement," *JWSE* 30, no. 3 (March 1925): 79; BLI, *A Sixteen Year Record of Achievement*, p. 80.

On property owners' suits, see *Chicago Herald and Examiner*, March 8, 1924; *Chicago Evening Post*, March 8, 1924; *Chicago Tribune*, Sept. 28, 1923. On court judgments, see *Chicago Herald and Examiner*, May 6, 1926; *Chicago Evening Post*, May 7, 1926, Jan. 7, 1927; CPC, *Proceedings* 6 (18th Annual Report of the CPC for the Year 1927): 1375.

On improvement blocked by property owners, see *Chicago Tribune*, May 4, 1924.

67. CPC, Executive Committee, *Proceedings*, Jan. 25, 1918, p. 934; *Chicago Tribune*, June 25, 1923 (editorial); *Chicago Commerce*, Jan. 25, 1930, 52.

68. *Chicago Commerce*, Oct. 25, 1919, 10; City Council, *Journal . . . 1917–1918*, pp. 2405, 2506; CBPE, *The City Bond Issue to Be Voted Upon February 22, 1921*

(Chicago, 1921), p. 1; *Chicago American,* May 22, 1924, Nov. 17, 1925; *Chicago Tribune,* May 22, 1924, Nov. 24, 1925; *Chicago Commerce,* Jan. 2, 1926, 12; CPC, *Proceedings,* 29th meeting (June 27, 1927): 1348; *Laws of Illinois, 55th General Assembly, 1927* (Springfield, 1928), pp. 226–27.

69. CPC, *West Side Superhighways,* p. 202. Quotation is from CPC engineer Hugh Young, "South Water Street Improvement," p. 75. See also Eugene S. Taylor (manager of the CPC), "Chicago's Superhighway Plan," *National Municipal Review* 18, no. 6 (June 1929): 393.

On the number of autos entering Chicago in 1925, see Major George A. Quinlin, "Vehicular Traffic in Chicago and Cook County," *JWSE* 30, no. 4 (April 1925): 163.

70. McClintock, *Report and Recommendations,* p. 14.

71. *Chicago Tribune,* Nov. 7, 1918, Nov. 1, 3, 1926; CBPE, *A Protest Against the Proposed New County Road Tax* (Chicago, 1923), p. 5; *Chicago Commerce,* Oct. 30, 1926, pp. 12–13; *Journal of the House of Representatives of the State of Illinois,* 56th General Assembly (March 5, 1929), p. 9; *Laws of Illinois,* 56th General Assembly, pp. 625–31; City Council, *Journal . . . 1929–1930,* April 18, 1929, p. 94.

72. Chicago Regional Planning Association, "Planning Progress in the Region of Chicago" (Chicago, Oct. 1927) (typescript in Municipal Reference Library, unpaged), pp. 3, 4, 7, 8, 9; *Chicago Commerce,* March 10, 1923, p. 20, May 5, 1923, pp. 11–12; *City Club Bulletin* 17, no. 11 (March 17, 1924): 42; Robert Kingery, "Highway Planning in the Chicago Region," *JWSE* 35, no. 1 (Feb. 1930): 14–15, 16–22.

73. *Chicago Tribune,* July 15, 1925; *Chicago Daily News,* Nov. 19, 1925; South Park Commissioners, *Annual Report . . . 1924,* p. 18. *Chicago Commerce,* Aug. 6, 1927, p. 8.

74. George A. Damon, "The Influence of the Automobile on Regional Transportation Planning," American Society of Civil Engineers, *Transactions* 88 (1925): 1132. See also "Attacking the Traffic Problem," *American City* 137, no. 22 (June 18, 1924): 257; Harland Bartholomew in Arthur S. Tuttle, "Increasing the Capacity of Existing Streets," in American Society of Civil Engineers, *Transactions* 88 (1925): 238–39.

75. Burnham and Bennett, *Plan of Chicago,* pp. 92–94; "An Ordinance of the City Council of the City of Chicago, Nov. 20, 1863" (in CSL Archives, box 12, doc. 79).

76. *Chicago Journal,* Feb. 7, 1927 (the quotation). The preference of motorists for boulevards is apparent from McClintock, *Report and Recommendations,* Appendix B, pp. 253–74. See also CBPE, *The Bond Issues to Be Voted Upon February 24, 1931* (Chicago, 1931), p. 17; *Chicago Tribune,* May 5, 1929; Kingery, "Highway Planning," pp. 22–23.

77. McClintock, *Report and Recommendations,* pp. 125–26; "Street Traffic Control," *American City,* 28, no. 2 (Feb. 1923): 215; C. E. DeLeuw, "Street Traffic Control," *JWSE* 34, no. 9 (Sept. 1929): 506.

78. *Chicago Evening Post,* Nov. 7, 1922; City Council, *Journal . . . 1922-1923,* pp. 994–95, 1303–4, 1549–50, 1875. On the CPC effort to rationalize the through-street system, see Taylor, "Relation of Transportation to the Chicago Plan," pp. 116, 121, 124–25; *Chicago Tribune,* June 5, 1924. The quotation in the text is from Elkins, "Private Auto Traffic," p. 104.

79. CLT, *Proceedings* 91 (May 4, 1923): 132.

80. Foster, *Streetcar to Superhighway,* p. 94.

81. [Chicago] Department of Superhighways, *A Comprehensive Superhighway*

Plan for the City of Chicago (Chicago, 1939), map. See also Carl Condit, *Chicago, 1930–1970: Building, Planning and Urban Technology* (Chicago, 1974), p. 45 n. 3 and p. 233.

82. *Chicago Journal*, Aug. 9, 1923; *Chicago Herald and Examiner*, April 28, 1926; Edward Bennett, *The Axis of Chicago* (Chicago, 1929), pp. 15, 17, 31, 39–43; CPC, *West Side Superhighways* (Chicago, 1929), pp. 13, 15, 51–69, 174, 212, 219.

83. *Chicago Commerce*, Aug. 13, 1927, p. 8; Sept. 14, 1929, p. 12; Carl Condit, *Chicago, 1930–1970*, pp. 23–25.

84. On the development of the superhighway in Chicago, see Hugh E. Young, "Ten Mile, $60 Million Express Motor Highway Proposed for Chicago," *American City* 38, no. 3 (March 1928): 92; CPC, "Summary of Plans for Avondale Superhighway" (typescript in Chicago Municipal Reference Library); CPC, *Proceedings*, 29th meeting (June 27, 1927): 1343. On the Detroit plan, see Foster, *Streetcar to Superhighway*, pp. 78–80.

85. Subcommittee on Two Level Streets and Separated Grades of the Committee on Traffic Regulation and Public Safety of the City Council of the City of Chicago, *A Memorandum and Preliminary Report with Reference to Elevated Through Highways for the Chicago Metropolitan Area* (Chicago, 1928), unpaged, p. 4; Subcommittee on Two Level Streets, *Further Preliminary Report*, (1929), unpaged, pp. 1–2; Committee on Traffic and Public Safety . . . , *Limited Ways for the Greater Chicago Traffic Area* (Chicago, 1932), pp. 29–30.

86. On Chicago's vehicle tax, see *Chicago Commerce*, Oct. 27, 1928, p. 11; Jacob Viver, "Report of Investigation of Urban Aspects of the Problem of Highway Finance," in National Highway Research Board, *Proceedings of the Fifth Annual Meeting* (Washington, D.C., 1926), p. 220.

Illinois had no motor fuel tax until 1929. Through 1931 its tax rate per vehicle remained among the nation's lowest (National Industrial Conference Board, Inc., *Taxation of Motor Vehicle Transportation* [New York, 1932], pp. 136, 155–56).

Massen's statement is in Committee on Traffic and Public Safety, *The Greater Chicago Traffic Area: A Preliminary Report on Major Traffic Facts of the City of Chicago and the Surrounding Region* (Chicago, 1932), p. 11.

87. Subcommittee on Two Level Streets, *Memorandum and Preliminary Report*, pp. 12–13; Committee on Traffic and Public Safety, *The Greater Chicago Traffic Area*, p. 8; Committee on Traffic and Public Safety, *A Limited Way Plan for the Greater Chicago Traffic Area: A Physical and Financial Program for Construction in the City of Chicago and the Surrounding Region*, (Chicago, 1933), pp. 44–59.

88. Subcommittee on Two Level Streets, *Memorandum and Preliminary Report*, p. 12; CLT, *Proceedings*, 134 (Jan. 27, 1937): 69 of meeting.

89. Cook County Board of Election Commissioners, Election Returns and Ballot Propositions, April 5, 1927, April 10, 1928, Nov. 6, 1928, Nov. 5, 1929 (in Chicago Municipal Reference Library).

90. *Chicago Tribune*, Nov. 2, 3, 1918, March 30, 1919; CPC, *South Water Street Must Go* (Chicago, 1917), p. 54; *Chicago Daily News*, Sept. 29, 1923; *Chicago Herald and Examiner*, March 4, 1927; *Chicago Journal*, March 4, 1927. The Chicago Bureau of Public Efficiency, good-government "watchdog" group, proved most wary of unnecessary bond issues. See, for example, CBPE, *The City Bond Issues to Be Voted Upon June 5, 1922: Vote "No" on Both Propositions* (Chicago, 1924), p. 4.

91. Examples of political identification with the Chicago Plan include BLI, *The*

Chicago Plan by the Board of Local Improvements (Chicago, n.d.); BLI, *A Sixteen Year Record of Achievement*, esp. pp. viii–x; A Historian, *Big Bill the Builder: A Chicago Epoc* (Chicago, 1927); campaign advertisement for William E. Dever in *Chicago Herald and Examiner*, April 1, 1927. Citations are merely indicative.

CPC spokesmen praise politicians in Charles H. Wacker, *The Michigan Avenue Extension* (Chicago, 1918), p. 7; CPC, *Proceedings* 5 (23rd meeting): 117; CPC, *Ten Years' Work*, pp. 1041–43.

92. CBPE, *The Bond Issues to Be Voted Upon November 4, 1924* (Chicago, 1924), p. 5; CBPE, *Tax and Bond Issue Propositions to Be Voted Upon Nov. 6, 1923* (Chicago, 1923), pp. 10–11; CBPE, *The Bond Issues to Be Voted Upon February 24, 1931* (Chicago, 1931).

93. CBPE, *The Bond Issues to Be Voted Upon November 24, 1924*, p. 5.

94. *Chicago Journal*, Dec. 27, 1924; Fixmer, *Data on Overloading of Trucks*, p. 16.

95. CBPE, *The Bond Issues to Be Voted Upon June 5, 1922* (Chicago, 1922), p. 4; *Chicago Tribune*, Nov. 4, 1928, Dec. 31, 1930 (on the alleged rebate).

96. On Chicago tax rates, see Francis X. Busch to William E. Dever, Feb. 11, 1926 (in Dever, *Papers*, box 6). On complaints of high taxation, see *City Club Bulletin* 11, no. 44 (Nov. 3, 1919): 215; *Chicago Journal*, Dec. 21, 1924. On the 1926 bond issues see CBPE, *Bond Issues to Be Voted Upon April 13, 1926* (Chicago, 1926), p. 13.

97. Herbert D. Simpson, *The Tax Racket* (Chicago, 1930), pp. 34–35, 64–65.

98. Ibid., p. 125. Homer Hoyt found in 1933 that Loop property values in 1926 had constituted one-fifth of those of the entire city (*100 Years of Land Values in Chicago* [Chicago, 1933], pp. 336–37).

99. A report of one tax-fixing scandal is in *Chicago Daily News*, April 10, 1926. On public apathy, see Simpson, *Tax Racket*, p. 126.

100. Simpson, *Tax Racket*, pp. 122, 124–25, 127.

101. Ibid., pp. 121, 135, 185; *Chicago Commerce*, March 8, 1930, p. 16; Liberal Club of Chicago, *High Cost of Government*, pp. 8–11; Herbert D. Simpson, "Chicago Complex," *Atlantic Monthly* 146, no. 10 (Oct. 1930): 544–46; *Business Week* (Aug. 27, 1930), pp. 24–25.

102. John O. Rees, *A Study of Real Estate Tax Delinquency in Chicago and Cook County* (Chicago, 1938), p. 4.

103. *Chicago Daily News*, Nov. 7, 1928; *Chicago Journal*, Nov. 16, 1928.

104. *Chicago Commerce*, Jan. 19, 1929, p. 30, Oct. 26, 1929, p. 12, Jan. 31, 1931, p. 36; *Chicago Daily News*, Nov. 4, 5, 1928; *Chicago Tribune*, Nov. 4, 1929.

105. Hardy, "American Privatism," p. 230.

106. *Chicago Tribune*, April 2–9, 1928; *Chicago Daily News*, April 7, 9, 11, 1928; *Chicago Evening Post*, April 6, 1928. *The Chicago Journal* (April 9, 1928) urged voters not to stop the Chicago Plan merely in order to hurt Thompson. The *News* and the *Tribune*, however, called for defeat of the bond issues as a rebuke to Thompson and Faherty.

107. *Chicago Tribune*, June 6, 1930. In November, 1930, the Civic Federation, the Citizens Association, CAC, the Governor's Relief Council, and the State Street Council supported $23 million in bond issues as weapons with which to combat the Depression (*Chicago Tribune*, Nov. 2, 1930). The CBPE, however, argued in 1931 that "not even the prospect of providing work for the unemployed justifies voting for bond issues which are otherwise undesirable and unnecessary" (*The Bond Issues to Be Voted Upon February 24, 1931*, p. 6).

108. On the successful efforts of bankers to force cuts in city expenditures, see "Chicago Broke," *Outlook and the Independent* 154, no. 5 (Feb. 5, 1930): 214; Herbert Gosnell, *Machine Politics: The Chicago Model* (Chicago, 1937), pp. 13, 17, 25. Mayor Anton Cermak's inaugural speech, reflecting his determination to send to the voters "no questionable or ill-considered" bond issues is in City Council, *Journal . . . 1930–1931*, April 27, 1931, pp. 89–91.

109. McClintock, *Report and Recommendations*, p. 101. In the original, the final sentence is italicized.

110. CLT, *Proceedings* 99 (Dec. 4, 1925): 443.

111. "Traffic Control in Chicago," *Roads and Streets* 65, no. 12 (Dec. 1926): 340.

112. *Chicago Commerce*, Feb. 6, 1919, p. 10, June 5, 1920, pp. 34–35; *Chicago Daily News*, May 2, Dec. 1, 1923.

113. *Chicago Commerce*, Jan. 31, 1924, p. 10, Nov. 14, 1925, p. 11; City Council, *Journal . . . 1925–1926*, pp. 1587–88. The CAC paid the cost of the study.

114. Miller McClintock, *Street Traffic Control* (New York, 1925), pp. 137–84.

115. Martin, "Obstruction of Street Traffic," p. 150; *Chicago Tribune*, Nov. 1, 1921; South Park Commissioners, *Annual Report . . . 1924*, p. 10. CAC opposition to traffic towers is reported in *Chicago Commerce*, June 11, 1921, p. 12, Jan. 14, 1922, p. 29. On the Chicago Motor Club's approval of synchronized traffic signals and opposition to this approach from the Chicago Surface Lines and city council engineer R. F. Kelker, Jr., see Eikins, "Private Auto Traffic," pp. 106–7. See also, R. F. Kelker, Jr., "Capacity of Roadways, *JWSE* 31, no. 4 (April, 1926): 141.

On the Michigan Avenue installation, see "Traffic Improvements Now Under Way in Chicago," *Motor Age* 42, no. 17 (Oct. 26, 1922): 30; *Chicago Commerce*, June 30, 1923, p. 10.

116. *Chicago Daily News*, Sept. 25, 1924; *Chicago Commerce*, Aug. 9, 1924, p. 58; *Chicago Herald and Examiner*, July 21, 1925; *Chicago Journal*, July 27, 1925.

117. William Canning, "A Study of the Traffic Problem and Traffic Control," *Roads and Streets* 68, no. 2 (Feb. 1928): 100; McClintock, *Report and Recommendations*, p. 186. The descriptions in this paragraph of the text are consistent with those of McClintock, but are largely the present author's conjecture based on interviews with former CSL staff engineer E. J. McIlraith (Dec. 14, 1973) and former CSL engineer Louis Traiser (Feb. 4, 1975).

118. McClintock, *Street Traffic Control*, pp. 131–43.

119. Chicago Surface Lines used advertisements to popularize traffic control. Examples are in *Chicago Commerce*, July 26, 1924, p. 10, Sept. 13, 1924, p. 33, Oct. 7, 1925, p. 1, Jan. 2, 1926, p. 1.

120. Interview with E. J. McIlraith, Dec. 15, 1973.

121. Ibid., Dec. 15, 16, 1973; Henry M. Lews, "Metropolitan Traffic Control," *Roads and Streets* 65, no. 5 (May 1926): 278.

122. *Chicago Commerce*, March 13, 1926, p. 7; *Electric Traction*, 22, no. 2 (Feb. 1926): 57–60; *Chicago Tribune*, Feb. 2, 7, 1926.

On the elimination of left turns, see *Chicago Daily News*, July 1, 1925.

123. *Chicago Commerce*, March 13, 1926, p. 7, May 1, 1926, p. 10; *Chicago Tribune*, Feb. 2, 1926.

124. Miller McClintock, "Street Traffic Bibliography" (mimeographed, Erskine Bureau for Street Traffic Research, Cambridge, Mass., 1933), p. 221; McClintock, "Street Traffic Control and Electric Railway Operation," in American Electric

Railway Association, *Proceedings* (New York, 1927), p. 28.

All other sources (outside the electric railway industry) credit either the Chicago Association of Commerce, the Police Traffic Squad, Public Works Commissioner A. A. Sprague, or city Traffic Engineer Leslie Sorenson. See *Chicago Commerce*, March 13, 1926, p. 7; *Chicago Post*, Feb. 4, 1926; *Chicago Tribune*, Feb. 7, 1926; *Roads and Streets* 68, no. 1 (Jan. 1928): 45.

On the timing of lights outside the Loop by CSL engineers, see CSL Traffic and Scheduling Department, *Annual Report . . . 1928*, p. 2; *Annual Report . . . 1929*, p. 3.

125. See, for example, E. J. McIlraith, *Address . . . Before the Canadian Electric Railway Association Toronto, Canada, Thursday, June 7, 1928* (n.p., n.d.), p. 5.

126. R. F. Kelker, Jr., in CLT, *Proceedings* 99 (Dec. 4, 1925): 503. On the CAC's longtime position, see *Chicago Commerce*, June 11, 1921, p. 12.

127. Young, "Day and Night Storage," p. 45; CLT, *Proceedings* 99 (April 12, 1925): 443.

128. Mr. Vorhees of Brentano's bookstore in CLT, *Proceedings* 99 (Dec. 4, 1925): 509. See also CLT, *Proceedings* 99 (Dec. 1, 1925): 353.

129. *Chicago Commerce*, Jan. 2, 1926, p. 1; CLT, *Proceedings* 99 (Dec. 1, 1925): 357–60, 368, 373.

130. McClintock, *Report and Recommendations*, table 29 (p. 159). The McClintock Survey was clearly expected to produce the determining evidence on this subject. See CLT, *Proceedings* 99 (Dec. 4, 1925): 469, 505.

131. *Chicago Tribune*, June 17, 21, 1926; *Chicago Commerce*, June 19, 1926, p. 1.

132. *Chicago Herald and Examiner*, Jan. 18, 21, 1927; *Chicago Daily News*, Jan. 18, 1927.

On the passage of the December 1927 ordinance, see City Council, *Journal . . . 1927–1928*, Dec. 14, 1927, p. 1569. The CAC's story of the background of and struggle for this ordinance is in *Chicago Commerce*, Dec. 17, 1927, p. 11. The fine was $1.

133. *ERJ* 71, no. 5 (Feb. 4, 1928): 188–90; Robert H. Naw, "No Parking, A Year and More of It," *American City* 40, no. 3 (March 1929): 85–88; *Chicago Commerce*, Jan. 14, 1928, pp. 16–17.

134. Aldis and Company, Papers (folder marked "1928–1939," first item, at Chicago Historical Society).

135. Miller McClintock, "Movement of Traffic a Vital Community Problem," in American Electric Railway Association, *Proceedings*, 1935, p. 101. The parking ban was possible only when off-street parking had become adequate. In 1927 Chicago had storage space for 14.961 automobiles in public and private garages and lots and on unused streets. The curb capacity of the Loop before the parking ban was 1,157 cars (*Chicago Commerce*, Dec. 24, 1927, p. 12).

136. City Council, *Journal . . . 1925–1926*, pp. 2093, 2485–88; *Journal . . . 1926–1927*, pp. 5728–35.

137. *Chicago Commerce*, March 20, 1926, p. 9; McClintock, *Report and Recommendations*, pp. 195–98; *Annual Reports . . . 1925–1931*, esp. *Annual Report . . . 1929*, p. 12.

138. McClintock, *Report and Recommendations*, pp. 197–98; *Chicago Commerce*, Dec. 10, 1927, p. 22; City Council, *Journal . . . 1927–1928*, p. 946.

139. *Chicago Tribune*, Nov. 23, 1923; *Chicago Daily News*, Nov. 1, 1928; *Chicago Commerce*, Jan. 26, 1929, p. 17, Jan. 25, 1930, p. 40.

140. Citizens' Police Committee, *Chicago Police Problems*, p. 154; J. Rowland Bibbins, "The Growing Transportation Problem of the Masses: Rails or Rubber or Both?" *National Municipal Review* 18, no. 8 (Aug. 1929): 520; *Automobile Industries* 58, no. 3 (Jan. 21, 1928): 78; *Roads and Streets* 70, no. 2 (Feb. 1930): 62.

141. On conflicting traffic rules, see City Council, *Journal . . . 1930–1931*, p. 4168. On the failure of offenders to appear in court, see *City Club Bulletin* 23, no. 7 (Feb. 17, 1930), p. 33; *Chicago Commerce*, Nov. 20, 1929, p. 15.

142. Committee on Traffic and Public Safety, *Limited Ways For the Greater Chicago Traffic Area*, pp. 24–26; BOSE, *24th Annual Report . . . 1931*, exhibit X (author's calculation).

143. McClintock, *Report and Recommendations*, p. 21; DPW, "Automobile Storage in the Central District" (mimeographed report, Sept. 1, 1936, in Chicago Municipal Reference Library).

144. C. E. DeLeuw, "Street Traffic Control," *JWSE* 34, no. 9 (Sept. 1929): 517.

145. McClintock, *Report and Recommendations*, p. 42.

146. McLellen & Junkersfield, Inc., *Report on the Transportation in the Milwaukee Metropolitan District to the Transportation Survey Committee of Milwaukee* (n.p., 1928), 1; 160–61, 168–71.

147. McClintock, *Report and Recommendations*, p. 22.

148. Werner W. Schroeder, *Metropolitan Transit Research Study* (Chicago, 1955), p. 38. More detail on employment and ridership trends are in Evan McIlraith, Papers, vol. 1 (unpaginated), in the possession of the author.

149. Chicago Surface Lines, "Average Daily Receipts by Months, 1926–1946," in McIlraith, Papers, vol. 1.

150. On the end of traffic control innovation after 1930, see George Barton, *Street Transportation in Chicago: An Analysis of Administrative Organization and Proceedings* (Evanston, 1948), pp. 24–26. On the tendency of improved streets to create improved traffic jams, see William S. Canning, "Report of the Committee on Traffic Regulation in Municipalities: One Way Streets," in National Highway Research Council of Highway Research Board, *Proceedings of the 18th Meeting* (Washington, D.C., 1938), p. 337; Chicago Park District, *9th Annual Report . . . 1944*, p. 31; Richard C. Fencl, "Jackson Boulevard: An Analysis of Traffic Operations," in National Research Council of Highway Research Board, *Proceedings of 30th Annual Meeting* (Washington, D.C., 1950), pp. 297–98, 303; Claiborne Pell, *Megalopolis Unbound* (New York, 1966), p. 76.

151. Miller McClintock, "How to Make a Community Traffic Survey," *American City* 35, no. 6 (Dec. 1926), p. 773.

152. McClintock, "How to Make a Community Traffic Survey," p. 772; R. F. Kelker, Jr., "City Traffic Problems," *Engineers and Engineering* 44, no. 4 (April 1927): 94.

Chapter 6

1. *Chicago Tribune*, Feb. 12, 1924.

2. Clarence Darrow to Donald Richberg, June 15, 1930, in Donald Richberg Papers, box 5, envelope 55, Chicago Historical Society.

3. Endorsements by various groups are reported in Chicago Association of Commerce, *Approval of the Traction Ordinance*, June 3, 1930 (copy at Chicago Historical Society); *Chicago Tribune*, June 16, 17, 20, 21, 22, 26, 28, 1930; CLT, *Proceedings* 119, (May 5, 1930): morning session, pp. 53, 96–98, 103, 107–8; *City Club Bulletin* 23, no. 24 (June 16, 1930).

4. Chicago City Council, *An Ordinance Providing for a Comprehensive Unified Local Transportation System for the City of Chicago and Metropolitan Area*, (passed by the City Council of the City of Chicago on May 19, 1930) (Chicago, 1930), pp. 2–3, 16–18 (hereafter *1930 Ordinance*).

5. Paul H. Douglas, "Chicago's Persistent Traction Problem," *National Municipal Review* 18 (Nov. 1929): pp. 669–75; CLT, *Proceedings* 114 (March 22, 1929): afternoon session, p. 43.

6. *1930 Ordinance*, pp. 2, 12–24.

7. Ibid., pp. 3–4, 12–13, 18.

8. CLT, *Proceedings* 38 (Jan. 21, 1913): 385; 39 (Feb. 13, 1913): 259 ff.; *Chicago Daily News*, Feb. 13, 1913.

9. *Chicago Daily News*, Feb. 2, 1923.

10. CLT, *Proceedings* 81 (Jan. 13, 1921): 74 and 81 of meeting; BOSE, *37th Annual Report*, exhibit T; Chicago Transit Authority, *Historical Information*, p. vii (calculations by the author). See also R. F. Kelker in Chicago City Club, *City Club Bulletin* 15, no. 19 (May 8, 1922): 75. On extensions, see CLT, *Proceedings* 90 (Dec. 14, 1922): 56–65. Insull's statement is in *Chicago Tribune*, Jan. 12, 1922.

11. Letter of Evan McIlraith to author, Nov. 28, 1974. On the causes of crowding, see CLT, *Proceedings* 81 (Jan. 13, 1921): 80–81; 87 (May 4, 1922): 3; *Chicago Daily News*, May 14, 1921, Jan. 17 (letters), Feb. 1, 2, 1923; William Reid, Commissioner of Public Service, *Annual Reports of the Department of Public Service, 1916–1920* (Chicago, 1921), pp. 10 and 14.

The Illinois Commerce Commission required that the Surface Lines' scheduling department submit evidence for the necessity of each schedule change, thus slowing the process of adapting car schedules to traffic and ridership (interview with Leroy Dutton [scheduler, later head of scheduling for the Chicago Transit Authority], Nov. 21, 1974).

12. Walter Jackson, "The Sale of the Ride," in *Papers Presented to the Illinois Electric Railway Association*, May 15, 1921, p. 7; *Electric Railway Journal* 63, no. 11 (Sept. 3, 1924); McDonald, *Insull*, pp. 243–45; *Chicago Tribune*, Oct. 29, 1923.

13. On elevated extensions, see Chicago Rapid Transit, *What to See and How to Go* (Chicago, 1926) (a flier with route information and public relations material, copy in Chicago Historical Society); *Chicago Daily News*, April 1, 1925; *Chicago Journal of Commerce*, Feb. 1, 1924; *Chicago Commerce* (April 4, 1925), pp. 17, 19. See also advertisements in *Chicago Daily News*, March 1, April 8, 1926, for "Zelosky's Westchester"—in a suburban area along a rapid transit extension. Each advertisement is headed: "Where the "L" goes, profit follows."

Insull explains his motives for these extensions in Samuel Insull, *Chicago's Future* (Chicago, 1924), pp. 5–7; Samuel Insull, *Rapid Transit Development for Chicago: Delivered before the Chicago and Cook County Bankers' Association, Union League Club, February 21, 1924* (n.p., n.d.), p. 3; Samuel Insull, *Memoirs*, pp. 191–93 (in Samuel Insull Papers, Cudahey Library, Loyola University, Chicago).

On the failure of the rapid transit extensions to attract or create patronage, see

Ralph J. Burton, "Mass Transportation in the Chicago Region: A Study of Metropolitan Government" (Ph.D. diss., University of Chicago, 1939), pp. 10-14.

Dever's statement is in *Chicago Tribune*, Dec. 12, 1924.

14. As will be recalled from Chapter 1, the Morgan syndicate began buying into Chicago City Railways in 1903. Chicago City's professional transit manager, T. E. Mitten, was replaced by banker Leonard Busby in 1910, and Busby became head of the CSL in 1914. Blair's relationship with Insull is discussed in Chapter 3.

15. McIlraith interview, Dec. 12, 1973. *Electric Railway Journal* 64, no. 14 (April 7, 1923): 616; A. N. Marquis Company, *Who's Who in the Midwest* (Chicago, 1949), pp. 841, 1044.

16. *Chicago Tribune*, Feb. 9, 1923; *Chicago Daily News*, Feb. 12, 1923; *Chicago Herald and Examiner*, Feb. 14, 20, 1923; *Chicago Economist*, Feb. 10, 1923.

17. Condit, *Chicago, 1910–1929*, pp. 235–36. Condit also cites the lack of subways. On new cars, see CTA, *Historical Information*, p. vii. On schedule changes, see table marked "CSL Schedule Changes, 1/1/23–3/1/24" in Dever Papers, box 4, folder 35, Chicago Historical Society.

18. *ERJ* 56, no. 7 (April 14, 1920): 337; 62, no. 17 (Oct. 27, 1923): 758; 67, no. 21 (May 22, 1926): 908; Chicago Surface Lines, *Chicago Surface Lines Yearbook, 1925– 1926* (Chicago, 1926), p. 9.

19. The number of cars scheduled as of January 31, and the number of total passengers carried each year fluctuated as follows:

Cars operating, 1–31 (*Above or below previous year*)	*Total passengers carried* (*compared with previous year*)
1922 over 1921: + .01%	+ .02%
1923 over 1922: + .02%	+ .08%
1924 over 1923: + .07%	+ .02%
1925 over 1924: − .01%	+ .02%
1926 over 1925: + .04%	+ .02%
1927 over 1926: = .03%	+ .04%

Data are from BOSE, *37th Annual Report*, exhibit T, and tables headed "Chicago Surface Lines: Scheduled cars as of January 31, 19—," bound with records of car assignments in the Office of Traffic and Scheduling, Chicago Transit Authority. System-wide data do not reflect the placement of cars or riders.

20. Chicago Surface Lines, Traffic and Scheduling Department, *Annual Report . . . January 31, 1924*, pp. 1–4, 7–8, 12; *Annual Report . . . 1925*, pp. 2–3; *Annual Report . . . 1926*, pp. 1, 3.

21. CLT, *Proceedings* 95 (Jan. 19, 1925): 583; *Chicago Daily News*, March 24, 1925; Chicago Surface Lines, Traffic and Scheduling Department, *Annual Report . . . 1929*, p. 1; *Annual Report . . . 1928*, p. 1.

22. American Transit Association, *Fare Structures in the Transit Industry* (New York, 1933), pp. 154, 198.

23. See all CLT meetings between Dec. 8 and Dec. 18, 1924, for example, CLT, *Proceedings* 94 (Dec. 18, 1924): 369. See also, for example, CLT, *Proceedings* 101 (July 13, 1926): 1083–84. Business, real estate, and neighborhood group lobbying for extensions can be found in almost every volume of the CLT's *Proceedings* between 1922 and 1930.

24. CLT, *Proceedings* 92 (Dec. 25, 1924): 333–37; 92 (March 7, 1925): 537; 101

(Dec. 14, 1931): 21, 34; *Chicago North Side Citizen*, Nov. 12, 1924; *Chicago Journal*, Nov. 2, 1925; *Chicago Daily News*, June 4, 1928.

25. William E. Dever, *Special Message of the Honorable William E. Dever, Mayor, Concerning Chicago's Local Transportation Problem* (submitted to the City Council of the City of Chicago, Oct. 22, 1924) (Chicago, 1924). See map.

26. BOSE, *Report on Certain Operating Features of the Chicago Surface Lines*, July 14, 1926.

27. See tables and maps in BOSE, *26th, 27th,* and *32nd Annual Reports.*

28. CLT, *Proceedings* 92 (Dec. 7, 1923): 113. Leroy Dutton, in his interview with the author, recalled that 100 persons per car was an accepted maximum through-out his career with the Surface lines.

On new cars, see BOSE, *23rd Annual Report*, p. 102; *ERJ* 73, no. 12 (March 23, 1929): 473.

29. On average fares between 1917 and 1932, see American Transit Association, *Fare Structures*, pp. 183–84. The average fare per ride (including transfer rides) was 3.85¢ in 1927. CSL's operating ration was 79.44 in 1927, as against a national average for surface companies of 77.2% (BOSE, *24th Annual Report*, exhibits W, X; Bureau of the Census, *Census of Electrical Industries, 1927* [Washington, D.C. 1931], p. 111).

30. *ERJ* 68, no. 6 [Aug. 7, 1926]: p. 235.

31. *ERJ* 69, no. 9 (Feb. 26, 1927): 368.

32. These efforts were enmeshed in the struggle between the city and the state public utilities commission, and policy toward busses was in part a reflection of this struggle. For the sake of brevity, this aspect of the local history of the bus is not dealt with except as it affected policy during the 1920s.

33. Poem is in undated clipping in Emily Larned, *Scrapbook* (unpaginated, Chicago Historical Society) (date inferred from context).

Quotation is in unidentified clipping in Larned, *Scrapbook*, April 11, 1915. See also, for example, *Chicago Tribune*, May 1, 6, June 13, 1915; *Chicago American*, March 22, 1915.

34. See correspondence between Larned and a variety of park board officials and politicians in Larned File, Chicago Historical Society. The bus company's strug-gles may be traced through *ERJ* 43, no. 10 (March 7, 1914): 555; DPS, *1st Annual Report* (1914), pp. 79–81; *ERJ* 51, no. 4 (Jan. 26, 1918): 98; 55, no. 22 (May 29, 1920): 1110–11; 56, no. 25 (Dec. 18, 1920): 1258. See also an untitled Chicago Motor Bus Company booklet, with the Larned *Scrapbook*, Chicago Historical Society. The photo is in *Chicago Herald and Examiner*, March 26, 1917. The final couplet is in an undated clipping in Larned, *Scrapbook.*

35. Petition in *Chicago Herald and Examiner*, March s, 1913. See also *Chicago American*, March 22, 1915; *Chicago Evening Post*, March 25, 1915; *Chicago Tribune*, March 17, 1915. See also a variety of unidentified clippings in Larned, *Scrapbook.*

36. *Bus Transportation* (hereafter *BT*), 1, no. 2 (Feb. 1922): 117; 1, no. 9 (Sept. 1922): 472–73.

37. *BT* 1, no. 1 (Jan. 1922): 31; Illinois Commerce Commission, *Opinions and Orders*, vol. 2 (1922–1923): In the Matter of the Chicago Motor Coach Company Relative to a Certificate of Convenience and Necessity to Operate Motor Buses on the South Side of the City of Chicago, Case 6066 (consolidated with other cases), pp. 67–68.

38. CLT, *Proceedings* 92 (March 21, 1924): 624–34.

39. Illinois Commerce Commission, *Opinions and Orders*, vol. 18 (1938–1939), City of Chicago v. A. A. Sprague *et al.* . . . Case 22981 (*Twenty-First Supplemental Order*), pp. 1128–32; Chicago Railways Company, *Corporate Record* 5 (Nov. 13, 1935): 395; CLT, *A Comprehensive Local Transportation Plan for the City of Chicago, November 22, 1937* (Chicago, 1937), fig. 2.

40. On the separate class of service constituted by Chicago busses, see *BT* 1, no. 12 (Dec. 1922): 629; CLT, *Proceedings* 90 (Dec. 8, 1922): 25; 98 (June 12, 1925): 476.

On the CMC's Outer Drive express service, see *Mass Transportation* 31, no. 8 (Aug. 1935): 231–32. On the percentage of traffic on each line, see charts in *McIlraith Collection*, volume labeled "Traffic" (in the possession of this author).

41. On Hertz's early career, see *Motor Coach Age* 24, no. 3 (Mar. 1972): 7–8; *BT* 2, no. 3 (Feb. 1923): 126.

42. John Ritchie (president of CMC), *The Place of the Motor Coach in a General Comprehensive Transportation Plan for the City of Chicago* (Chicago, 1923), p. 3.

43. *BT* 7, no. 7 (Aug. 1928): 424–25; *ERJ* 75, no. 10 (Sept. 15, 1931): 546–60; letter of Evan McIlraith to author, April 11, 1974.

On the Chicago Motor Coach Plan for complete bus substitution, see CLT, *Proceedings* 104 (Dec. 10, 1926): 4–8 of meeting; *ERJ* 68, no. 5 (Dec. 18, 1926): 1105–6; *BT* 1, no. 6 (Jan. 1927): 51–52. McIlraith's statement is in a letter to this author, Jan. 26, 1974.

44. Snell, "American Ground Transport"; David J. St. Clair, "The Motorization and Decline of Urban Public Transit, 1935–1950," *Journal of Economic History* 41 no. 3 (Sept. 1981): 579–600, esp. pp. 595–600. (St. Clair does not necessarily subscribe to Snell's hypothesis.)

45. For Kelker's position see CLT, *Proceedings* 90 (Dec. 8, 1922): 2–14; 104, (Dec. 22, 1926): 48–49. On the incompatibility of busses with subways, see CLT, *Proceedings* 104 (Dec. 10, 1926): 58.

46. CLT, *Proceedings* 117 (March 17, 1930): 219; 117 (April 24, 1930): 451–57. Quotation is in *Chicago Tribune*, July 3, 1924.

47. On transit building and job creation, see Henry D. Capitain, *The New Traction Ordinance* (Chicago, 1918), p. 18; Citizens' Committee for the Unification Traction Ordinance, "Why?" (undated leaflet in Hooker Collection); *Chicago Tribune*, Oct. 28, 1918; *Chicago Daily News*, Oct. 18, 1918; *Chicago Evening Post*, Nov. 1, 1918. In 1923 the president of Chicago Motor Coach estimated that a bus system required an investment of $1.25 per dollar of gross revenue against $4.00 per dollar of gross for streetcar systems (*BT* 2, no. 8 [Aug. 1923]: 387). See also Werner Schroeder, *Metropolitan Transit Research Study* (Chicago, 1955), p. 186, for post-1945 Chicago data.

48. The story is highly complex—an extreme example of the way in which state regulation played havoc with city policy. See the sources cited in the author's dissertation ("Mass Transit, the Automobile and Public Policy in Chicago, 1900–1930" [University of Illinois, 1976], pp. 608–11).

49. McIlraith to this author, Nov. 28, 1974.

50. Chicago Transit Authority, *Historical Information*, p. vii; "Taking Inventory of Vehicles, Track and Bus Routes," *Transit Journal* (1939) (reprinted in *Traction Heritage* 4, no. 6 [Nov. 1971]: 6).

On the positive effects (on ridership) of bus substitutions in other cities, see

272

Walter Shaw, "Interim Report of Walter Shaw" (in Receivership Record of Chicago City Railways, Northwestern Transportation Library, Evanston, Illinois), dated Sept. 26, 1938, pp. 39–52.

Chicagoans asked for buses instead of streetcars or elevated estensions in CLT, *Proceedings* 94 (Dec. 18, 1924): 330; 94 (Dec. 19, 1924): 378; 94 (Dec. 22, 1924): 399; 120 (May 7, 1930): 106; 135 (March 9, 1937): 101. See also Chicago Railways Company, *Annual Report for the Year Ending January 31, 1928* (Chicago, 1928), p. 6.

51. CLT, *Proceedings* 102 (Traction Conference Subcommittee) (Sept. 17, 1926): 82–83.

52. Paul R. Leach, comp., "Chicago's Traction Problem"; *Chicago Daily News*, Reprints, no. 17 (1925), pp. 17–20 (copy in Chicago Historical Society); *Chicago Tribune*, Dec. 19, 1924.

A typical company position paper issued under Blair's name is Henry A. Blair, *Plan for a Unified Transportation System for the City of Chicago* (Chicago, 1927).

53. See remarks by Walter Fisher in CLT, *Proceedings* 54 (May 27, 1914): 7 P.M. session, pp. 109–10.

On street railway overcapitalization in general, see Committee on Public Utilities of the Chamber of Commerce of the United States, *Conference on Street Railways*, 4th sess., June 13, 1919 (Washington, D.C., 1919), esp. pp. 14–15, and Bureau of the Census, *Electrical Industries*, (Washington, 1917) p. 53, table 54.

54. Arthur Stowe Dewing, *The Financial Policy of Corporations* (New York, 1922), pp. 321 ff.; H. M. Addinsell, "Public Service Bonds," *AAAPSS* 88 (March 1920): 63.

55. For a table of these securities, see *The Fitch Bond Book* (1917), pp. 98, 210–12.

56. Under modern conditions, long-term bonds must yield a higher return than those which return the investor's money to his use in a few years. Before 1930, this rule was reversed except in times of great uncertainty, e.g., 1916–1917 (George Makiel Burton, *The Term Structure of Interest Rates: Expectations and Behavior Patterns* [Princeton, 1966], pp. 6–7). Prime capital bonds maturing in a 10-year period paid 4.25% in 1910, 6.11% in 1920, and 4.40% in 1930. Thirty-year bonds paid 3.80% in 1910, 5.10% in 1920, and 4.40% in 1930. On inflation, see Dewing, *Financial Policy*, pp. 1138–39.

57. On fares, costs, and length of haul, see Addinsell, "Public Service Bonds," pp. 64–66; Dugald C. Jackson and David J. McGrath, *Street Railway Fares: Their Relation to Length of Haul and Cost of Service* (New York, 1917), pp. 2–8.

58. On cost of labor and equipment, see FERC, *Proceedings* (Washington, D.C., 1920), pp. 112, 388, 399–400; *ERJ* 53, no. 13 (Sept. 28, 1918): 589; CTA, Employee Relations Department, "History of Labor Negotiations: Wages and Working Conditions, CTA and CSL" (mimeographed, Chicago, 1972), p. 1.

The fare hike controversy may be traced through *ERJ* 55, no. 5 (Jan. 31, 1920): 265; 56, no. 20 (Nov. 13, 1920): 1034–35; 58, no. 22 (Dec. 3, 1921): 1009; 50, no. 12 (March 25, 1922): 544; 59, no. 14 (April 8, 1922): 615; 50, no. 22 (June 2, 1922): 910.

59. BOSE, *37th Annual Report*, exhibit T; *Chicago Tribune*, Jan. 27, 1922; *Chicago Daily News*, Jan. 17, 1923; *Chicago Commerce*, May 8, 1920, pp. 9–10, May 22, 1920, pp. 15, 26, 28; Chicago City Railway Corporate Record, 5: 214, 222–23, 288, 315 (in CSL Archives); *ERJ* 55, no. 19 (May 8, 1920): 939–41; 55, no. 20 (May 15, 1920): 999–1000.

60. CLT, *Proceedings* 95 (Jan. 19, 1925): 475.

61. Quotation is in CLT, *Proceedings* 102 (Sept. 17, 1926): 47 of meeting.
Sentiments against referenda and for terminable permits are in CLT, *Proceedings* 102 (Sept. 10, 1926): 14; Lobdell, *Chicago's Traction Problem in 1927*, pp. 6–7; Howard R. Traylor to William E. Dever, Feb. 1, 1926, in Dever, Papers, box 5, folder 46.

62. Barrett, "Mass Transit . . . ," pp. 583, 586–89, 628–31.

63. On Insulls influence, see Chicago City Club, *City Club Bulletin* 23, no. 26 (June 30, 1930): 63; Walter L. Fisher, *Analysis of the Traction Ordinance: A Report to the Honorable Judge James H. Wilkerson, Judge of the U.S. District Court* (Chicago, 1930), pp. 21–22; *Chicago Herald and Examiner* June 25, 1930. Insull's own position is stated in Samuel Insull, "Chicago's Future" (an address delivered at the 42nd Annual Meeting of the Chicago Real Estate Board, Dec. 4, 1924) (Chicago, n.d.), p. 11.

64. Quotation is in CLT, *Proceedings* 78 (Aug. 1, 1918): 2:30 P.M. session, p. 30.
On Insull's role in 1918, see *Chicago Commerce*, Sept. 26, 1918; CLT, *Proceedings* 100 (Jan. 11, 1926): 216, 218.

65. *Chicago Tribune*, May 20, 1923; Leach, *Chicago's Traction Problem*, p. 13; Samuel Insull, *Public Utilities in Modern Life*, pp. 65, 289, 312–14.

66. On Insull's role in the compromise (including the arrangement of half-a-million in "grease"), see McDonald, *Insull*, p. 261. On the plan's financial structure, see Halsey, Stuart and Company, Reorganization Managers, *Reorganization Plan and Agreement of Chicago Local Transportation Company* (Chicago, Dec. 1, 1930), pp. 1, 7, 13, 22–23, and charts opposite pp. 28, 43–35. On the subway, see *1930 Ordinance*, pp. 5–6.

67. McDonald, *Insull*, pp. 237–39, 261; *Chicago Commerce*, May 29, 1926, p. 38; *Chicago Tribune*, July 11, 12, Sept. 6, Nov. 23, 1924; *Chicago Journal of Commerce*, Feb. 17, 1923.

68. CLT, *Proceedings* 88 (Subcommittee on Traction Program) (June 7, 1922): 29 of meeting.

69. *Chicago Commerce* (Oct. 10, 1918), p. 7.

70. The 1925 valuation was twice the market value of the companies' securities. See Edwin Lobdell, *Chicago's Transportation Problem at the Close of 1925* (Chicago, Dec. 12, 1925), pp. 1–4; *Chicago Tribune*, Jan. 27, 1925; William E. Dever to Raymond Robbins, Dec. 22, 1924 in Dever, Papers, box 5, folder 38; CLT, *Proceedings* 115 (Sept. 10, 1929): 374–86.

71. Representative of the Western Society of Engineers in CLT, *Proceedings* 92 (Feb. 21, 1924): 512–13; James Rowland Bibbins, "City Building and Transportation," *JWSE* 25, no. 12 (Aug. 20, 1920): 435–36. See also George C. Sikes, "The Eternal Traction Question," p. 4.
On the 1930 extensions, see CLT, *Proceedings* 116 (Oct. 15, 1929): 252–54; 120 (May 10, 1930): 251–87.

72. Insull quoted in *Chicago Commerce* (Apr. 10, 1920), p. 11; Samuel Insull, "The Engineer's Influence in Public Utilities," *JWSE* 25, n. 10 (May 20, 1920): 358–60; J. Rowland Bibbins, "The Economic Future of Transportation Utilities," *JWSE* 25, no. 1 (Jan. 1920): 10.

73. On Thompson's candidate for WSE president, see *Chicago Daily News*, May 3, 1921. On 24-hour engineers, see CLT, *Proceedings* 90 (Nov. 23, 1922): 15 of session. On the "Expert Fees" padding charge, see *Chicago Tribune*, July 9, 1921; *Chicago Daily News*, Dec. 16, 1921 (editorial); *Chicago Journal*, Aug. 23, 1923.

74. Leroy Dutton and C. M. Traiser, Surface Lines traffic and scheduling

experts, both agreed on Kelker's competence and ability to move with the political tide in their interviews with this author. Both men had many occasions to witness Kelker's testimony before city and state bodies.

On Kelker's consulting work for other cities and the consistency of his support for street railways and faith in their future, see Foster, *Streetcar to Superhighway*, pp. 52, 200.

75. For the biographies of Young and D'Esposito, see Marquis, *Who's Who in Chicago, 1931*, pp. 1080, 253. A number of Young's public statements are cited in Chapter 5, notes 7, 21, 56, 90.

76. On Parson's achievements, see *Webster's Biographical Dictionary*.

77. *1916 Report*, pp. 9–14, 165–69, and plate 8.

78. Ibid., pp. 245, 251–52, and Bion Arnold in CLT, *Proceedings* 65, (Dec. 27, 1916): 66. Later plans adopting the same principle include Kelker, *A Comprehensive Plan for a Unified Local Transportation System for the City of Chicago* fig. 3; Chicago City Council, *An Ordinance Providing for a Comprehensive Municipal Local Transportation System . . .* (Chicago, 1925), pp. 60–62; *1930 Ordinance*, pp. 35–38, 41–43.

79. *1916 Report*, p. 145.

80. Ibid., pp. 43–45.

81. Ibid., pp. 41, 42.

82. *Chicago Commerce*, Sept. 26, 1918, pp. 11–12; *Chicago Tribune*, Aug. 14, 1918; *ERJ* 52, no. 3 (July 20, 1918): 125; CLT, *Proceedings* 78 (Aug. 5, 1918): 10 A.M. session, p. 143 of meeting.

83. The initial trustees are listed in CLT, *An Ordinance . . . Reported to the City Council August 8, 1918* (Pamphlet 914), *Authorizing the Chicago Traction Company to Construct, Operate and Maintain a System of Local Transportation in the Streets, Alleys and Public Ways of the City of Chicago* (hereafter, *1918 Ordinance*), pp. 2–3.

84. *Chicago Tribune*, Nov. 7, 1918; City Council, *An Ordinance Providing for a Comprehensive Municipal Local Transportation System, Passed by the City Council of the City of Chicago, February 27, 1925 to be Voted on by the People at a Special Election April 7, 1925* (hereafter *1925 Ordinance*) (Chicago, 1925), pp. 1, 5.

85. Quotation is from Edward E. Gore (CAC president) in CLT, *Proceedings* 96 (Feb. 6, 1925): 199.

On the incompetence of the city to manage public utilities, see *Chicago Journal of Commerce*, Nov. 14, 1923 (editorial); *Chicago Tribune*, May 29, 1923 (editorial), Feb. 21, 1924; *Chicago Evening Post*, July 3, 1923 (editorial); *Chicago Daily News*, Aug. 22, 1923.

86. Sikes, "Chicago's Eternal Traction Question," pp. 3–4.

87. Tomaz Deuther in CLT, *Proceedings* 69 (March 20, 1917): 46–48; Tomaz Deuther, "Letter to the Honorable Mayor Thompson and the Members of the Chicago City Council, August 10, 1918," *Chicago Daily News*, Oct. 29, 30, 1918; Greater Chicago Federation, *Looking Backwards: Reminiscences on Traction* (Chicago, n.d. [1918]; unpaginated, Hooker Collection); The Better Transportation League, "To Those Who Helped Defeat the Recent Traction Ordinance" (circular letter dated Jan. 27, 1919, in Hooker Collection).

88. See below, this chapter, page 201.

89. *Chicago Commerce*, Oct. 7, 1922, 5, 7, 37.

90. Ibid., p. 3.

91. CLT, *Proceedings* 87 (May 10, 1922): 2–5 of meeting; 88 (June 2, 1922): 53–66 of meeting; *Chicago Commerce*, June 10, 1922, pp. 3–4.

92. Ibid., p. 4; CLT, *Proceedings* 90 (Nov. 23, 1922): 37, 43.

93. *Chicago Commerce*, May 26, 1923, p. 17.

94. Samuel Insull, *Rapid Transit Development for Chicago* (Chicago, n.d. [1924]); all Chicago newspapers, Feb. 22, 1924.

95. Marshall Field and Company declared for the idea in 1926 (*ERJ* 67, no. 11 [March 3, 1926]: 460). The idea gathered support thereafter (*Chicago Journal*, Nov. 22, 1926; *Chicago Daily News*, Jan. 4, 1927). The *1930 Ordinance* (p. 5) included provisions for the building of CBD subways by special assessment—thus signifying acceptance of the idea by the most important elements of the political and business communities.

For Deuther's position, see his *First Issue of Civic Questions*, 1st ed. (1924), pp. 36–37, 58, 62; 3d ed. (1925), pp. E, M; and CLT, *Proceedings* 114 (March 25, 1929): 370; 119 (May 5, 1930): 10.30 A.M. session: 53.

On the 1928 referendum, see *Chicago Tribune*, April 6, 1928.

96. Douglas, "Chicago's Persistent Traction Problem," p. 670.

97. CLT, *Proceedings* 117 (April 24, 1930): 541.

98. Most Notably *Hammer* v. *Dagenhart* (247 U.S. 251 [1918]), which eviscerated federal child labor legislation.

99. In the Supreme Court of the United States, October Term, A.D. 1919, *City of Chicago, Plaintiff in Error,* v. *Thomas E. Dempsey*, Chairman of the Illinois Public Utilities Commission, Defendant in Error, *Reply Brief on Motion to Dismiss Writ of Error*, pp. 1, 2–6.

100. In the Matter of the Petition of the Chicago Stage Company Relative to a Certificate of Convenience and Necessity, Case 6642, Illinois Public Utilities Commission, *Opinions and Orders*, vol. 5 (Jan. 18, 1918), pp. 16–72. On the involved story of Thompson's "effort" to regain the power of regulation for the city, see Barrett, "Mass Transit," pp. 409, 428–29, 430, and attendant notes.

101. CLT, *Proceedings* 117 (Mar. 6, 1930): 36 of meeting.

102. On the issue of valuation and its implications in other cities at the end of World War I see FERC, *Proceedings . . .* (Washington, D.C., 1920), pp. 255, 407–8, 573, 773, 815, 832, 1516, 1861–62, 2105, 2111, 2116.

103. *Chicago Commerce*, Jan. 31, 1921, p. 24.

104. The Commerce Commission's position can be found in In the Matter of the Petition of the Chicago Railways Company, *et al.*, Relative to Street Car Fares, Case 9357, Illinois Public Utilities Commission, *Opinions and Orders*, vol. 6 (Aug. 6, 1919), pp. 1067–68. In the Matter of the Petition of the Chicago Railways Company *et al.*, Operating as the Chicago Surface Lines, Relative to Street Railway Rates, Case 9357, November 5, 1920, ibid., 7: 105–9; *Chicago Tribune*, Nov. 24, 1921, Jan. 10, April 29, 1922.

The federal decision is *United Railways and Electric Company* v. *West*, 280 U.S. 324, 50 Superior Court 123 74: 390, in *U.S. Supreme Court Reports* 74 (Law Edition, Oct. term, 1929).

105. See note 50, this chapter. On the rationale for valuing utility properties as "going concerns," see Dewing, *Financial Policy of Corporations*, p. 264, n. B.

106. CLT, *Proceedings* 102 (Sept. 10, 1926): 14 of meeting.

107. *Chicago Journal*, Nov. 2, 1918.

108. A good account of the dramatics involved is Roberts, "Portrait of a Robber Baron," pp. 344–72.

109. On progressive hopes for regulation, see, for example, Clyde Lyndon King, *The Regulation of Municipal Utilities* (New York, 1912), esp. pp. 20–24. Quotation is from Charles Merriam, *Chicago: A More Intimate View of Urban Politics* (New York, 1929), p. 22.

110. McDonald, *Insull*, p. 178. Also Merriam, *Chicago*, pp. 113–14.

111. "The Chicago Election," *Outlook* 109 (April 21, 1915): 901–2; Merriam, *Chicago*, pp. 183, 185; Alex Gottfried, *Boss Cermak of Chicago: A Study of Political Leadership* (Seattle, 1962), pp. 80–82.

112. *Chicago Tribune*, July 27, 29, 1919; *ERJ* 53, no. 18 (May 3, 1919): 891; CLT, *Proceedings* 89 (July 31, 1922): entire meeting.

113. Referendum results, as reported in Chicago newspapers, were as follows:

Year	Proposition	Vote	Number voting as % of population
1914	Comprehensive Independent Subway	380,092	15.7%
1918	Comprehensive Private System	453,692	17.3
1925	Comprehensive Municipal System	556,772	18.0
1930	Comprehensive Private System	382,527	11.3

114. *1918 Ordinance*, pp. 27–30, 40.

115. The best arguments for the ordinance are in Capitain, *The New Traction Ordinance.*

116. *Chicago Commerce*, Nov. 7, 1918, p. 7; *Chicago Journal*, Nov. 2, 1918. On Dunne's anomalous position, see *Chicago Tribune*, Oct. 30, 1918 (editorial). A convenient summary on Thompson's behavior during World War I is in the *New York Times*, June 26, Sept. 5, 1917. A list of aldermanic opponents and supporters is in *Chicago Tribune*, Nov. 3, 1918. Other supporters are discussed in *Chicago Daily News*, Aug. 28, 1918.

117. The story is summarized in *ERJ* 52, no. 10 (Sept. 7, 1918): 429.

118. Greater Chicago Federation, "Looking Backwards"; William Hale Thompson, "To the Voters of Chicago"; *Chicago Daily News*, Oct. 30, 1918; *Chicago Journal*, Nov. 1, 2, 4, 6, 1918; *Chicago Tribune*, Nov. 1, 3, 1918; *Chicago Daily News*, Oct. 29, 1918.

119. *Chicago Tribune*, Nov. 3, 1918. See also, for example, *Chicago Evening Post*, Nov. 2, 4, 1918; *Chicago Daily News*, Oct. 26, 1918; *Chicago Defender*, Nov. 2, 1918; Citizens for the Unification Traction Ordinance *Why?*

120. *Chicago Daily News*, Oct. 29, 31, 1918; *Chicago Tribune*, Nov. 1, 3, 1918; *Chicago Herald and Examiner*, Oct. 31, Nov. 1, 3, 1918.

121. *Chicago Journal*, Nov. 6, 1918.

122. *Dziennik Zwiazkowy*, Aug. 20, 1918.

123. George Hooker in *Chicago Journal*, Nov. 3, 1918. Election returns from *Chicago Tribune*, Nov. 7, 1918.

124. Ibid.

125. William Hale Thompson, *The Thompson Plan for People's Ownership and Operation of the Street Railway Systems at a 5¢ Fare* (submitted to the City Council in Special Session, Sept. 9, 1919, by Mayor William Hale Thompson) (Chicago, n.d.), p. 8.

126. Ibid., esp. p. 19.

127. *ERJ* 54, no. 16 (Oct. 18, 1919): 769; see also *Chicago Herald and Examiner*, Sept. 11, 1919.

128. Commission on Local Transportation (Thompson's own commission, not the CLT), *Report . . . to the Honorable William Hale Thompson, Mayor, December 31, 1920* (Dec. 1921). On the fate of the legislation connected with this plan, see *Chicago Daily News*, April 11, May 18, 20, 1921, Jan. 19, 1923; *Chicago Tribune*, June 19, Dec. 25, 1921.

129. Chicago City Club, *City Club Bulletin* 15, no. 19 (May 8, 1922): 76.

130. CLT, *Proceedings* 81 (Jan. 13, 1921): 16–17; *ERJ* 59, no. 1 (Jan. 7, 1922): 49.

131. *Chicago Daily News*, May 20 (editorial), May 22 (editorial), Dec. 22, 1921; *Chicago Tribune*, Sept. 10, 1919; *Chicago Evening Post*, Sept. 10, 1919 (editorial); *Chicago Journal*, Sept. 12, 1919 (editorial).

132. *Chicago Journal*, Feb. 12, 1919 (editorial); CLT, *Proceedings* 80 (Oct. 29, 1919): 72–73. Thompson never proposed to buy the Insull-owned rapid transit lines.

133. On the CSL acquiescence, see *Chicago Commerce*, June 18, 1921, p. 35.

On municipal ownership as a "bailout," see Delos F. Wilcox, "What Shall We Do with the Street Railways?" *American City* 20, no. 4 (April 1919): 334; Paul Douglas, "Seattle's Municipal Street Railway System," *Journal of Political Economy* 29, no. 6 (June 1921): 455–77. On Detroit, see Davis, "The City Remodeled," pp. 465–69. See also Morris Llewellyn Cooke, *Public Utility Regulation* (New York, 1924), p. 47.

134. Typed biography of Ulysses S. Schwartz with Schwartz, *Scrapbooks* (Chicago Historical Society); *Chicago American*, April 3, 1916: *Chicago Examiner*, April 4, 1916; *Chicago Tribune*, April 20, 1920; *Chicago American*, May 20, 1920. Schwartz' connections thus touched all of the most important groupings in Chicago political parties.

135. Schwartz' approach evolved between 1919 and 1921. See Ulysses S. Schwartz, *Memorandum of a Plan for the Settlement of Chicago's Traction Problem* (undated pamphlet, apparently 1919, in Hooker Collection); *Chicago Commerce*, May 17, 1919; *Chicago Daily News*, April 28, 1920; Ulysses S. Schwartz, *A Discussion of the Transportation Problem in Chicago, with a Financial Plan for City Acquisition of the Property and the Operation Thereof* (Chicago, 1921), esp. pp. 16–18.

136. *Chicago Evening Post*, Jan. 16, 1921. See also (CLT, *Proceedings* 82 [Jan. 17, 1921]: 3–4, 9). Schwartz ran for a seat on the Cook County Board in 1922 on a platform of "no tax boosting" (*Chicago Tribune*, Sept. 28, 1922).

137. CLT, *Proceedings* 83 (Dec. 22, 1921): 4–15; 86 (Feb. 9, 1922): 71–88; 86 (Subcommittee on Traction Plans) (March 16, 1922); 4 of meeting.

On the reluctance of Rapid Transit representatives to discuss the plan, see *Chicago Evening Post*, Feb. 9, 1922.

138. *Chicago Tribune*, Jan. 16, 20, 1923; *Chicago Daily News*, Jan. 17, 1923. On Brennan's background, see *New York Post*, April 4, 1923.

139. *Chicago Daily News*, March 2, 1923; *Chicago American*, May 14, 1923; *Chicago Tribune*, March 5, 1923; *Chicago Herald and Examiner*, March 5, 1923.

140. *Chicago American*, March 19, 1923. See also *Chicago Tribune*, June 5, 1923.

141. Socialist Joseph Carr, quoted in *Chicago American*, Feb. 22, 1923.

142. On Dever's attempt to enforce "moral standards," see, for example, *Chicago Tribune*, April 8, 1925; *Chicago American*, May 30, 1923.

The "midwives" remark is in *Chicago Herald and Examiner*, April 3, 1925. Also Dever to Augusta Rosenwald, March 18, 1925, in Dever, Papers, box 5, folder 39.

On the Dever administration's problems with nontransit issues, in particular organized crime, see John M. Allswang, *A House for All Peoples* (Lexington, 1971), pp. 125, 134–35, 173–75.

143. *Chicago Herald and Examiner*, April 4, 1925.

144. *Chicago Daily News*, March 27, 1923.

145. *Chicago Commerce* (Feb. 24, 1923), pp. 13–14; *Chicago Evening Post*, July 3, 1923. See also, for example, *Chicago Tribune*, Feb. 12, 1923; *Chicago Daily News*, Feb. 27, 1923; *Chicago Tribune*, Aug. 23, 1923.

146. *Chicago Tribune*, July 3, 1923 (first quotation). Second statement is quoted in Deuther, *First Issue of Civic Questions*, p. j. Final statement is in William E. Dever, *Special Message of the Honorable William E. Dever, Mayor, Concerning Chicago's Local Transportation Problem* (submitted to the City Council of the City of Chicago, Oct. 22, 1924) (Chicago, n.d.), p. 11.

147. Ibid., p. 9; *Chicago Journal*, May 7, 1924; *Chicago Tribune*, May 14, Sept. 27, 1924.

148. Negotiations fill the CLT *Proceedings* between Nov. 18, 1924, and Feb. 20, 1925. For a summary, see *Chicago Tribune*, Jan. 27, 1925.

149. *1925 Ordinance*, pp. 1, 5.

150. Ibid., p. 7.

151. Dever's role in putting the ordinance through the city council is evident in the transcript of the debate (in CLT, *Proceedings* 96: 494–584; see esp. pp. 510–14). Among the plan's supporters were the All Chicago Council (CLT, *Proceedings* 96 [Feb. 6, 1925]: 198–99), the Chicago Association of Commerce (CLT, *Proceedings* 96 [Feb. 23, 1925]: 459–69), Edwin Lobdell, an influential broker dealing in traction securities (Lobdell to Dever [telegram], Feb. 28, 1925, in Dever, Papers, box 28, folder 39); Bion J. Arnold (*Chicago Tribune*, March 28, 1925), Harold Ickes (*Chicago Daily News*, March 28, 1925), Walter L. Fisher (*Chicago Daily News*, March 31, 1925; *Chicago Tribune*, April 1, 1925), 350 ethnic, neighborhood, and business groups (*Chicago Daily News*, April 6, 1925), and Brennan's Democratic organization (*Chicago Tribune*, April 4, 1925).

On the CSL's position, see *Chicago Commerce*, Feb. 23, 1925, p. 31; Chicago City Railway, *Annual Report, 1925*, p. 4; *ERJ* 65 no. 12 (March 21, 1925): 481. On Insull's backing of the ordinance, see *Chicago Tribune*, April 6, 7, 1925.

152. On opposition to the ordinance, see *Chicago Herald and Examiner*, April 6, 1925; *Chicago Journal*, April 1, 8, 1925; *Chicago American*, April 6, 1925; *Chicago Daily News*, March 30, 1925; *Chicago Tribune*, March 22, 1925. These citations are only indicative.

Election results are from *Chicago Daily News*, April 8, 1925.

153. Lobdell, *Chicago's Transportation Problem . . . 1925*, p. 1; Patrick Lucy (of Chicago Rapid Transit), "Transportation in Chicago," *JWSE* 33, no. 11 (Nov. 1928):

567. See also Henry F. Bruner, "How Chicago Is Attempting to Solve Its Traction Problem," *Harvard Business Review* 9, no. 4 (July 1931): 459.

154. *1925 Ordinance*, p. 7.

155. Ibid., pp. 5–6.

156. *Chicago Journal*, June 25, Aug. 27, 30, 1923, April 4, 1924, March 3, 1925; *Chicago Tribune*, July 5, 1923; *Chicago Herald and Examiner*, July 6, 1923 (cartoon), April 2, 1925.

157. *Chicago Tribune*, April 8, 1925.

158. On the city council, the only municipal ownership spokesman was Ald. Wiley Mills, once a member of Mayor Dunne's "radical" school board and a single-taxer. See CLT, *Proceedings* 96 (Feb. 11, 1925): 361–62, 378; *Chicago Daily News*, Jan. 4, 1927. See also Detroit Plan Traction Purchase Club, *The Municipal Ownership of Chicago's Street Railways* (Chicago, 1927); Carl D. Thompson, *The Chicago Traction Problem and Its Solution* (Chicago, 1928).

159. Dever's statement is in CLT, *Proceedings* 98 (May 25, 1925): 228–30; Schwartz' statement is in *Chicago Daily News*, Dec. 31, 1925. For Ald. Jacob Arvey's statement, see CLT, *Proceedings* 99 (Jan. 4, 1926): 712–14.

On Thompson's 1927 campaign, see *Chicago Journal*, Dec. 11, 1926; "A Historian," *Big Bill, the Builder*, rear cover.

Wilcox's position is in *City Club Bulletin* 19, no. 9 (March 1, 1926): 53.

160. A summary of the various plans before the council in 1927 is in *ERJ* 69, no. 6 (Jan. 22, 1927): 177–78. See also *Statement of Principles and Outline of a Financial Plan with Draft of an Ordinance for a Unified Transportation System for Chicago* (prepared under the Direction of the Subcommittee on Traction Settlement of the CLT of the City Council of the City of Chicago, Ald. E. Frankhauser, Chairman, Sept. 22, 1928).

161. U. S. Schwartz, *The Chicago Traction Problem: A Discussion Before the City Club of Chicago, November 18, 1935* (n.p., n.d.); Francis X. Busch, "The Public Relations of Transportation," *ERJ* 75, no. 11 (Oct. 1931): 567–68.

Melvin G. Holli and Peter d'A. Jones, *Biographical Dictionary of American Mayors* (Westport, 1981), p. 101. Dever's Commissioner of Public Works, A. A. Sprague, became a trustee for the Chicago Rapid Transit during its receivership.

162. See Chicago Surface Lines advertisements in *Chicago Commerce*, Nov. 7, 1925, p. 1, June 16, 1928, p. 1, May 21, 1927, p. 1, July 14, 1928, p. 1. Protective committees for the surface companies were headed by chief executives of the city's largest banks (*ERJ* 67, no. 5 [Jan. 30, 1926]: 215). CSL's receivership judge was James Wilkerson, former head of the State Utilities Commission.

163. "Chronology of the Chicago Traction Situation" (anonymous typescript in the John Crerar Library, Chicago, c. 1931), pp. 10–14. See also *Chicago Tribune*, Aug. 6, 20, 1929.

164, See, for example, CLT, *Proceedings* 117 (March 8, 1930): 72–78; *Chicago Tribune*, May 20, 1930.

165. *Chicago Tribune*, June 16, 1930.

166. CLT, *Proceedings* 120 (May 7, 1930): 75.

167. On the housing industry, see *Commerce* (successor to *Chicago Commerce*) 30, no. 12 (Jan. 1934): 115. On general unemployment, which reached higher levels in 1927–28 than in any year since 1921, see State of Illinois, *Thirteenth Administrative*

Report of Directors of Departments Under Civil Administrative Code for the Year July 1, 1929 to June 30, 1930 (Springfield, 1930), p. 1310, fig. 5.

168. CLT, *Proceedings* 115 (July 1, 1929): 198.

169. CLT, *Proceedings* 117 (April 20, 1930): morning session, pp. 256–58.

170. "Chronology," p. 27. On the promise that the ordinance would help to revive the city's economy see, for example, *Chicago Herald and Examiner*, June 4, 1930; *Chicago Daily News*, June 25, 1930; *Chicago Defender*, June 21, 1930 (editorial); *Chicago Commerce*, June 28, 1930, p. 27. Citations are merely indicative.

171. *Chicago Tribune*, July 2, 1930.

172. Insull's utility empire collapsed in the Depression. He was tried for mail fraud and ultimately acquitted in a jury trial presided over by Judge James H. Wilkerson. See McDonald, *Insull*, pp. 275–333. Wilkerson's behavior was, according to McDonald, "coldly impartial" (p. 324).

Epilogue

1. C. F. Kettering, "Prosperity and the Automobile Industry," in American Electric Railway Association, *Proceedings, 1928* (New York, 1928), p. 91.

2. E. J. McIlraith, "Costs and Competition in Street Use," *ERJ* 75, no. 11 (Oct. 1931): 569.

3. Chicago City Council, Committee on Traffic and Public Safety, *Limited Ways*, pp. 21–25, 29–31.

4. Transfer studies are in McIlraith, Papers, vol. 2. On two-car trains, see CTA, *Historical Information*, pp. 10, 17, 23, 12, 28. On the congested west side line, see CLT, *Proceedings* 95 (Feb. 23, 1924): 525. During the 1930s the most congested lines were also west side crosstowns and diagonals (Milwaukee and Western Avenues); see Phillip Harrington, R. F. Kelker, and Charles E. DeLeuw, *A Comprehensive Local Transportation Plan for the City of Chicago* (Chicago, 1937), pp. 46–47. On neighborhood business districts, see Proudfoot, "Outlying Business Centers."

5. On the spread of population, see maps in CPC, *Master Plan of Residential Land Use of Chicago* (Chicago, 1943), p. 22, and Chicago Area Transportation Study, *Final Report*, (Chicago, 1960), 2: 18. The exceptions are the city's far southwest and far south sides. On city and suburban populations, see *1939 Superhighway Plan*, p. 23.

6. Studies of Carnegie Steel and Swift and Company employees are in McIlraith, Papers, vol. 2. Data for 1916 are from *1916 Report*, pp. 238–39.

7. *Chicago American*, May 20, 1920.

8. On company tax and public benefit payments, see Schwartz, *The Chicago Traction Problem*, p. 8. Profits are estimated from BOSE, *37th Report*, pp. 86–87.

9. Donald F. Davis, "Mass Transit and Private Ownership: An Alternative Perspective on the Case of Toronto," *Urban History Review* 3 (Feb. 1979): 60–98. See also McShane, *Technology and Reform*, pp. 131–35.

10. CLT, *Proceedings* 137 (Feb. 2, 1939): 52; *Transit Journal* 79, no. 9 (Sept. 1935): 281; 79, no. 12 (Dec. 1935): 433–34; 80, no. 3 (March 1936): 103; 85, no. 11 (Nov. 1941): 441; *Chicago Tribune*, June 30, 1941, Feb. 21, 1942, April 2, 27, 1943.

11. Public Works Administration, Department of the Interior, *In re the Application of the City of Chicago for P.W.A. Aid to Start a Chicago Subway System* (Washington, D.C., Sept. 29, 1938), pp. 73, 199–203; H. M. Waite to Harold Ickes,

Sept. 23, 1938, in CSL, *Archives*, box 6, folder 32; CLT, *Proceedings* 135 (May 12, 1937): 10–11.

On the streetcar as an obstacle to traffic see Jarvis Hunt, *A Plan for the Subways and New Rapid Transit Lines of Chicago* (Chicago, 1917), p. 5; Committee on Traffic and Public Safety, *Limited Ways*, p. 13.

12. *Chicago Tribune*, April 13, May 28, June 1, 3, 1945; *Mass Transportation* 43, no. 2 (Feb. 1947): 46; 43, no. 5 (May 1947): 188–89; *Laws of Illinois*, 64th General Assembly, 1945, pp. 1170–87.

13. Werner Schroeder, *Metropolitan Transit Research Study* (Chicago, 1955), p. 181.

14. On the RTA, See *Laws of Illinois*, 78th General Assembly, 1973 2: 3192–3246; *Chicago Tribune*, March 15, 19, 21, 1974; *Chicago Sun Times*, March 11, 1972; *Motor News*, Nov. 1973; *Chicago Defender*, March 16, 1974.

15. On suburban complaints, see *Chicago Tribune*, Aug. 1, 2, 4, 1982. On black concerns, see *Chicago Sun-Times*, Nov. 3, 4, 1979; *Chicago Metro News*, Feb. 17, 1979; *Chicago Defender*, April 14, 1979; *Chicago Tribune*, Jan. 17, 1981; *The Chicago Journal* 6, no. 12 (July 7, 1982): 6–7. Quotation in text is from *The Reader* 11, no. 4 (Oct. 23, 1981): 3.

16. *Chicago Tribune*, June 7, 11, 17, 1981, April 30, May 21, 1982; *Chicago Sun-Times*, June 11, Aug. 25, 1981.

17. On fares and wages, see *Chicago Sun-Times*, March 3, May 24, 1981; *Chicago Tribune*, May 20, June 12, July 7, 1981. On the "need" for higher fares, see U.S. Comptroller General, *Soaring Transit Subsidies Must Be Controlled* (Washington, D.C., Feb. 26, 1981), pp. 8–25, 31–41. The nexus between rising wages, fares, and subsidies is too complex to be discussed here.

On the man without carfare, see *Chicago Tribune*, Jan. 16, 1982.

18. *Chicago Sun-Times*, Nov. 23, 1980, Feb. 9, 1982.

19. See, for example, Samuel P. Hays, "The Politics of Reform in Municipal Government in the Progressive Era," *Pacific Northwest Quarterly* 55 (Oct. 1964): 157–69; Hays, "The Changing Political Structure of the City in Industrial America," *Journal of Urban History* 1 (Nov. 1974): 6–34; Kenneth Fox, *Better City Government: Innovation in American Urban Politics, 1850–1937* (Philadelphia, 1977), esp. pp. 63–65; Carl V. Harris, *Political Power in Birmingham, 1871–1921* (Knoxville, 1977), pp. 270–71; Joseph L. Arnold, "The Neighborhood and City Hall: The Origins of Neighborhood Associations in Baltimore, 1880–1911," *Journal of Urban History* 6, no. 1 (Nov. 1979): 3–30. A good survey is David C. Hammack's "Problems in the Historical Study of Power in the Cities and Towns of the United States, 1800–1960," *American Historical Review* 83 (April 1978): 323–49.

20. For a variety of views, see Mark Foster, "City Planners and Urban Transportation: The American Response, 1900–1940," *Journal of Urban History* 5, no. 3 (May 1979): 365–67; Kenneth T. Jackson, "Race, Ethnicity and Real Estate Appraisal: The Home Owners Loan Corporation and the Federal Housing Administration," *Journal of Urban History* 6, no. 4 (Aug. 1980): 419–52; Robert Caro, *The Power Broker* (New York, 1975); Robert Goodman, *After the Planners* (New York, 1977).

21. The city began to develop a superhighway planning bureaucracy of its own with the creation of a permanent Committee on Traffic and Public Safety, which worked to centralize the planning of urban highways. See the Committee's *Limited Ways: A Plan for the Greater Chicago Traffic Area* (Chicago, 1933), esp. pp. 7–9. The

later position of the CPC in the highway planning process is exemplified by the discussion of alternative highway plans in City of Chicago, Department of Superhighways, *A Comprehensive Superhighway Plan for the City of Chicago* (Chicago, 1939), pp. 37–40.

Hugh E. Young directed the WPA's land use survey during the later 1930s. See Federal Works Agency, WPA (Illinois), Official Program no. 165-1-54-151 (3), "Basic Records, Tables, Maps, and Reports" (n.d., mimeographed, Chicago Municipal Reference Library). A summary of the resultant material, with the CPC's definition of "blight," is in Chicago Plan Commission, *Master Plan of Residential Land Use in Chicago* (Chicago, 1943).

22. CLT, *Proceedings* 66 (Dec. 27, 1916): 65 of meeting.

23. See, for example, Alan D. Anderson, *The Origins and Resolutions of an Urban Crisis: Baltimore, 1890–1930* (Baltimore, 1977).

24. On the role of the debate over "resource allocation" in American urban history, see, for example, Zane Miller, "Scarcity, Abundance, and American Urban History," *Journal of Urban History* 4, no. 2 (Feb. 1978): 131–55; Sam Bass Warner, *The Private City: Philadelphia in Three Periods of Its Growth* (Philadelphia, 1972); Jon C. Teaford, *The Municipal Revolution in America: Origins of Modern Urban Government, 1650–1823* (Chicago, 1975).

25. For comparison, see Brownell, *Urban Ethos in the South*, pp. 116–23, and his "Symbol of Modernity."

26. A recent treatment of the early parks movement in Glenn E. Holt, "Private Plans for Public Spaces: The Origins of Chicago's Park System," *Chicago History* 7 (Fall 1979).

27. City of Chicago, Department of Superhighways, *A Comprehensive Superhighway Plan for the City of Chicago* (Chicago, 1939), p. 25.

28. Martin, *Enterprise Denied*, esp. pp. 352–67. Mallach, "Origins of the Decline," p. 14.

29. On Chicago fares compared with those of other cities, see *Mass Transportation* 35, no. 10 (Oct. 1939): 303–4.

Bibliographical Note

Because the primary literature concerned with planning and local transportation in Chicago is extensive, it is possible to provide only a guide to the locations of the most important bodies of material. Fortunately, anyone wishing to wade into this sea of largely unorganized data will find that it is, at least, easily accessible.

The papers of several political figures involved in the events discussed in this work are to be found in the Chicago Historical Society. The papers of William Dever, covering Dever's mayoral term, contain important communications within the administration as well as a useful scrapbook of newspaper clippings. The papers of Alderman Ulysses S. Schwartz are in the archives of Gottlieb, Schwartz, and Company, in the Chicago Historical Society. There is also a valuable collection of newspaper material left behind by Mayor Edward F. Dunne, as well as several scrapbooks on local transportation issues in the 1890–1906 period, catalogued as the Ambler Scrapbooks.

Other important and useful collections at the Chicago Historical Society include the archives of the Chicago Surface Lines (containing corporate records, minute books, stock ledgers, and some early correspondence), the papers of the Citizens' Association of Chicago, the papers and reports of the City Club of Chicago, and the papers of Donald Richberg (useful for the later period) and Emily Larned (on early motorbus politics).

The most significant collection of materials on early city planning, transit planning, and transit politics is that made by Hull House resident George Hooker between 1900 and 1920. The collection can be found in several locations within the Regenstein Library at the University of Chicago. The largest body of materials bears the call number HE 4491/C41 c5.

Another valuable collection, concerned with finance and politics as well as transit policy, is the papers of Samuel Insull, at the Cudahey Memorial Library, Loyola University, Chicago. Those interested in the legal and political maneuvering of the later period may wish to consult the Receivership Records of the surface and elevated companies in the library of the Northwestern University Transportation Institute. These papers also contain some useful transit data.

Transit and traffic engineers left behind few papers relating specifically to Chicago. The large collection kept by Bion Arnold was, unfortunately, lost after Arnold's death. Useful information can be found, however, in the reports of the Chicago Surface Lines Department of Traffic and Scheduling which are kept by the Chicago Transit Authority. A small collection of transit use data, especially useful for the period after 1930, was left by Evan McIlraith and is currently in the possession of the author. In addition, Miller McClontick and the staff of the Erskine Institute created a valuable bibliography of street traffic and regulation studies, a copy of which is in the Regenstein Library.

Most useful for a study of planning and policymaking are the records of the various bodies involved. The papers of Danial Burnham, at the Chicago Art Institute, are helpful for a study of the background of the *Plan of Chicago*. On the implementation of the plan, the printed *Proceedings* of the Chicago Plan Commission are indispensable. On both mass transit and traffic policy, the basic source is the *Proceedings* of the Chicago City Council Committee on Local Transportation. Over 150 unindexed volumes of the typed records of this committee are in the Chicago Municipal Reference Library in the Chicago City Hall. These records, which also contain material on the taxi and trucking industries, are as rewarding as they are difficult to use. The Municipal Reference Library also contains a variety of reports and documents from city agencies concerned with transportation, including the annual reports of the Board of Supervising Engineers and the Park Board reports.

Also available at the Municipal Reference Library are all of the transit and traffic engineering reports made for Chicago during the period with which we are concerned. Many of these reports, some of which are important planning documents, are available in other libraries as well. The studies of transportation systems in other parts of the country—cited in the text—are important as a context for the Chicago experience.

Essential sources for the study of traffic, transit, and planning issues in Chicago (and in the nation) are a variety of journals. For general urban concerns, the most important are *American City, Municipal Journal and Engineer*, and scattered issues of the *Annals of the American Academy of Political and Social Science*. For street traffic and street building, see *Roads and Streets, Automobile Trade Journal, Motor Age, Engineering News-Record, Engineering and Contracting*, and the *Transactions* of the American Society of Civil Engineers. The most important journals covering the transit industry include the *Electric Railway Journal*, the *Street Railway Review* (later, *Electric Traction*), and *Bus Transportation*. Also important are two publications of the American Electric Railway Association: *AERA* (a monthly magazine) and the Association's annual *Proceedings*. These, and the *Journal of the Western Society of Engineers*, offer perspectives on auto and city-planning as well as mass transit issues. The best single source on Chicago traffic problems and policy is *Chicago Commerce* (the journal of the Chicago Association of Commerce). Like all the other journals, it represents a distinct point of view, but it is generally reliable.

Essential for those seeking a broader knowledge of the process by which the city was planned and built during the twentieth century are two works by Carl Condit: *Chicago, 1910–1929: Building, Planning and Urban Technology* (Chicago, 1973) and *Chicago, 1930–1970* . . . (Chicago, 1974). If a reader is desirous of more detailed information on the political aspects of mass transit policy, three good sources await him: Samuel Wilbur Norton, *Chicago Traction: A History Legislative and Political* (Chicago, 1907); Ralph E. Heilman, "Chicago Traction: A Study of the Efforts of the Public to Secure Good Service," *American Economic Association Quarterly* (3d ser.) 9, no. 2 (New York, 1908); and Werner Schroeder, *Metropolitan Transit Research Study* (Chicago, 1955), which contains excellent transportation use data and a summary history. Finally, for those interested in the equipment and operations of surface mass transit, Alan R. Lind's *Chicago Surface Lines: An Illustrated History* (Park

Forest, 1974, 1980) offers not only photographs, but an excellent narrative which makes it much more than a "hobby" book.

Other recent, scholarly work on transportation and city planning has been cited in the notes; space does not permit a listing of these works here, but the author wishes to acknowledge his debt to the scholars in this rapidly growing field.

Index

Accidents: automobile, 132, 148; mass transit, 11, 18, 224
Adams, Charles Francis, 39
Addams, Jane, 39
All Chicago Council, 188–89
Altgeld, John Peter, 27
Alvord, John: 1904 report of, on street paving, 51–52
Amalgamated Transit Workers' Union, 194–95
Arnold, Bion J.: character of, 95–96; commuting habits of, 25; on engineering, 183; and flat fares, 29, 217; on mass transit service and profits, 119; and 1907 ordinances, 40–41, 83; 1902 Chicago study by, 29–30, 72, 155; and 1916 Commission, 105, 185; on over capitalization, 120–21; and politicians, 91–92, 95; as progressive engineer, 95–96; on public opinion, 95, 215; on regulation, 92; on subways, 67–68, 93; on taxes for mass transit, 115; on teamsters, 50; on traffic segregation, 46, 64
Arvey, Jacob, 189, 205–6
Automobile clubs, 133–34. *See also* Chicago Motor Club; Motor clubs
Automobile industry, 131, 133–34
Automobiles: accommodation of, by city, 133–34, 141; and boulevards, 147; and CBD, 54, 57, 131–32, 158–61, 209, 213; cost of ownership, 58, 140, 259; early class connotations of, 58–61, 62, 72; and dispersion of commerce, 131; and dispersion of population, 139–40, 210; improvements in, 141; and mass transit, comparable roles of, 4–5, 101–3, 106–7, 111, 113; and mass transit service, 132, 141, 157–58, 159, 161, 212–13; and neighborhood business, 159; ownership of, 62, 130–32, 140–41, 162, 209; and parks, 71;

recreational use of, 146; speed of, in traffic, 141, 161; uses of, 140–41, 209–10. *See also* Accidents; Parking; Planning; Journey to work; Street improvements; Taxes; Traffic regulation

Baker, Ray Stanard, 58
Baltimore, journey to work in, 162
Bankers: control of surface transit companies by, 170; and regulation, 204–5; and transit policy, 179–83, 201
Bemis, E. W., 95
Bentley, Arthur, 32, 44
Bicycles, 71
Blair, Henry, 97, 170
Board of Supervising Engineers, Chicago Traction (BOSE), 40, 65, 68–69, 89, 94–95, 117, 186
Bond issues, city, 150–53
Boston: local government and mass transit in, 4, 186; special assessments in, 153; street improvement spending in, 143; subways in, 24; transportation policy in, and Chicago transit policy, 28, 31–32, 186; unification of transit service in, 83
Boulevards, 51, 70–72, 143, 147
Brennan, George, 194
Brinkerhoff, Henry, 126
Brownell, Blaine, 141–42
Burnham, Daniel, 5, 45, 66, 80, 149, 150
Bus companies. *See* Busses
Busby, Leonard, 63, 170, 201
Business, hours of: and mass transit crowding, 99, 117; proposed staggering of, 65; and street traffic congestion, 55
Business interests: and mass transit policy, 38, 186–87, 200, 203–4; and traffic regulation, 131. *See also* Central Business District interests; Neighborhood interests
Busse, Fred A., 38